Dendrobium Orchids of Australia

Dendrobium Orchids of Australia

Walter T. Upton

Houghton Mifflin Australia

First published in 1989 by
Houghton Mifflin Australia Pty Ltd
PO Box 289, Ferntree Gully, Victoria 3156
112 Lewis Rd, Knoxfield, Victoria 3180, Australia

National Library of Australia
Cataloguing-in-Publication entry:

Upton, W. T. (Walter T.), 1922–
Dendrobium orchids of Australia.

Bibliography.
Includes index.
ISBN 0 86770 052 1.

1. *Dendrobium* – Australia. 2. Orchid culture – Australia. I. Title.

635.9′3415

Designed by Tom Kurema
Typeset in Australia by Solo Typesetting, South Australia
Printed in Hong Kong by South China Printing Company (1988) Limited

Contents

Foreword

The genus *Dendrobium* has a special place among the orchids native to Australia. The genus is so large as to be taxonomically unwieldy, but taxonomists have not yet given universal approval to a break up into smaller genera; rather, they have treated it in sections. Several sections are represented in the Australian species, ranging from those (mainly southern) which have their greatest development within Australia, to those which have developed in areas to the north and north-west of Australia. Of the orchids generally present in Australia, *Dendrobium* has the largest number of non-terrestrial species, many of which have large, colourful or showy flowers, making it natural that hybridists of Australian orchids have concentrated largely on this genus.

Wal Upton is well known throughout Australia and beyond as a fine grower of orchids. Ably assisted by Jill Upton, he has in recent years directed his energies and expertise to the growing of Australian orchids. He has taken great pains to accumulate a stock of fine plants of *Dendrobium* species and, using these as parents, has proceeded to follow a carefully thought out programme of hybridization. His reward can be found in the R. H. S. registrations in his name, on the show benches, in the collections throughout Australia and New Zealand to which he has contributed, and now in the publication of this book.

All the recent name changes are here and the book is taxonomically up to date. A mooted revision of the Australian species of *Dendrobium* would not alter the substance of the work. Commencing with the foundation of the genus, the author gives descriptions of all the species in Australia and, characteristically, reveals all his cultural methods and techniques of hybridization. The text is complemented by the author's line drawings and colour photographs.

As the only complete reference to *Dendrobium* in Australia this work will be of great value to all orchid lovers, be they specialist or amateur.

L. J. Lawler
Atherton
1988

Acknowledgements

This book could not have been produced without the active assistance of my wife, Jill, and I thank her for her tolerance and encouragement as well as her help in deciphering and typing the manuscript.

I am greatly indebted to Mr Leonard J. Lawler for reading through and checking the entire manuscript; for his understanding and interest; and for providing most useful advice. He has materially contributed to the accuracy of this work.

My sincere thanks to Mr Mark A. Clements for giving me access to his recent research of the many species of the genus *Dendrobium*, therefore ensuring that this book is taxonomically up to date.

I would also like to thank numerous friends who have assisted over the years by the loan of plants and in various other ways. In particular I would like to thank Mr S. C. Clemesha, Mr M. J. Corrigan, Mr M. Harrison, Mr H. Hirsch, Mr G. Mobbs and Mr J. Nield.

With much joy I remember the many friends and colleagues who have joined with me in the pleasure of an excursion into the beautiful 'Australian bush', and the total enjoyment of seeing orchids in their native habitat.

Walter T. Upton
Elanora Heights
NSW, Australia
1988

Introduction

The *Dendrobium* genus of orchids is not only one of the largest but one of the most popular with enthusiasts throughout the world. Each year more people become aware of the beauty of the *Dendrobium* orchid.

The Australian members of the genus *Dendrobium* are highly regarded and much sought after in orchid collections. In Australia and New Zealand the genus plays a major role in horticultural developments and in collections. Orchid societies hold monthly competitions or annual shows and classes for native orchids are frequently won by a *Dendrobium* species or hybrid. Many home gardens have a native *Dendrobium* growing somewhere.

Enthusiasts realize that the superior plants are those produced by line breeding of species. In many instances hybrids are a great improvement on species from the native habitat as the most worthwhile characteristics of different species are brought together to produce superior clones. Modern *Dendrobium* hybrids are also easier to cultivate than species from the wild.

Australian native orchids are protected plants and illegal collectors are heavily fined. As most native *Dendrobium* species are readily available for sale, there is no real need to collect them from the wild. Further, Australia is a signatory to the Convention on International Trade in Endangered Species of Wild Fauna and Flora (CITES), making it necessary to obtain authority to export or import plants. This agreement does not ban commercial trade but seeks to prevent over-exploitation and other threats to the world's plants. Flasks of artificially produced hybrids are not subject to these regulations and can be obtained in Australia from a few reputable native *Dendrobium* orchid enthusiasts.

Native Australian members of the genus *Dendrobium* are confined to a narrow belt along the east coast (Fig. 1), from Tasmania to the tip of Cape York Peninsula, and to a narrow strip along the extreme north coast of the Northern Territory and Western Australia. Within this area are also found the main forests.

Australian topography is unique. The continent was formed in early geological times and has not been subject to recent forces such as mountain-building upheavals, with the result that the land was worn smooth in comparison to other continents. Only 5 per cent of the land is over 615 m high. It is the world's flattest continent and, next to Antarctica, the world's driest one.

On the east coast there is a narrow coastal strip varying from 1·5 to 3·5 km wide in part of southern NSW, to about 160 km wide in southern Queensland. To the west of this coastal strip is the Great Dividing Range, which is around 900 m high. This range is of sufficient height to influence the rainfall and, as Australia lies within the south-east trade winds, moisture-laden air is brought in from the Pacific Ocean. The warm East Australia Current coming from the tropics flows southward along the east coast and further stimulates rainfall by increasing the water vapour in the atmosphere. These climatic conditions on the east coast are ideal for the genus *Dendrobium*. West of the Great Dividing Range, rainfall is lower and native *Dendrobium* orchids are rarely found.

The northern coastal strip of the continent, where *Dendrobium* species also occur, is an area of summer monsoonal rains. Heavy rains occur during 3 or 4 months, and the rest of the year is hot and dry.

Rainfall figures and average temperatures for some Australian cities and towns are shown in tables 1 and 2. July is the month with the lowest average temperature in all parts of the continent. The highest average temperature in the south is in January and February, and in the north December.

Although native *Dendrobium* plants grow in only a restricted area of Australia, the necessary conditions can be provided in orchid houses throughout the world if care and thought is given to creating a similar environment. *Dendrobium* orchids are extremely diverse, and while many microclimates in an orchid house may be necessary to accommodate them all, it is possible to grow a great many fairly easily by establishing a basic, compatible environment.

My first interest in *Dendrobium* orchids began in the early 1950s and has increased continually since. By the late 1950s I was producing my first hybrids. This book began when I started drawing native orchids in 1960. The fascination of these most special orchids totally absorbed me, and I trust this book will help and encourage many enthusiasts, giving them some of the pleasure I have had in writing it.

Table 1 Average rainfall in some Australian cities and towns (mm)

	Darwin	Cairns	Rockhampton	Brisbane	Sydney	Melbourne	Hobart	Adelaide	Perth
Jan.	393	421	179	167	100	48	46	20	8
Feb.	330	422	187	162	113	50	42	21	11
Mar.	258	460	110	145	126	53	44	24	20
Apr.	103	264	58	88	132	59	52	45	46
May	14	110	41	70	124	58	52	69	125
June	3	72	58	69	129	50	57	72	186
July	1	39	41	57	106	49	56	67	175
Aug.	2	42	22	48	81	49	52	62	139
Sept.	13	43	29	48	70	59	56	51	81
Oct.	50	50	45	75	80	67	60	44	56
Nov.	124	98	61	94	78	59	60	31	21
Dec.	244	203	112	129	80	58	56	27	14
Annual total	1535	2224	943	1152	1219	659	633	533	882

Table 2 Average maximum and minimum temperatures of some Australian cities and towns (°C)

	Darwin	Cairns	Rockhampton	Brisbane	Sydney	Melbourne	Hobart	Adelaide	Perth
Jan. (av max)	32.3	31.5	31.4	25.4	21.4	19.9	16.7	22.4	23.2
July (av min)	23.3	16.7	8.6	15.5	13.6	9.9	8.1	11.8	13.1

1

The genus *Dendrobium*

The genus *Dendrobium* was established by Olaf Swartz in 1799 in *Nova Acta Regiae Societatis Scientiarum Upsaliensis* Vol. 6. The generic name, *Dendrobium*, has been conserved although two earlier names, *Ceraia* Lour. (1790) and *Callista* Lour. (1790), had been established. When Swartz created the genus he also named 19 species of *Dendrobium*.

In *Taxon* 418 1979, Holttum, Brieger and Cribb gave reasons why *D. crumenatum* Sw. was inappropriate as the type species of the genus and proposed *D. moniliforme* (Lindley) Sw. as the type species. *D. crumenatum* was placed on the list of types of conserved names in 1930 after the Cambridge Congress. The reasons given for the retypification were that *D. crumenatum* was not included in the paper in which Swartz established the genus, whereas *D. moniliforme* was; and that *D. crumenatum* is not in the section *Dendrobium* whereas *D. moniliforme* is.

D. linguiforme Sw. was the first Australian species to be included in the genus *Dendrobium* and was admitted to the genus by Swartz in a third paper on the genus published in *Kongl. Vetensk. Acad. Nya Handl.* in 1800.

The name *Dendrobium* comes from the Greek words *dendron*, meaning tree, and *bios*, meaning life. The name meant 'living on a tree'—referring to the epiphytic habit of the genus.

World distribution

The *Dendrobium* genus occurs from India and Sri Lanka, through Burma then eastward to southern and eastern China; from Korea to Japan and south through the Ryukyu Archipelago to Taiwan; from Thailand, Laos, Vietnam and Cambodia through Malaysia to the Philippines; and from Indonesia, eastward to Papua New Guinea then South to Australia and New Zealand and east through Melanesia to Polynesia.

Characteristics of the Australian members of the genus *Dendrobium*

Plants vary in size from less than 1 cm to well over 5 m; epiphytic or lithophytic with the exception of *D. lobbii* which is terrestrial. Growth sympodial, rhizomes of few to several internodes or indistinct stems clustered together, creeping, erect or pendulous; wiry or fleshy (pseudobulbous) throughout or fleshy in part only; either leafy throughout or one to few at top of the pseudobulbous stem only. *Leaves* jointed at the base; very variable in shape, from linear to almost round to terete; and from thin to thick and leathery.

Inflorescences racemose, terminal or lateral from upper nodes; one to many flowered. Flowers generally small to large. (Fig. 2 shows a typical flower of *D. kingianum.*) *Sepals* almost equal, short and broad to long, almost filiform. Lateral sepals joined to the column foot at their bases to form a distinct mentum. (In the two Norfolk Island forms of *D. macropus*, the column foot is absent.) *Petals* mostly narrower and thinner than sepals and frequently shorter. (Broader in *Phalaenanthe* section and, if *D. johnsoniae* is accepted as an Australian native, this is also broader.)

Fig. 1 Distribution of the genus *Dendrobium* in Australia.

Fig. 2 Most common of the Dendrobes, *D. kingianum* showing parts of the flower

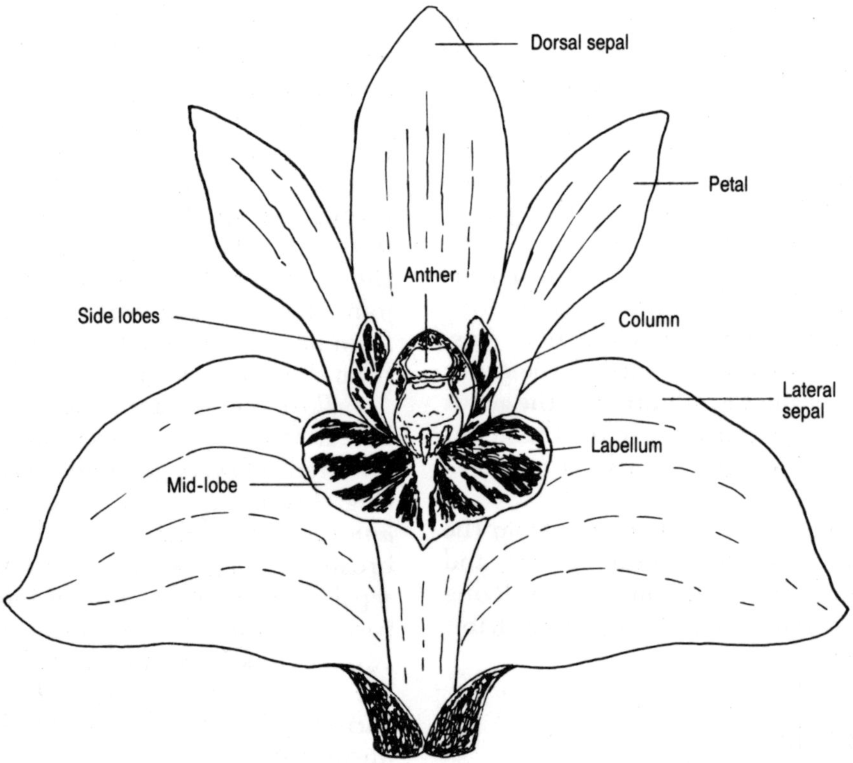

Dendrobium kingianum Bidw.

Fig. 3 shows some variants in the flowers. A, B, C, D, E & F shows the different sepal and petal shapes together with the different size and shape of the mentum. G, H, I & J shows the column and labellum and some of the various methods by which the sepals and labellum are joined to the column foot. *Labellum*, Fig. 4 shows some variants in labella. It will be seen that it is more or less three lobed, (*D. cancroides* Fig. 4J and *D. smilliae* Fig. 4D indistinctly and, *D. stuartii* Fig. 4E is not three lobed) joined to column foot at base (Fig. 3H) and occasionally to part of sides, thus forming a spur (Fig. 5C); lamina usually with longitudinal keels or ridges (callus), very distinct in Fig. 4A, E, G, H, J, K & M, or with a disc or plate as in Fig. 4B, C, I & L. It will be noticed in Fig. 4D *D. smilliae* that the ridge is across the labellum instead of parallel with it. In the Norfolk Island forms of *D. macropus* the labellum is indistinguishable from the other petals. (peloric)

Column, Fig. 5 shows some variants in the columns. Column with a distinct foot basally, which is often quite long although

Fig. 3 Some variants in the flowers of the genus *Dendrobium* in Australia

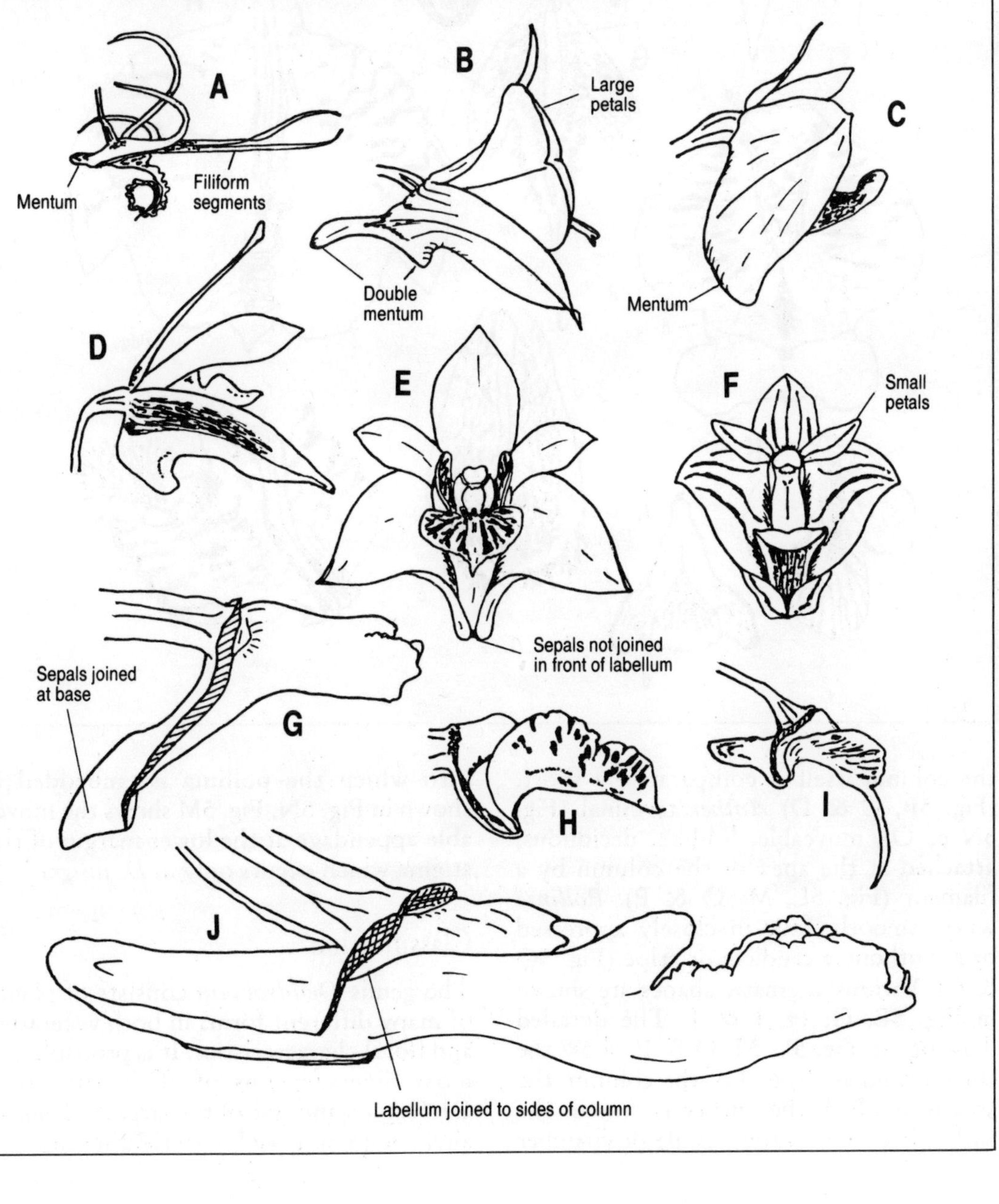

A *D. teretifolium*
B *D. bigibbum*
C *D. malbrownii*
D *D. adae*
E *D. kingianum*
F *D. lichenastrum*
G Column of *D. antennatum*
H Column and labellum of *D. speciosum*
I Column and labellum of *D. affine*
J Column and labellum of *D. lobbii*

Fig. 4 Some variants in the labella of species of the genus *Dendrobium* in Australia

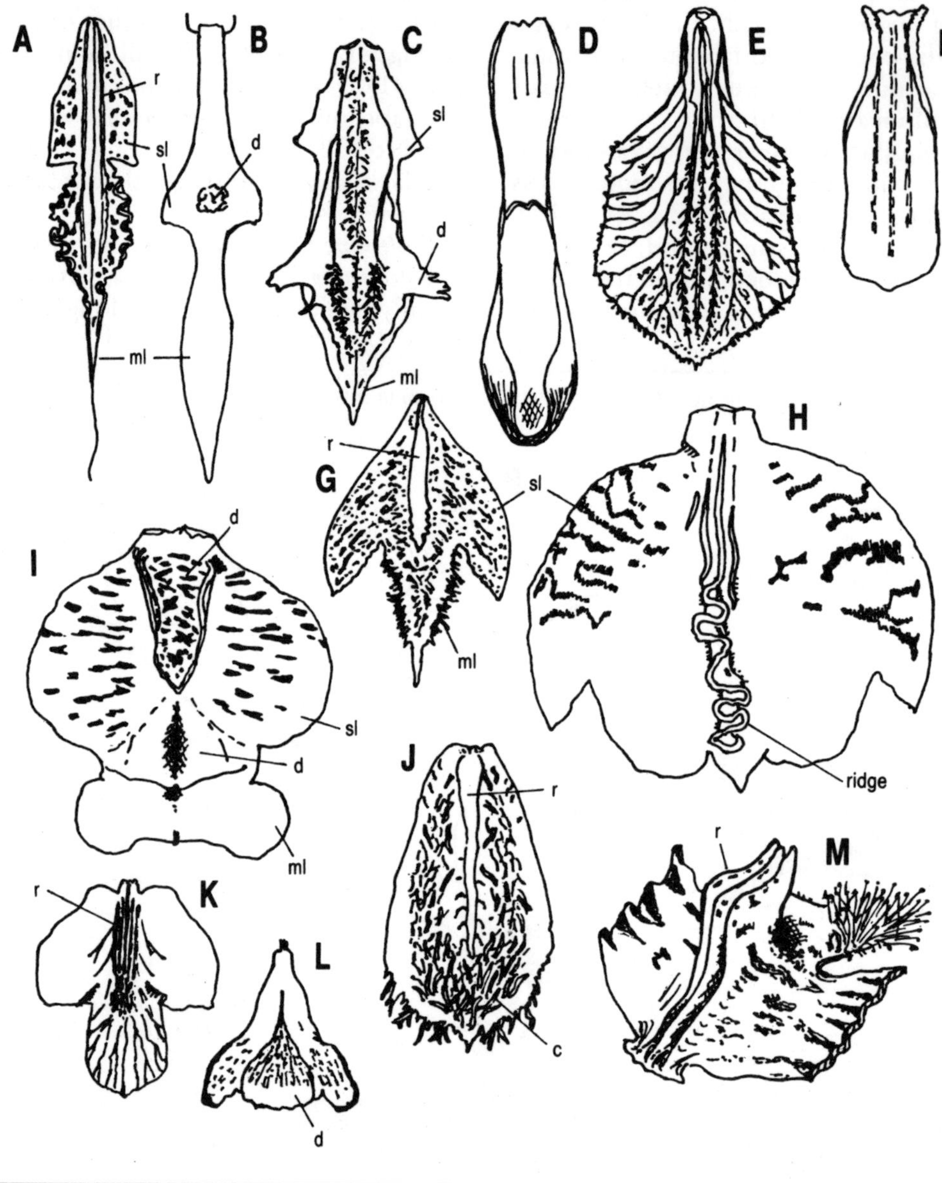

A *D. teretifolium.* Three ridges, frilly and large mid-lobe.

B *D. moorei.* Small square disc.

C *D. insigne.* Outer ridges of disc, extensive, small side lobes.

D *D. smillieae.* Indistinct side lobes.

E *D. stuartii.* No side lobes, three ridges.

F *D. lichenastrum.* No noticeable side lobes, three ridges.

G *D. baileyi.* Single ridge, mid-lobe ciliate.

H *D. aemulum.* Three ridges, centre one very crenulate.

I *D. agrostophyllum.* Two large discs.

J *D. cancroides.* Single ridge, indistinctly trilobate, ciliate.

K *D. bigibbum.* Five ridges.

L *D. schneiderae.* Large triangular disc at front.

M Basal portion of labellum of *D. fleckeri.* Three raised tooth-like ridges.

sl = side lobe
r = ridge
ml = mid-lobe
c = cilia
d = disc

the column itself is comparatively short. (Fig. 5B, C & D) *Anther* terminal (Fig. 5N & O), moveable, lid-like, deciduous, attached at the apex of the column by a filament (Fig. 5L, M, O & P). *Pollinia* waxy, smooth, four in closely appressed pairs without a caudicle or stipe (Fig. 5Q & R). Various stigmatic shapes are shown in Fig. 5E, G, H, I & J. The detailed drawings in Fig. 5L, M, O & P, show the clinandrium at the end of the column, the area into which the anther is seated, the anther bed. The anther locule or chamber into which the pollinia is embedded is shown in Fig. 5N. Fig. 5M shows the moveable appendage at the lower margin of the stigma which occurs only in *D. insigne.*

Classification

The genus *Dendrobium* consists of plants of many different forms in both vegetative and floral characteristics. It is probably the most diverse genus of all in the order *Orchidales*, and one of the largest numerically, comprising well over 1000 species.

Fig. 5 Some variants in the columns of species of the genus *Dendrobium* in Australia

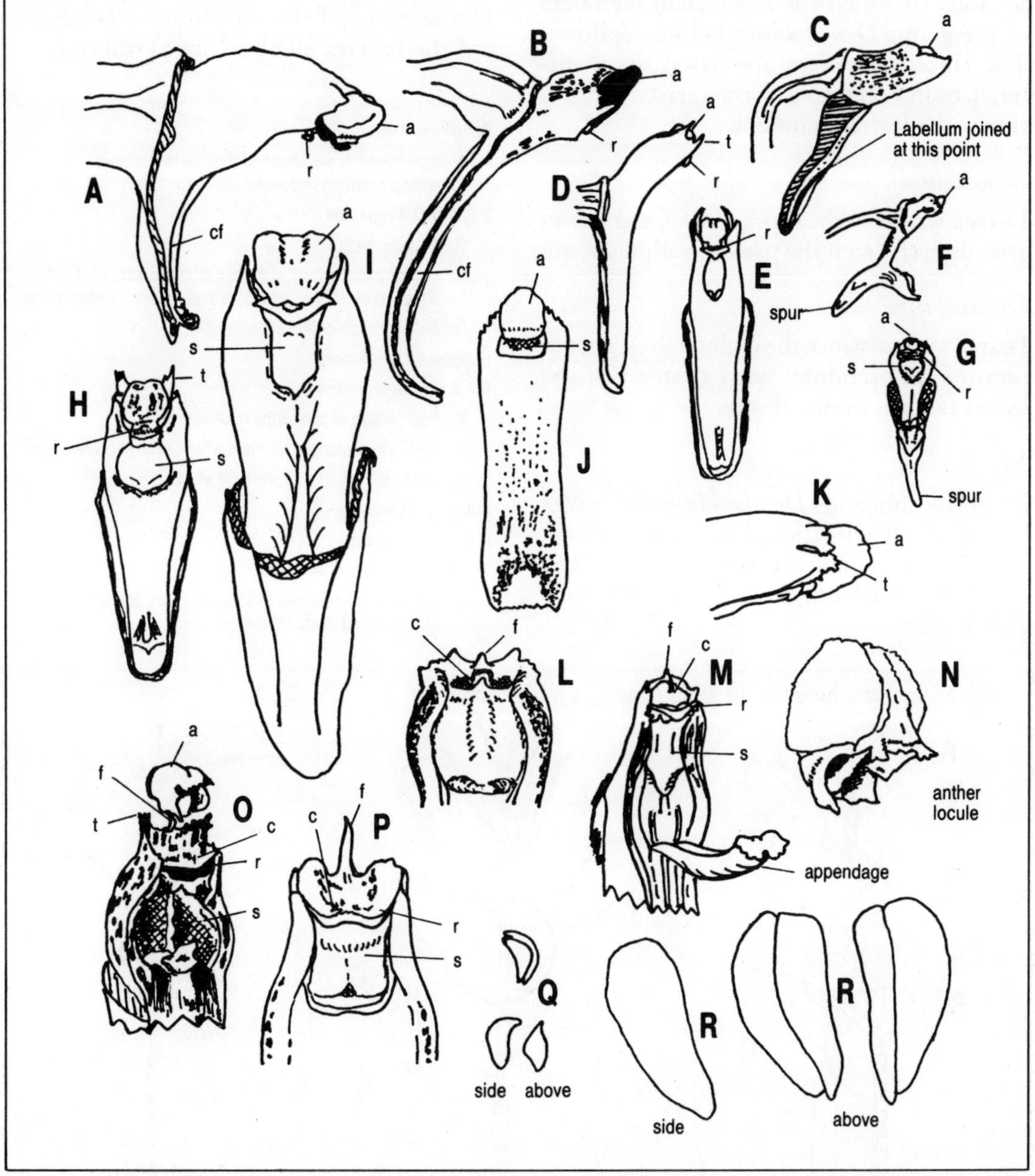

A *D. luteocilium*
B *D. malbrownii*
C *D. discolor*. Labellum joined to base of column foot to form a nectary
D & E *D. striolatum*
F & G *D. bigibbum*
H *D. pugioniforme*
I *D. lobbii*
J *D. toressae*
K *D. cancroides*
L *D. mirbelianum*
M *D. insigne* showing appendage from base of stigma
N Anther cap of *D. stuartii*
O *D. baileyi*
P *D. rigidum*
Q Pollinia of *D. toressae*
R Pollinia of *D. bigibbum*

a = anther
r = rostellum
s = stigma
t = column teeth
f = filament
c = clinandrium
cf = column foot

With such diversity and large numbers, this genus has proved a great problem to the taxonomist and many reviews have been undertaken. The most recent one by Dr F. G. Brieger has been subject to much criticism and is unacceptable to a large number of leading botanists. This revision, as it applies to the genus *Dendrobium* in Australia, is shown in Appendix 4.

Schlechter's (1982a) treatment, in the main, is used here. He divided the genus into four subgenera and each of these subgenera into sections giving a total of forty-one sections in all; since then four of Schlechter's original sections have been raised to generic rank. In Australia there are 54 species and seven named natural hybrids. In Norfolk Island there is one species of two forms (they are both peloric forms). On Lord Howe Island there is one species plus one subspecies of a Norfolk Island species. In New Zealand there is one species. No representatives of the subgenus *Rhopalobium* are found in the area and many of the sections are also not represented.

A brief description of the subgenera, and sections to which the Australian members of the genus *Dendrobium* belong, follows. The three subgenera are based upon differences in vegetative characteristics. Fig. 6 shows the leaf attachment.

Athecebium

Leaves without sheathing bases, the leaves arise directly from the pseudobulbous stem.

Dendrobium

Leaves with distinct sheathing bases, sheath remains on pseudobulbous stem when the leaves fall off. Stems fleshy.

Xerobium

As in the subgenus *Dendrobium* the leaves have distinct sheathing bases that remain on the stem when the leaves fall off; the tufted stems are not fleshy but tough, wiry and slender.

Athecebium

The sections of the subgenus, *Athecebium*, and the species involved are as follows.

Subgenus *Athecebium*

A *D. linguiforme* Leaf attached directly to rhizome. Sheathing bract on rhizome.

B C & D *D. kingianum*

- **B** Aerial growth.
- **C** Part of aerial growth enlarged to show leaf join. Leaf attached directly to pseudobulbous stem, no sheathing.
- **D** Basal portion of leaf showing join.

Subgenus *Dendrobium*

E F & G *D. antennatum*

- **E** Top portion of pseudobulbous stem.
- **F** Part of pseudobulbous stem enlarged to show leaf join.
- **G** Basal portion of leaf, showing sheathing base.

Subgenus *Xerobium*

H & I *D. malbrownii*

- **H** Top portion of wiry stem.
- **I** Part of stem enlarged to show leaf join. Sheathing base of leaf pulled away from stem.

Fig. 6 Leaf attachment in the genus *Dendrobium* in Australia

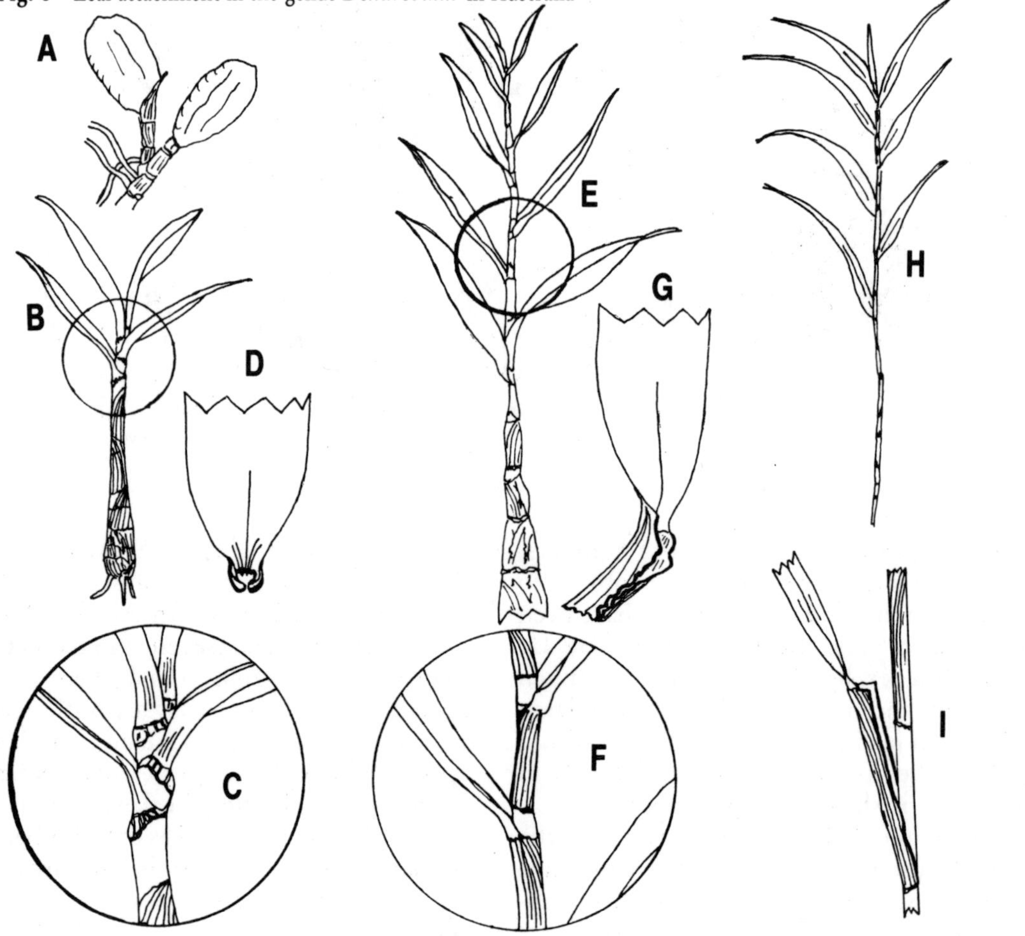

Section 1 ***Rhizobium***

Plants in this section have thick, fleshy, terete, leathery leaves. Pseudobulbous stems are hard to distinguish. I find it difficult to accept Dockrill's treatment in his *Australian Indigenous Orchids* of dividing the fleshy-leaved dendrobiums of Australia into three sections. I prefer to have them all within the one section but split into three separate groups, even then the boundaries are hard to distinguish clearly and definitely. Based on my examination of a large number of plants of *D. lichenastrum* and *D. toressae*, I do not think they fit here at all satisfactorily. Certainly the way the capsule of *D. lichenastrum* opens is different from other species of the genus *Dendrobium* in Australia.

Although this may sound contradictory, if I were to accept any one of Brieger's new genera, it would be *Dockrillia*, as the species under discussion are so distinctly different from the other Australian species of *Dendrobium*. But it would be necessary to examine members occurring in Papua New Guinea or elsewhere, before coming to such a conclusion.

Group one

Those with oval, oblong or furrowed semi-terete leaves: (Dockrill's Section 1) *D. cucumerinum*, *D. linguiforme*, *D. wassellii*.

Group two

Those with elongate, terete or sub-terete to ovate-acuminate leaves (Dockrill's Section 3): *D. bowmanii*, *D. calamiforme*, *D. fairfaxii*, *D.* sp. aff. *fairfaxi*, *D. mortii*, *D. pugioniforme*, *D. racemosum*, *D. rigidum*, *D. schoeninum*, *D. striolatum*, *D. teretifolium*.

Group three

Those with prostrate leaves that are oblong, ovate to almost circular and terete. Smaller in all parts than Group One or Group Two (Dockrill's Section 2): *D. lichenastrum*, *D. prenticei*, *D. toressae*.

Section 2 ***Monophyllaea***

Short pseudobulbous stems of one to few internodes bearing one to two thin leaves: *D. carrii*, *D. monophyllum*, *D. schneiderae*.

Section 3 ***Dendrocoryne***

Pseudobulbous stems in tufts and with several nodes (except some short cane forms of *D. kingianum*). Two to several leaves near the apex; labellum not generally held close to the column; keels of the disc not usually high (*D. fleckeri* and *D. falcorostrum* are exceptions): *D. adae*, *D. aemulum*, *D. falcorostrum*, *D. fleckeri*, *D. jonesii*, *D.* sp. aff. *jonesii*, *D. kingianum*, *D. macropus* (Norfolk Island), *D. macropus* subsp. *gracilicaule*, *D. macropus* subsp. *howeanum* (Lord Howe Island), *D. moorei* (Lord Howe Island), *D. speciosum*, *D. tetragonum*.

Section 4 ***Latouria***

Plant similar to section *Dendrocoryne*. Labellum held close to the column; keels of disc quite high; petals of *D. johnsoniae* much larger than sepals; *D. bifalce*, *D. johnsoniae*.

Dendrobium

The sections of the subgenus *Dendrobium* and the species involved are as follows.

Section 5 ***Phalaenanthe***

Flowers with a double mentum: *D. affine*, *D. bigibbum*.

Section 6 ***Eleutheroglossum***

Flower with incurved conical mentum: *D. fellowsii*.

Section 7 ***Spatulata*** **(Syn.** ***Ceratobium*****)**

Plant similar to section *Phalaenanthe*. Petals longer and narrower than sepals and mostly twisted: *D. antennatum*, *D. carronii*, *D. canaliculatum*, *D. discolor*, *D. johannis*, *D. mirbelianum*, *D. nindii*, *D. trilamellatum*.

Section 8 ***Dendrobium***

Stems fleshy, inflorescence occurs on defoliated stem; mentum short; lip ciliate: *D. stuartii*.

Section 9 ***Trachyrhizum***

Stems slender and leafy; inflorescence usually short and from nodes along the stem; labellum with reasonably large side and mid-lobes and quite complex: *D. agrostophyllum*.

Section 10 *Pedilonum*

Inflorescence short and compact on defoliated stems; labellum partly joined to column foot to form a spur: *D. smillieae.*

Xerobium

The sections of the subgenus *Xerobium* and the species involved are as follows:

Section 11 *Grastidium*

Inflorescence usually two-flowered, lasting less than a day: *D. baileyi*, *D. cancroides*, *D. luteocilium*, *D. tozerensis*.

Section 12 *Dichopus*

Habit similar to section *Grastidium*. Unusual moveable appendage on the lower margin of the stigma; flowers lasting 1 day: *D. insigne*.

Section 13 *Monanthos*

Inflorescences one-flowered, lasting longer than 1 day; flowers mostly inverted: *D. malbrownii*.

Section 14 *Conostalix*

Habit similar to section *Grastidium*. Inflorescences one- to two-flowered; base of labellum joined to the sides of the column foot to form a spur: *D. lobbii*.

The Genus *Dendrobium* in Australia

Leaves without sheathing bases	Leaves with distinct sheathing bases: Stems fleshy	Leaves with distinct sheathing bases: Stems not fleshy but wiry
Subgenus—ATHECEBIUM	Subgenus—DENBROBIUM	Subgenus—XEROBIUM
Section 1 Rhizobium Group One: *cucumerinum, linguiforme, wassellii* Group Two: *bowmanii, calamiforme, fairfaxii,* sp. aff. *fairfaxii, mortii, pugioniforme, racemosum, rigidum, schoeninum, striolatum, teretifolium* Group Three: *lichenastrum, prenticei, toressae*	**Section 5 Phalaenanthe** *affine, bigibbum*	**Section 11 Grastidium** *baileyi, cancroides, luteocilium, tozerensis*
Section 2 Monophyllaea *carrii, monophyllum, schneiderae*	**Section 6 Eleutheroglossum** *fellowsii*	**Section 12 Dichopus** *insigne*
Section 3 Dendrocoryne *adae, aemulum, falcorostrum, fleckeri, jonesii,* sp. aff. *jonesii, kingianum, macropus, macropus* subsp. *gracilicaule, macropus* subsp. *howeanum, moorei, speciosum, tetragonum*	**Section 7 Spatulata** *antennatum, carronii, canaliculatum, discolor, johannis, mirbelianum, nindii, trilamellatum*	**Section 13 Monanthos** *malbrownii*
Section 4 Latouria *bifalce, johnsoniae*	**Section 8 Dendrobium** *stuartii*	**Section 14 Conostalix** *lobbii*
	Section 9 Trachyrhizum *agrostophyllum*	
	Section 10 Pedilonum *smillieae*	

2

History of the genus *Dendrobium* in Australia

If you look at the history of the botanical collection of orchids in Australia, it becomes obvious that species of the genus *Dendrobium* were not among the first recorded collections. But, when considering the early exploration of the area, who knows? They may still have been the first orchids seen by a European. One thing is certain: they were used by the Aborigines for many purposes and for hundreds of years before any European set foot in Australia.

Who then was the first European to see an Australian *Dendrobium*? In the 16th century, trading between Portugal and the East Indies and Timor was established—surely some of the Portugese sailors made the relatively easy crossing of the Timor Sea to the north coast of Australia. Willem Janszen, captain of the Dutch East India Company vessel *Duyfken* is stated to be the first European to land on Australian soil; this was on the western side of Cape York Peninsula—the date sometime around March 1606.

About 9 December 1606, Luis de Torres sailed through Torres Strait and probably sighted the north of Cape York. In October 1616, Dirk Hartog, captain of the *Eendracht*, landed on the west coast of Australia (no dendrobes would be seen on that landfall). In 1623, vessels from Amboina visited the coast of the Gulf of Carpentaria. In 1642, Abel Tasman sailed well to the west of Australia and then south-east to the west coast of Tasmania (no dendrobes would be seen on that west coast) and again, in 1644, he sailed around the Gulf of Carpentaria looking for a southern sea passage.

In 1688, the pirate ship *Cygnet* visited the north-west coast of Australia; this is the first recorded British contact. William Dampier was a crew member, and described the landing in his book *A New Voyage Around the World* published in 1697. Dampier returned to Australia in command of H.M.S. Roebuck in 1699, sailing up the coast of Western Australia.

After this, very little exploration near the areas where dendrobes grow took place until 1770. Captain James Cook, on the 20 April 1770, became the first European to reach the east coast, and it was then that the first real scientific botanizing was carried out in Australia. Sir Joseph Banks and Dr Daniel Solander accompanied Captain Cook when they landed at Botany Bay. Charles Moore (1893) in his *Handbook of the Flora of New South Wales*, states 'within an area of a few miles, these naturalists, during a week's stay, collected a considerable herbarium of new plants, which were described and named on their return to England, and formed the nucleus of the Banksian Herbarium in the British Museum'. No dendrobiums were included.

Leaving Botany Bay, Captain Cook travelled northward and, on the 18 June 1770, after striking a reef, he beached his ship the *Endeavour* near the mouth of a river, which he named the Endeavour River. The area became a port and was named Cooktown. While repairs were taking place, about 6 weeks, Banks and Solander spent many hours searching the surrounding areas for plants, and it was here that the first official record of collection of an Australian *Dendrobium* was made:

D. discolor was collected near the Endeavour River by Banks.
D. canaliculatum was collected by Banks and Solander on the Endeavour River.
D. rigidum was collected on the Endeavour River by Solander.

It would seem therefore that one of these was the first *Dendrobium* collected by a European.

They were not, however, the first Australian dendrobiums to be named, that honour goes to *Dendrobium linguiforme*. The Swedish botanist, Olaf Swartz, founded the genus *Dendrobium* in 1799 when he described 19 species, and in 1800 he published *D. linguiforme*.

The second species to be named was *Dendrobium speciosum* by Sir James E. Smith in his *Exotic Botany* in 1804; on the type sheet is written 'Port Jackson, New South Wales'. It was apparently collected by Surgeon-General John White.

Robert Brown named five Australian dendrobiums in his *Prodromus Florae Novae Hollandiae et Insulae Van-Diemen* published in 1810; three of these were collected by Banks and Solander in 1770. (The extensive herbarium collected together by Banks was divided for identification between Banks and Brown.) The other two, namely *D. aemulum* and *D. teretifolium*, were collected by Robert Brown during his stay in Australia between 1801 and 1805. Brown was naturalist of the expedition that charted and circumnavigated the coast of Australia, under Captain Matthew Flinders, and was assisted by Ferdinand Bauer, a distinguished botanical draughtsman, and by Peter Good, a very intelligent and practical gardener.

Allan Cunningham, a Kew botanist, came to Australia in the capacity of King's botanist in 1816 and as a collector for Kew. He spent much time as a botanist on expeditions of the interior under John Oxley and subsequently with Captain (later Rear-Admiral) King exploring the western and northern areas of Australia. He returned to England in 1832 but came back as Colonial Botanist in 1837. In 1839 he named *D. pugioniforme* and *D. tetragonum*, plants he had collected earlier.

Some naming of Australian dendrobes was carried out by such distinguished botanists as John Lindley and Professor Heinrich G. Reichenbach who had never seen Australia and depended on specimens sent to them by collectors. *Dendrobium kingianum* was first described by John Lindley, but John Carne Bidwill suggested the name, in Edwards' *Botanical Register* in 1844. Bidwill, a botanist, came to Australia in September 1838 and did some botanizing while waiting for land on which to settle. In 1847 he was appointed Government Botanist and Director of the Sydney Botanic Gardens.

William Sharp Macleay, who named *D. cucumerinum*, was a naturalist particularly interested in entomology and marine fauna. He was a fellow of the Royal Society and Trustee of the Australian Museum, the eldest son of Alexander Macleay, Colonial Secretary of NSW.

German-born Baron Sir Ferdinand Jakob Heinrich Mueller named ten Australian dendrobes. He arrived in Australia in 1847, aged 22 years. In 1852 he became Government Botanist for Victoria and held this position until his death in 1896. Walter Hill, who named *D. nindii* in 1874, was Queensland Colonial Botanist and the Director and founder of the Brisbane Botanic Gardens.

Although Robert David Fitzgerald named few dendrobiums (*D. falcorostrum*, one variety and one form of *D. discolor* and one form of *D. teretifolium*), his contribution to Australian orchidology was tremendous. His splendid work *Australian Orchids*, in which 17 dendrobiums are depicted, is a lasting monument of his great ability. Fitzgerald was Deputy Surveyor General in Sydney.

Towards the end of the 19th century, Frederick Manson Bailey named six *Dendrobium* species, the natural hybrid *D. x delicatum* and numerous varieties. Bailey was Colonial Botanist of Queensland and the author of the six volume *Flora of Queensland*.

After Bailey named his last *Dendrobium* in 1890 (actually he named it as a *Bulbo-*

phyllum and it was later moved to the genus *Dendrobium* by Dockrill), there was a gap of 37 years before further plants were named. In 1927 Rev. H. M. R. Rupp named *D. tenuissimum*, although this species later proved to be a synonym of *D. mortii*.

The list below gives the species of *Dendrobium* occurring in Australia by the year they were named or by the year they were found in Australia if they had been found elsewhere and named previously. *Dendrobium* species are still being found. It was only in 1982 that Lavarack and Cribb named *D. carronii*, although in this particular case the orchid had been known for quite some time.

In recent years Dr P. S. Lavarack, Senior Botanist of the National Parks and Wildlife Service of Queensland, has led several expeditions into various areas of the Cape York Peninsula. These have been extremely fruitful, not only because several new orchids have been discovered, including two new dendrobiums, but also for their contribution in the advancement of knowledge of the orchids occurring in Australia. The expeditions have been jointly funded by the Australian Orchid Foundation and the Queensland National Parks and Wildlife Service.

Alick W. Dockrill, the author of *Australian Indigenous Orchids*, can be said to be the greatest authority on Australian orchids alive today. He has named only one *Dendrobium* species, but he has done much work within the genus and his researching has made this complex group of plants more readily understandable.

Naming of the species of *Dendrobium* occurring in Australia

Date named	Name of orchid	Originally named by	
1800	*D. linguiforme*	O. Swartz	Possibly collected at Port Jackson
1804	*D. speciosum*	J. E. Smith	Collected at Port Jackson—possibly by Surgeon-General White
1810	*D. canaliculatum*	R. Brown	Collected by Banks and Solander near the Endeavour River in north Queensland in 1770
1810	*D. rigidum*	R. Brown	Collected by Solander near the Endeavour River 1770
1810	*D. discolor* (as *D. undulatum*)	R. Brown	Collected by Banks near the Endeavour River 1770
1810	*D. aemulum*	R. Brown	Collected by Brown 1801 to 1805
1810	*D. teretifolium*	R. Brown	Collected by Brown at Port Jackson 1801 to 1805
1839	*D. pugioniforme*	A. Cunningham	Found by Cunningham in the Illawarra district of NSW
1839	*D. tetragonum*	A. Cunningham	Found by Cunningham about 1820 near Moreton Bay
1841	*D. calamiforme*	C. Loddiges	—
1842	*D. cucumerinum*	W. S. Macleay ex Lindley	Found by Macleay at Brownlow Hill, near Camden, NSW
1845	*D. kingianum*	J. C. Bidwill	Found by Bidwill about 1839 on the Buckets, near Gloucester, NSW
1846	*D. schoeninum*	J. Lindley	Described from a plant imported by Loddiges
1852	*D. bigibbum*	J. Lindley	Collected from Mt Adolphus Island in Torres Strait by Dr Thomas. In cultivation at Kew since 1824
1855	*D. insigne*	H. G. Reichenbach (Blume)	First collected for Australia by Geoff Stocker on Saibai Island in 1974

Date named	Name of orchid	Originally named by	
1857	*D. striolatum*	H. G. Reichenbach	Described from a specimen grown by F. Strange in the greenhouses of Consul Schiller
1859	*D. macropus* subsp. *gracilicaule* (as *D. gracilicaule*)	F. Mueller	Collected by Hill at Moreton Bay, Queensland
1859	*D. monophyllum*	F. Mueller	Found by Hill in the woods at Moreton Bay, Queensland
1859	*D. mortii*	F. Mueller	Collected by Dr Beckler at the Hastings River, NSW
1865	*D. johannis*	H. G. Reichenbach	Collected in north Qld by John G. Veitch in 1863
1867	*D. smillieae*	F. Mueller	Found at Rockingham Bay, north Qld by J. Dallachy
1869	*D. lichenastrum* (as '*Bolbophyllum*')	F. Mueller	Collected by J. Dallachy near Townsville in 1869
1869	*D. moorei*	F. Mueller	Found by Fitzgerald on Lord Howe Island in 1869
1870	*D. fellowsii*	F. Mueller	Collected by Dallachy at Rockingham Bay, prior to 1870
1872	*D. fairfaxii*	F. Mueller and R. D. Fitzgerald	Collected at Mt Tomah by Fitzgerald in 1872
1873	*D. agrostophyllum*	F. Mueller	Collected by J. Dallachy near Townsville in 1873
1873	*D. bowmanii*	G. Bentham	From near Rockhampton
1873	*D. affine* (as *D. dicuphum*)	F. Mueller	Possibly by Mueller in 1855 to 56 in Northern Territory
1874	*D. baileyi*	F. Mueller	Found by Bailey in the mountains behind Rockingham Bay, north Qld
1874	*D. nindii*	W. Hill	Found by P. H. Nind near Innisfail in 1873
1876	*D. x superbiens*	H. G. Reichenbach	Sent to England by William MacArthur in 1876
1876	*D. falcorostrum*	R. D. Fitzgerald	Found by Fitzgerald on Mount Banda Banda near Macleay River
1878	*D. jonesii* (as *D. fusiforme*)	F. M. Bailey	Found by Bailey at Cardwell, Queensland
1881	*D. prenticei* (as *Bulbophyllum prenticei*)	F. Mueller	—
1883	*D. trilamellatum* (as *D. johannis* var. *semifuscum*)	J. J. Smith	Described as *D. johannis* var. *semifuscum* from plants cultivated by W. Bull
1884	*D. adae*	F. M. Bailey	Collected at Herberton
1884	*D. stuartii*	F. M. Bailey	Described from a plant sent to Brisbane Botanic Gardens by J. W. R. Stuart from near Herberton, north Qld before 1884
1884	*D. x delicatum*	F. M. Bailey	Collected by Benjamen Crow at Spring Bluff, near Toowoomba in 1884
1886	*D. fellowsii* (as *D. bairdianum*)	F. M. Bailey	Found by J. W. R. Stuart between Herberton and Mourilyan Harbour, north Qld

Date named	Name of orchid	Originally named by	
1886	*D. schneiderae*	F. M. Bailey	Collected by H. Schneider in the Darlington Range SE Queensland
1890	*D. toressae* (as *Bulbophyllum toressae*)	F. M. Bailey	Found in June 1889 by Archibald Meston and Bailey on Bellinden Ker Range
1898	*D. bifalce*	F. M. Bailey	As *D. breviracemosum*
1927	*D. mortii* (as *D. tenuissimum*)	H. M. R. Rupp	Found in 1925 by John Hobson and Rupp at Upper Allyn River, NSW
1934	*D. x grimesii*	C. T. White and V. S. Summerhayes	Found by Ken Wadsworth on the Atherton Tableland
1936	*D. racemosum*	W. H. Nicholls	Sent to Nicholls by H. G. Henson of Cairns, Qld
1937	*D. carrii*	H. M. R. Rupp and C. T. White	Found by White and T. Carr in September 1936 on Mount Spurgeon, north Qld
1937	*D. fleckeri*	H. M. R. Rupp and C. T. White	Found by Dr H. Flecker on upper Mossman River near Mount Spurgeon in 1935
1941	*D. mirbelianum* (as *D. wilkianum*)	C. Gaudichaud-Beaupre	*D. wilkianum* found by J. H. Wilkie at Babinda, north Qld in October 1941. Originally found by G. Bates and B. Kerns in 1934
1945	D. bifalce	J. Lindley	Collected by R. L. Hunter in the Portlands Roads area of north Qld
1945	D. x *gracillimum*	H. M. R. Rupp	Found by Mr and Mrs John Tucker in 1925 in forests near Paterson, NSW
1945	*D. luteocilium*	H. M. R. Rupp	Found by J. H. Wilkie at Babinda, north Qld in 1945
1947	*D. cancroides*	T. E. Hunt	Found by J. H. Wilkie near Babinda
1948	*D. lobbii*	J. E. Teijsmann and S. Binnendijk	Collected by Dr L. J. Brass in Iron Range area in 1948
1955	*D. antennatum*	J. Lindley	Found in the Massey Scrubb, near Coen, Qld by J. L. H. Wassell in 1955
1955	*D. x foederatum*	S. F. Goesling-St Cloud	Collected by J. Dyson-Holland at Aeroglen near Cairns in September 1954
1956	*D. x vinicolor*	S. F. Goesling-St Cloud	Collected by N. Whittaker, 40 km north of Cairns in August 1955
1963	*D. wassellii*	S. T. Blake	Collected by J. L. H. Wassell prior to 1959
1964	*D. x suffusum*	L. Cady	Collected by Alan Willows from North Brother Mountain, Laurieton, NSW, in September 1961
1967	*D. malbrownii*	A. W. Dockrill	Found by Malcolm Brown in the McIlwraith Range in March 1966
1977	*D. tozerensis*	P. S. Lavarack	Collected at Tozer's Gap by Lavarack and Gray in 1975
1982	*D. carronii*	P. S. Lavarack and P. J. Cribb	Previously known as the 'pink form' of *D. canaliculatum*

3

Descriptions of *Dendrobium* species in Australia

Dendrobium adae F. M. Bail.

Proc. Roy. Soc. Q/ld. 1:149 (1884)
Fig. 7, Plate 6

SYNONYMS

Callista adae (F. M. Bail.) Kuntze, *Rev. Gen. Pl.* 2:654 (1891)
Dendrobium palmerstoniae Schltr., *Repert. Spec. Nov. Regn. Veg. Beih.* 3:317 (1907)
Dendrobium ancorarium Rupp, *Q/ld. Nat.* 12:114, pl.6 (1945)
Tropilis adae (F. M. Bail.) Rauschert, *Feddes Repert.* 94:470 (1983)

DESCRIPTION

Originally collected at Herberton, north Queensland. Named in honour of Mrs Ada Stuart, the wife of J. W. R. Stuart after whom *D. stuartii* was named.

Pseudobulbous stems, up to 40 cm long, arise from a short creeping rhizome. Slender and furrowed, with two to four lanceolate leaves at the apex. Plant similar to *D. fleckeri* but stems generally longer and more robust.

Flowers about 2 to 2·5 cm in diameter, expand widely and are beautifully perfumed (similar to orange blossom) on warm days, but at night and on cloudy days remain half closed and scentless. Up to nine flowers (most frequently about three to four in cultivation) borne on a short raceme; vary considerably in colour from white to creamy-white to greenish-white through pale green to pale yellow and almost apricot. Sometimes petals are white and sepals pale yellow to pale green, with some forms pinky-brown on the rear of the sepals. Flowers turn to apricot as they wither. Labellum white to green, with faint reddish markings. Mid-lobe densely tomentose, but the cilia very short. Anterior margins of the mid-lobe folded under giving it a very succulent appearance. Column white to greenish-white and speckled with orange-red dots (almost in the form of an anchor in some clones) on the inside of the column foot. Individual flowers last from 15 to 20 days.

CHROMOSOME NUMBER

2n = 38 (K. Y. Lim, 1985).

FLOWERING PERIOD

Mainly July to October (mid winter to spring).

DISTRIBUTION

North-eastern Queensland, confined to eastern tropics between Annan River (15° 30′S) and Burdekin River (20°S), just south of Townsville.

HABITAT

Highland rainforests of north Queensland, where early morning mists and heavy dews are commonplace. From approximately 760 m to 1220 m. Mostly well above 760 m.

CULTURE

Requires good ventilation (possibly its most essential requirement), protection from dry winds, a sunny situation and high humidity. 50 per cent shade is ample, possibly a little more shade while being established. If well watered, grows happily on a block of hardwood or cork. If grown in pots, medium must be very open (fir-bark-type mix) and plant should dry out between waterings. Does not appear to be cold sensitive and can stand temperatures down to just below 0°C for very short periods.

Easy to grow and is quite common in collections in Sydney, NSW. Free-flowering.

Generally one forward growth is produced each year, but a well-established plant can be backcut to produce a specimen plant. If backcutting, leave three pseudobulbous stems in front of the cut and at least two behind to ensure the whole plant continues to grow. Pseudobulbous stems that no longer flower can be cut off to tidy up the plant.

IN HYBRIDIZATION

Appears to improve the upright habit of the plant and the raceme. Labellum shape dominant. Generally produces clear, clean colours. Progeny frequently perfumed. The few-flower habit is unfortunately dominant in most instances, although not so in *D.* Orana, which regularly carries 14 flowers.

Fig. 7 *Dendrobium adae* F. M. Bail. Collected S. Clemesha, Mt. Lewis, North Queensland, 1964

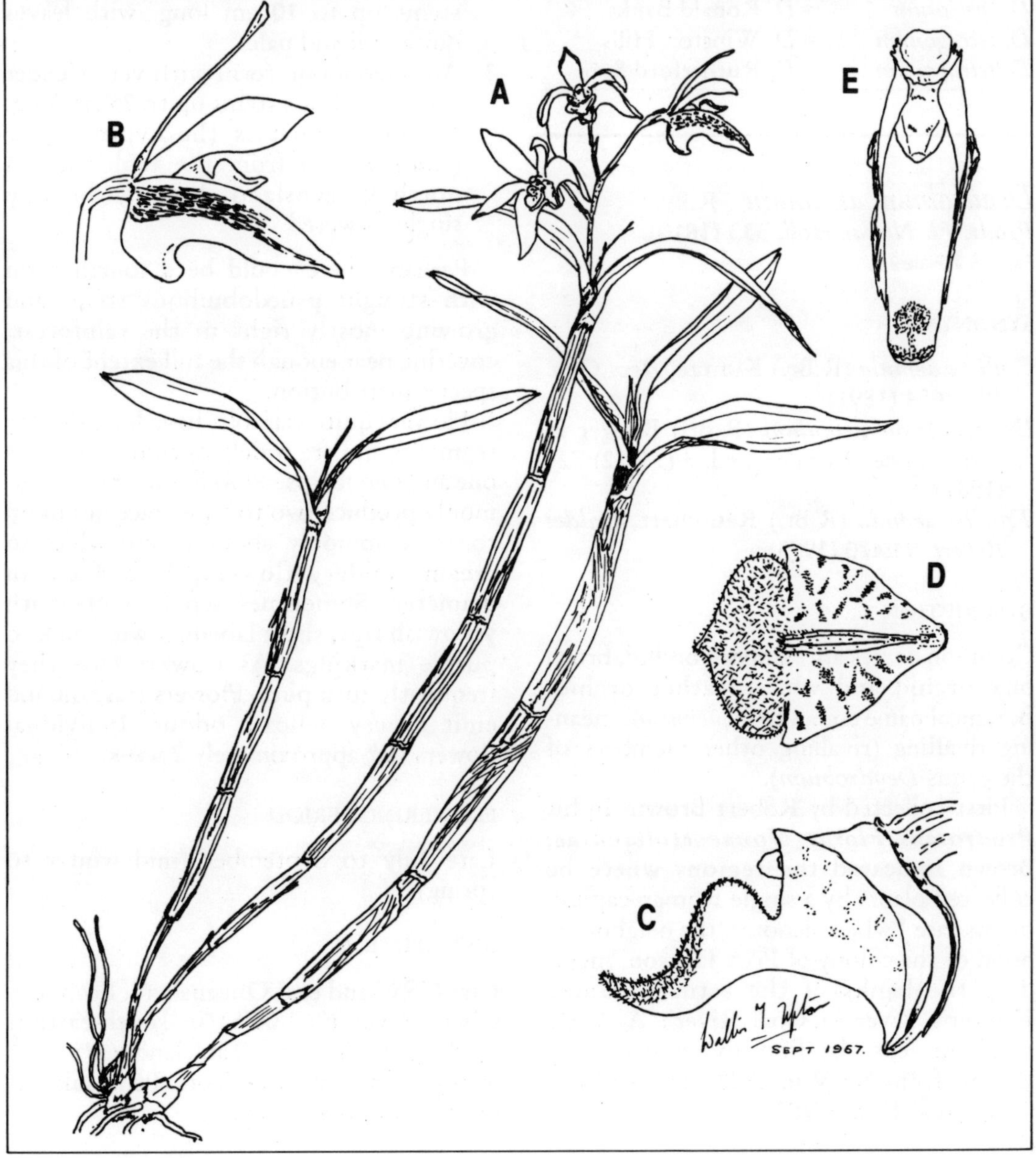

A Plant $^2/_3$X
B Flower, side-view 2X
C Column and labellum, side-view 4X
D Labellum, pressed flat 3X
E Column, front-view 4X

HYBRIDS

D. adae	X *D. x delicatum*	= *D.* Kevin Banks
D. adae	X *D.* Eureka	= *D.* Warooa
D. adae	X *D. falcorostrum*	= *D.* Leonie Jane
D. adae	X *D. macropus* subsp. *gracilicaule* (as *D. gracilicaule*)	= *D.* Gracia
D. adae	X *D. x gracillimum*	= *D.* Double Two
D. adae	X *D. kingianum*	= *D.* Tooloom
D. adae	X *D. monophyllum*	= *D.* Orana
D. adae	X *D.* Nunkumbil	= *D.* Whitehart
D. adae	X *D. speciosum*	= *D.* Yellow Venture
D. adae	X *D.* Susan	= *D.* Green Goddess
D. adae	X *D. tetragonum*	= *D.* Star Imp
D. Double Two	X *D. kingianum*	= *D.* Nathan Wall
D. Double Two	X *D. speciosum*	= *D.* Ronald Banks
D. Kevin Banks	X *D. tetragonum*	= *D.* Winston Hills
D. Star Imp	X *D. tetragonum*	= *D.* Rutherford Star

Dendrobium aemulum R.Br.

Prodr. Fl. Novae Holl. 333 (1810)

Figs 8 & 9, Plate 9

SYNONYMS

Callista aemula (R.Br.) Kuntze, *Rev. Gen. Pl.* 2:654 (1891)

Dendrocoryne aemulum (R.Br.) Brieger in Schltr., *Die Orchideen*, ed. 3 (11/12):724 (1981)

Tropilis aemula (R.Br.) Rauschert, *Feddes Repert.* 94:470 (1983)

DESCRIPTION

Common names are ironbark orchid, brush box orchid and white feather orchid. Botanical name comes from *aemulus*, meaning rivalling (rivalling other members of the genus *Dendrobium*).

First collected by Robert Brown. In his *Prodromus Florae Novae Hollandiae*, Brown indicated the regions where he collected plants by a single Roman capital, in this case J which denotes 'the neighbourhood of the colony of Port Jackson, including the banks of the estuary named Hunter's River or Coal River'. A. Cunningham sent the first living plants to Europe from NSW in 1823 and they first flowered at Kew in 1824.

An extremely variable species both in plant and floral form. Three vegetative forms exist, the first two are more common.

1. Those that chiefly grow on brush box trees (*Lophostemon confertus*), and grow in the rainforests or the more sheltered areas from southern Queensland to northern NSW. Rarely seen as far south as Sydney. Narrow pseudobulbous stems up to 25 cm long, with dark-green, shiny leaves.
2. Those that grow on ironbark eucalypts (*Eucalyptus* spp.) in open forest from southern Queensland to southern NSW. Short, stumpy, crowded pseudobulbous stems up to 10 cm long, with leaves quite dull and pale.
3. An uncommon form with very slender pseudobulbous stems up to 25 cm long. Favourite host is the cypress pine (*Callitris* sp.) from the tablelands of north Queensland. Frequently only single-flowered.

Perhaps there could be a fourth form with straight psuedobulbous stems and growing mostly right in the rainforest, covering near enough the full extent of this species distribution.

Flowers quite variable, have long slender segments and are widely opening. In forms one and two the pseudobulbous stems commonly produce two to three racemes of up to 14 (commonly about seven) white to cream, spidery flowers, 3 to 4 cm in diameter. Sometimes sepals white with yellowish tips; short labellum with pink to purple markings. As flowers fade they frequently turn pink. Flowers fragrant and emit a very delicate odour. Individual flowers last approximately 2 weeks.

FLOWERING PERIOD

Late July to September (mid winter to spring).

DISTRIBUTION

East NSW and east Queensland. From the Clyde River (36° 00′S) in south-eastern NSW to the Atherton Tablelands (17° 00′S) in south-eastern Cape York Peninsula. In both lowlands and ranges.

HABITAT

Given previously when mentioning the three vegetative forms. Generally grows pressed close to the trunk of the tree at varying heights. Not unusual to see many tiny seedlings lying flat against the trunk of the tree from almost ground level.

CULTURE

Grows best on hardwood or cork slabs in a sunny situation. During its active growing period enjoys a daily watering (with caution if grown in pots) and a weekly fertilizing with a product high in nitrogen, an NPK mixture with a 30:10:10 composition should be used (p. 186). After growth is completed, reduce watering, as plant commences its resting period. In winter hardly any water required. In early autumn a fertilizer high in potassium and phosphorus is beneficial for the production of flowers. (Nitrophoska is good.) Free-flowering.

Fig. 8 *Dendrobium aemulum* R.Br. Collected from near Springbrook, Queensland, on 3/9/1966. Brush Box tree form

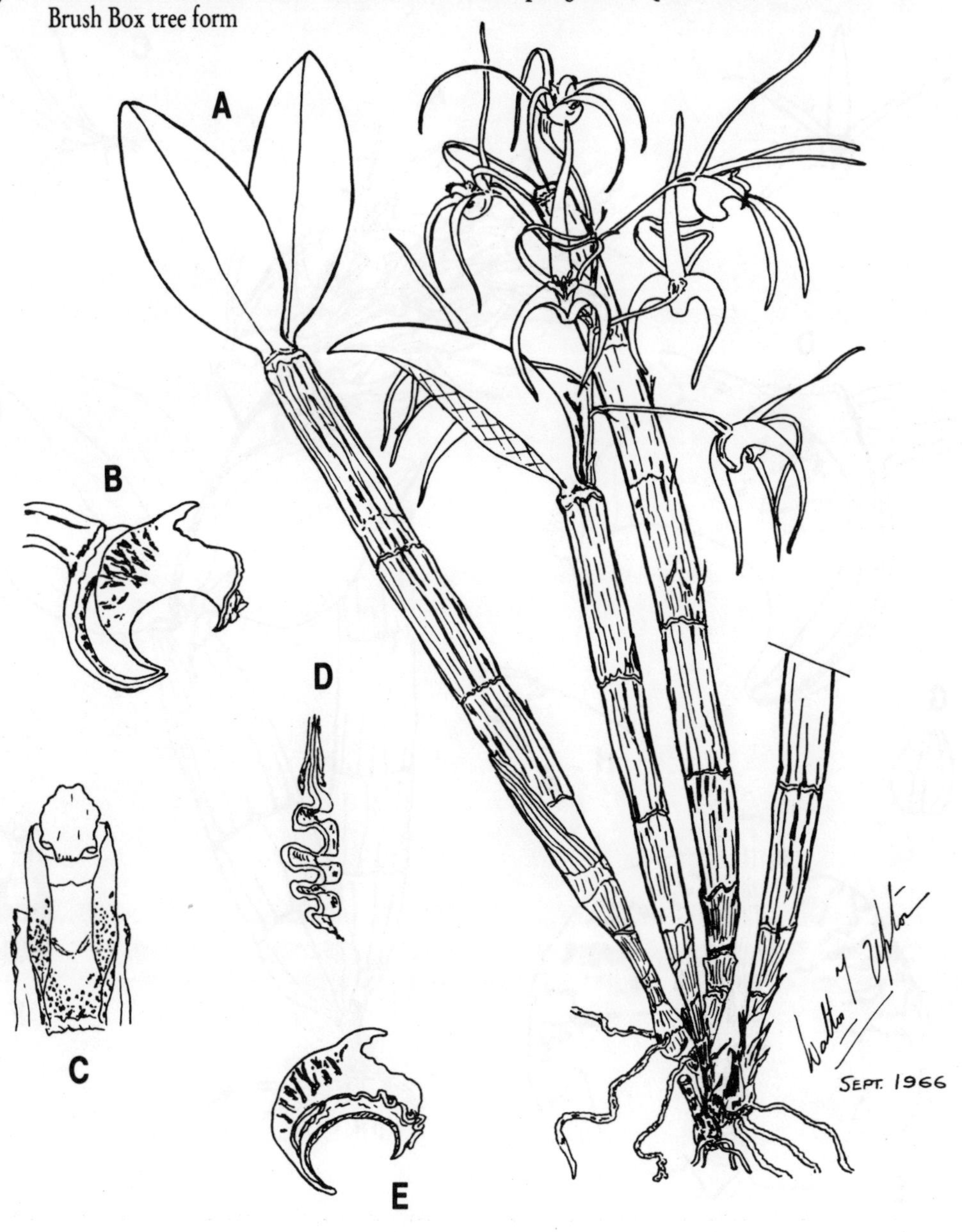

A Plant 1X
B Column and labellum, side-view 4X
C Column, front view 8X
D Callus (ridge) on labellum —
E Cross-section of labellum 4X

HYBRIDS

D. aemulum	X *D. kingianum*	= *D.* Emmy
D. aemulum	X *D. macropus* subsp. *gracilicaule* (as *D. gracilicaule*)	= *D.* Liela Gilbert
D. Emmy	X *D. speciosum*	= *D.* Craig Hewitt
D. Emmy	X *D.* Ellen	= *D.* Goblin
D. Goblin	X *D.* Penny Ann	= *D.* Minuet
D. Goblin	X *D. kingianum*	= *D.* King Goblin

IN HYBRIDIZATION

Little is known yet as only six hybrids have been named, but it seems to maintain its 'whispy' or feathery flower character, and its progeny appear to be free-flowering. Plant habit seems to be midway between the two parents.

Fig. 9 *Dendrobium aemulum* R.Br. Collected from near Nowra, NSW, 1980. Ironbark form

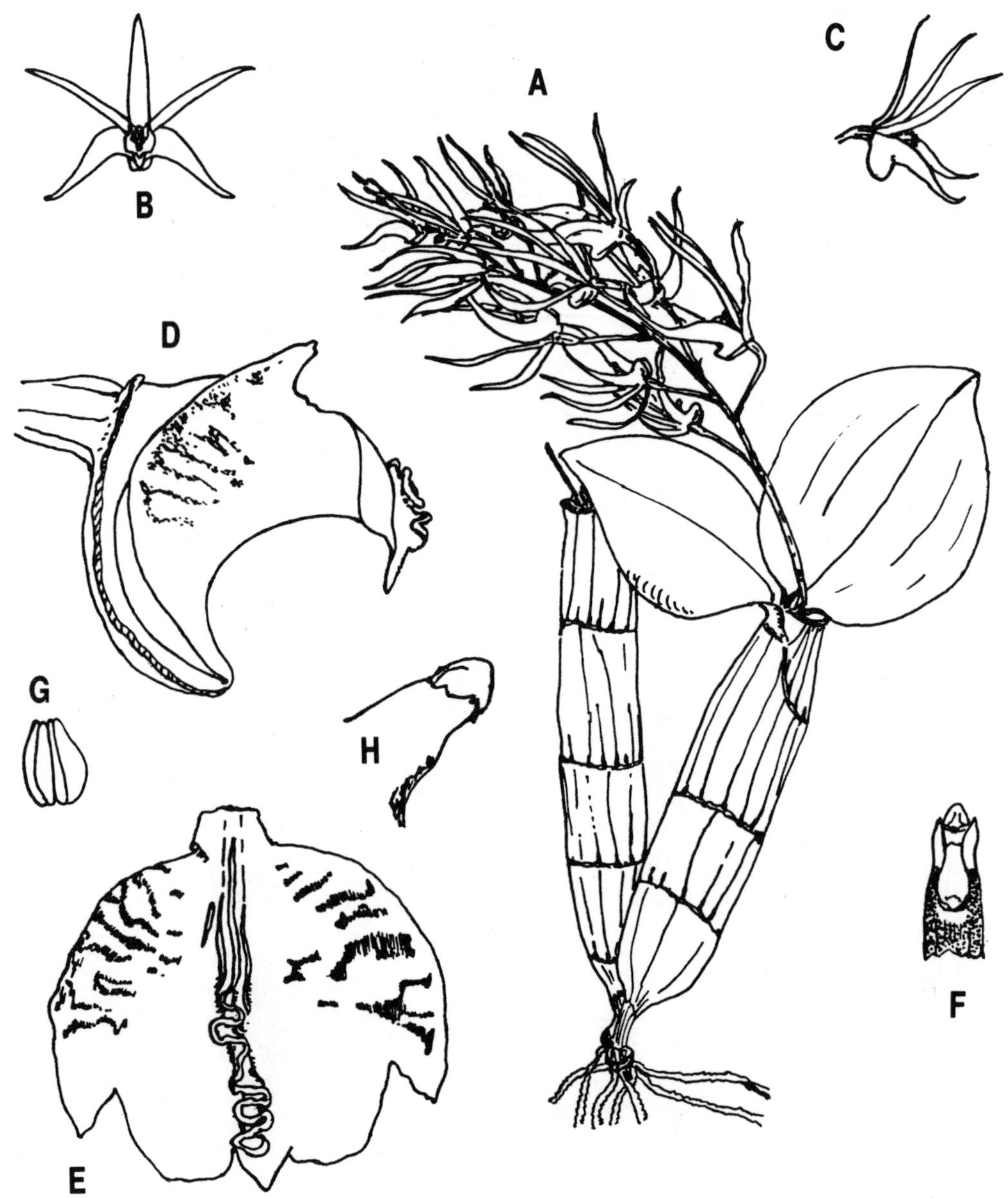

A Plant 1X
B Flower, front-view 1X
C Flower, side-view 1X
D Column and labellum, side-view 8X
E Labellum, pressed flat 8X
F Column, front-view 4X
G Pollinia greatly enlarged
H Top of column, side-view 8X

Dendrobium affine (Dcne) Steud.

Nomencl. ed. 2:489 (1840)

Fig. 10, Plate 7

BASIONYM

Onychium affine Dcne, *Nouv. Ann. Mus.* 3:365 (1836)

SYNONYMS

Dendrobium bigibbum sensu F. Muell., *Fragm. Phyt. Aust.* 6:119 (1868), non Lindley (1853) nom. invalid

Dendrobium dicuphum F. Muell., *Fragm. Phyt. Aust.* 8:28 (1873)

Dendrobium leucolophotum H. G. Reichb., *Gard. Chron.* ser. II, 18:582 (1882)

Callista dicupha (F. Muell.) Kuntze, *Rev. Gen. Pl.* 2:654 (1891)

Dendrobium urvillei Finet, *Bull. Soc. Bot. France* 50:372, t.11, f 1–9 (1903)

Dendrobium dicuphum F. Muell. var. *grandiflorum* Rupp et Hunt, *Proc. Linn. Soc. N.S.W.* 72:241 (1948)

Dendrobium dicuphum F. Muell., var. *album* Hort. ex E. C. Cooper, *Roy. Hort. Soc. Dict. Gdng.* 2:653 (1951)

DESCRIPTION

F. Mueller described *D. dicuphum* in 1868 under the name *D. bigibbum.* His first use of the name *D. dicuphum* was in 1873 and was based on the 1868 description. He cites 'Liverpool River', Gulliver as the type. As F. Mueller was the botanist on the North Australia Expedition led by A. C. Gregory in 1855 to 1856, it is hard to believe he did not see *D. dicuphum* then. In recent research by Dr Phillip Cribb and Dr P. S. Lavarack, it has been determined that *D. dicuphum* is a synonym of *D. affine* from New Guinea (Pers. Comm. Dr P. Cribb and M. Clements, 1986).

Pseudobulbous stems up to 50 cm high, generally 10 to 20 cm. In some areas this species is quite a dwarf plant, the mature pseudobulbous stems being between 10 and 13 cm high, often flowering from pseudobulbous stems only 15 mm high. Two to ten leaves, generally deciduous.

Up to 20 flowers, 2·5 to 5 cm in diameter, produced on racemes up to 50 cm long. Generally ten to 12 flowers and, very rarely, up to 60 flowers seen. Flowers widely opening and white or white with sepals greenish on the outside, or segments suffused at the base with deep-maroon markings. Labellum white, with deep-maroon markings on the side lobes. Occasionally whole flower is pink. Flowers long lasting.

FLOWERING PERIOD

March to August (autumn to winter).

DISTRIBUTION

From north-eastern Western Australia, the north of the Northern Territory, Melville, Bathurst and Groote Eylandt, some of the islands in the Gulf of Carpentaria and possibly in Queensland. All north of approximately 18° 00′S.

HABITAT

Generally grows on paperbark trees (*Melaleuca* spp.) overhanging watercourses, lagoons, etc.; in swampy areas; and in little pockets of rainforest near streams and inland springs where the light intensity is quite strong. It is subject to a long dry season. Another common host is *Cycas armstrongii.*

The climatic conditions under which this species grows consists of two definite seasons: the 'wet' season from late November to early May (the rainfall during this period is about 1300 mm); and the 'dry' season from May to early November (when almost drought conditions prevail and three consecutive dry months is not unusual). During the dry season the temperature range is from a maximum of approximately 32°C to a minimum of approximately 20°C. Not only does this species survive these tough conditions but grows and multiplies. It stores moisture in its pseudobulbous stems to see it through this dry period.

CULTURE

Slab culture suits best, if to be potted then as small a pot as possible and a very open mix. Very high light intensity necessary, so place high up in the greenhouse or, if outside, the maximum amount of light (just short of burning the leaves) is necessary.

Good air movement at all times, heavy watering while in full growth during the summer months. During the winter months no water whatsoever. Minimum winter temperature of 15°C.

IN HYBRIDIZATION

An excellent parent, producing quite compact plants with upright to arching racemes, generally carrying numerous flowers of very reasonable shape. Flowers long lasting. All hybrids to date have been registered under the species name *D. dicuphum.*

HYBRIDS

D. affine (as *D. dicuphum*)	X *D. bigibbum* (as *D. compactum*)	= *D.* Merritt Island
D. affine (as *D. dicuphum*)	X *D. canaliculatum*	= *D.* Judy Leroy
D. affine (as *D. dicuphum*)	X *D. discolor*	= *D.* Chieno
D. affine (as *D. dicuphum*)	X *D. fleckeri*	= *D.* Reg Burns
D. affine (as *D. dicuphum*)	X *D. johannis*	= *D.* Siswati
D. affine (as *D. dicuphum*)	X *D. nindii*	= *D.* Gresford
D. affine (as *D. dicuphum*)	X *D. tetragonum*	= *D.* Maron
D. Judy Leroy	X *D. antennatum*	= *D.* Lowana Tiki
D. Judy Leroy	X *D. bigibbum*	= *D.* Eleebana
D. Judy Leroy	X *D. canaliculatum*	= *D.* Dusky Girl
D. Judy Leroy	X *D.* Colin	= *D.* Jombock
D. Judy Leroy	X *D. johannis*	= *D.* Blue Tiki
D. Judy Leroy	X *D.* Star of Gold	= *D.* Andrew Upton
D. Judy Leroy	X *D. x superbiens*	= *D.* Gold Baby
D. Judy Leroy	X *D. tetragonum*	= *D.* Duke
D. Merritt Island	X *D. canaliculatum*	= *D.* Penang Angel

Fig. 10 *Dendrobium affine* (Dcne.) Steud.
From south of Darwin

A Plant 2/3X
B Flower, front-view 2X
C Flower, side-view 2X
D Column and labellum, side-view 2X
E Labellum, pressed flat 2X
F Top of column, front-view 4X
G Top of column, side-view 4X

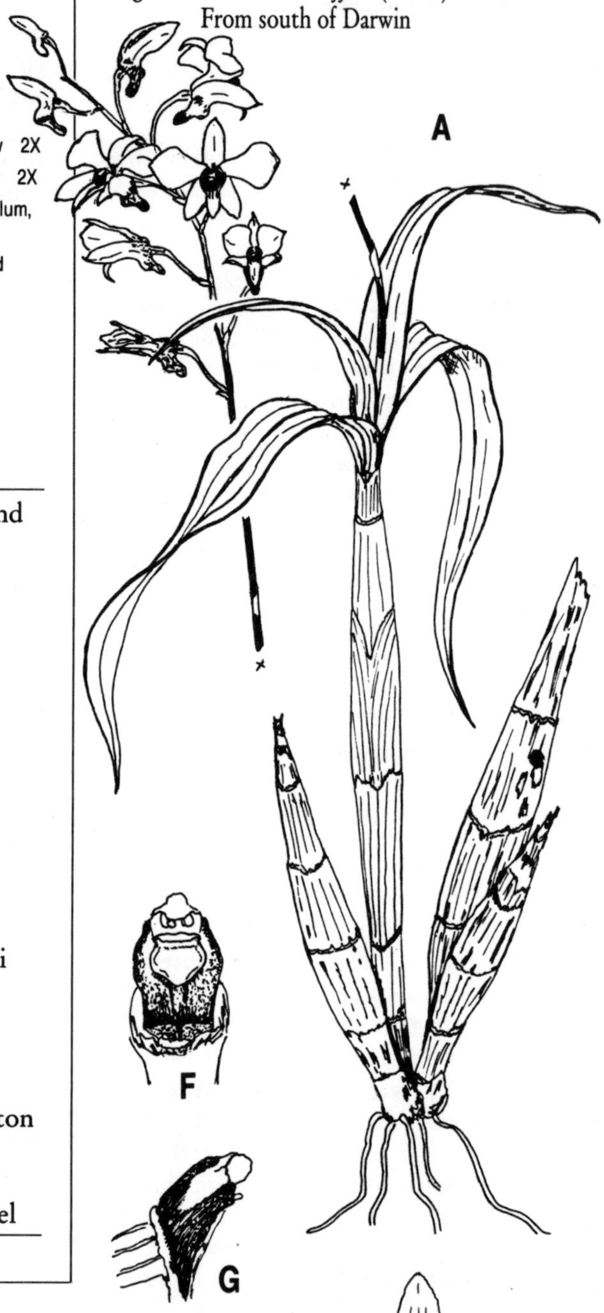

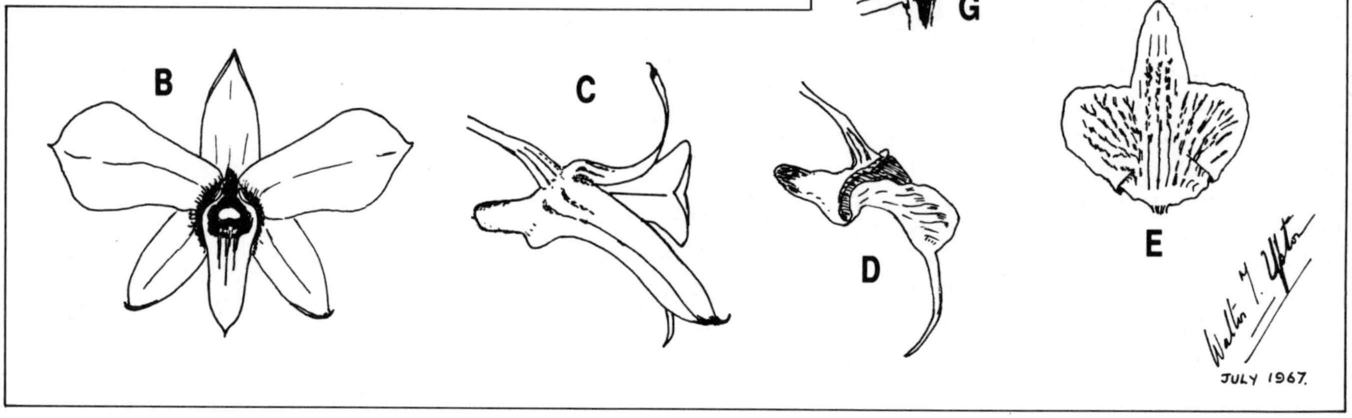

Dendrobium agrostophyllum F. Muell.

Fragm. Phyt. Aust. 8:28 (1873)
Fig. 11, Plate 11

SYNONYMS

Callista agrostophylla (F. Muell.) Kuntze, *Rev. Gen. Pl.* 2:654 (1891)
Dendrobium muellerianum Schltr., *Repert. Spec. Nov. Regn. Veg. Beih.* 3:316 (1907)
Trachyrhizum agrostophyllum (F. Muell.) Rauschert, *Feddes Repert.* 94:468 (1983)

DESCRIPTION

Named from *agrostis*, meaning grass, and *phyllon*, meaning leaf—from the grass-like leaves. Collected in 1873 by John Dallachy near Townsville.

Slender pseudobulbous stems up to 90 cm long (generally much less in cultivation) which, unlike the rest of the Australian dendrobiums, take 2 to 3 years to mature. Pseudobulbous stems well spaced, from 1 to 6 cm on the rhizome, the distance between them frequently increasing as the plant matures. Thin, narrow, lanceolate leaves, up to about 20, well spaced along the apical third to half, or even more, of the pseudobulbous stem.

Pseudobulbous stems flower for several years in succession, producing short racemes of two to ten (commonly about four in cultivation) pale to deep yellow, waxy flowers. Flowers fragrant, approximately 2 cm in diameter. Labellum yellow with purple-brown markings, mid-lobe very

Fig. 11 *Dendrobium agrostophyllum* F. Muell. Collected Mt. Lewis, North Queensland, 1964

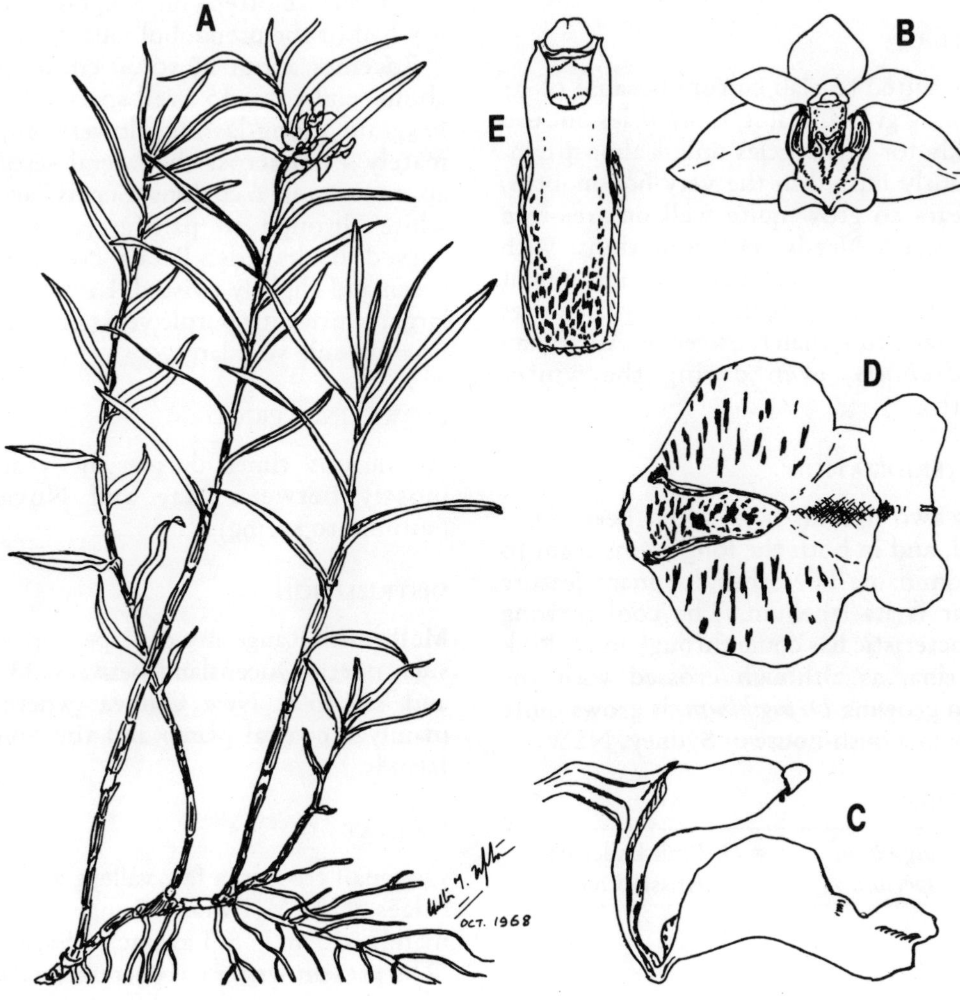

A Plant 1/3X
B Flower, front-view 2X
C Column and labellum, side-view 4X
D Labellum, pressed flat 4X
E Column, front-view 4X

short and broad giving the appearance of a double-lobed mid-lobe. Flowers inclined to droop a little but open widely, stay in good condition for up to 3 weeks.

FLOWERING PERIOD

August to October (winter to spring).

DISTRIBUTION

North-eastern Queensland between 15° 30′S and 20° 00′S (just south of Townsville).

HABITAT

In the upper branches or trunks of trees in the cool cloud forests at approximately 900 to 1300 m, occasionally in lower altitudes in gullies. In some localities grows in semi-exposed positions low down on small trees and rocks.

CULTURE

Most suited to slab culture because of its rambling growth habit. Cork dries out too quickly for this species unless almost continuously misted in the very hot months. Appears to grow quite well on tree-fern fibre slabs. Needs cool conditions, with high humidity and plenty of air movement together with high light intensity. Requires more moisture than the average Australian *Dendrobium*, even during the winter months.

IN HYBRIDIZATION

Only two hybrids have so far been registered, and in both the long stems seem to be dominant. The other dominant feature so far is its labellum. The cool-growing characteristic has come through in *D.* Pink Ballerina, as although crossed with the warm growing *D. bigibbum*, it grows quite easily in a bush-house in Sydney, NSW.

HYBRIDS

D. agrostophyllum	X *D. bigibbum*	= *D.* Pink Ballerina
D. agrostophyllum	X *D. speciosum*	= *D.* Aussie Ochre

Dendrobium antennatum Lindley

J. Bot. Lond. 2:236 (1843)

Fig. 12, Plates 10 and 212

SYNONYMS

Dendrobium d'albertisii H. G. Reichb., *Gard. Chron.* ser. II, 9:366 (1878)
Callista antennata (Lindley) Kuntze, *Rev. Gen. Pl.* 2:654 (1891)

DESCRIPTION

Originally found in Australia by J. L. H. Wassell in 1955 in the Massey Scrub near Coen. Name means with antennae, referring to the position of the petals.

Pseudobulbous stems of Australian plants usually up to 60 cm long, but a New Guinea plant of this species in my collection is over 130 cm tall. Approximately 12 narrow, ovate, quite leathery leaves, the upper surface often concave, borne on the top half of the pseudobulbous stem.

Racemes about 20 to 30 cm long with about eight to 15 well-spaced flowers. Fragrant, long-lasting flowers approximately 4 cm across the lateral sepals and approximately 6 cm long. Sepals vary from white through to pale green, generally curved backwards. Petals erect, mostly green and slightly twisted. Labellum quite large, white with purple veins or markings. Seed capsule very large.

FLOWERING PERIOD

At various times during the year, but mostly between May and November (autumn to spring).

DISTRIBUTION

McIlwraith Range area of Cape York Peninsula, north Queensland between 13° 35′S and 13° 55′S, New Guinea (where it is mainly a coastal plant) and the Solomon Islands.

HABITAT

On small trees in a few valleys and on the ridges of the McIlwraith Range at approximately 320 m. P. S. Lavarack (1980a) states 'The present writer has encountered plants of this species on three different field trips,

1 *D. striolatum* growing on rock in full sun

2 Rainforest, north Queensland, home of many species

3 *D. kingianum* on cliff overlooking deep ravine

4 These trees are host to *D. canaliculatum*

5 Typical habitat of *D. bigibbum*

6 *D. adae* 'Jill'

7 *D. affine*

8 *D. bigibbum* var. *bigibbum*

9 *D. aemulum*

10 *D. antennatum*

11 *D. agrostophyllum*

12 *D. calamiforme*

but always in similar habitats. It occurs on the steep, fast flowing streams that flow from the 800 m high McIlwraith Range to the coast. It is almost invariably found where the streams drop over a waterfall or series of steep rapids and where the rocky topography has forced a break in the canopy. The conditions then are exposed, with high light, but also high humidity. They are not plants of dark rainforests.'

CULTURE

Needs lots of light and high humidity. Minimum winter temperature of 15°C (60°F). Slab culture suits best; if potted needs a very open mix and frequent repotting—every 2 years minimum. Needs plenty of water when growing in the summer months; and very little during the winter rest.

IN HYBRIDIZATION

Plant form dominant. Generally good arching to upright racemes produced by progeny. Most hybrids to date have been within its own section (*Spatulata*) and flowers mostly midway between parents.

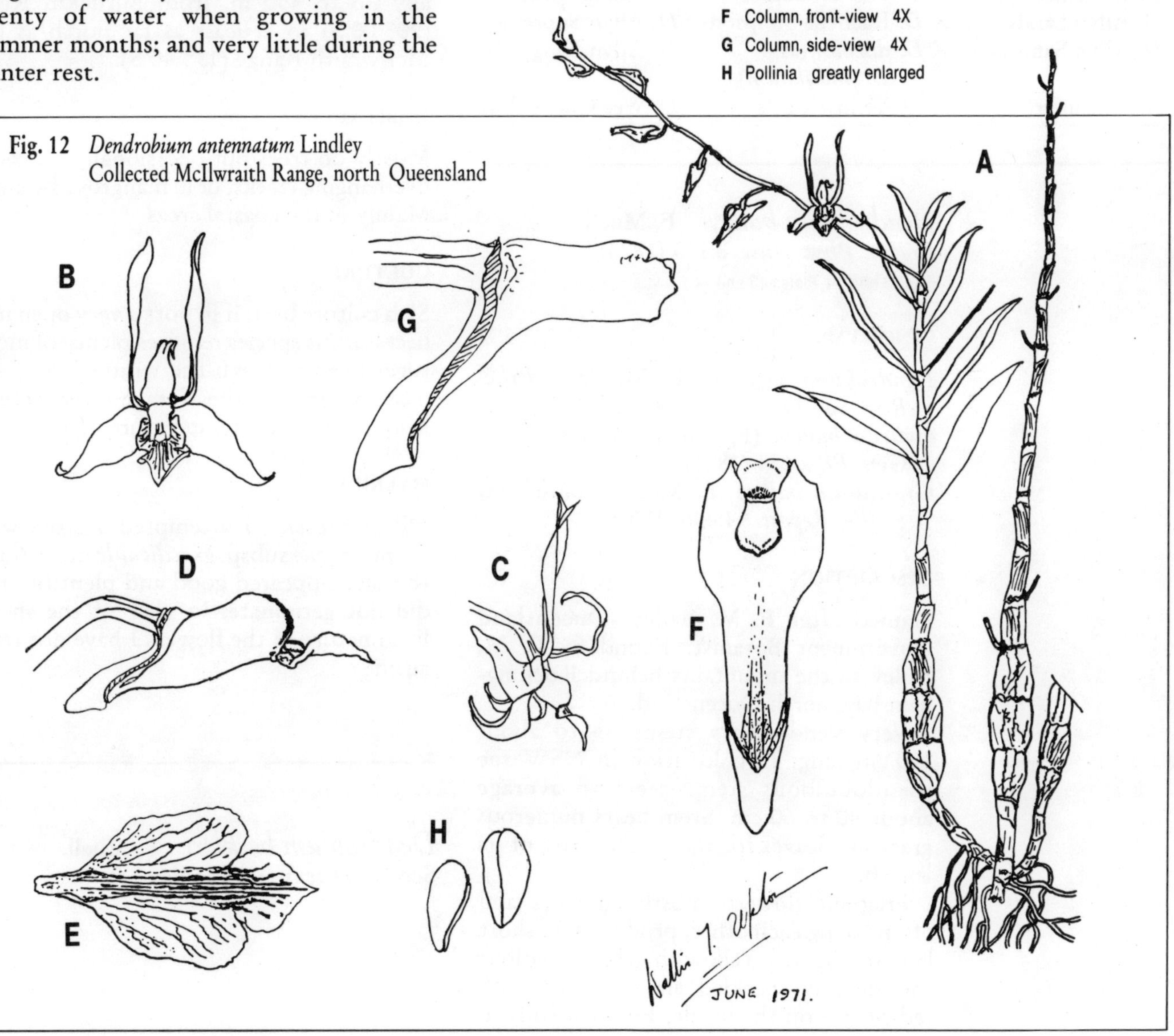

A Plant 1/3X
B Flower, front-view 1X
C Flower, side-view 1X
D Column and labellum, side-view 2X
E Labellum, pressed flat 2X
F Column, front-view 4X
G Column, side-view 4X
H Pollinia greatly enlarged

Fig. 12 *Dendrobium antennatum* Lindley
Collected McIlwraith Range, north Queensland

HYBRIDS

D. antennatum (as *d'albertisii*)	X *D. bigibbum*	= *D.* Albertine
D. antennatum (as *d'albertisii*)	X *D. canaliculatum*	= *D.* Lowana Nioka
D. antennatum (as *d'albertisii*)	X *D. discolor* (as *D. undulatum*)	= *D.* Verninha
D. antennatum (as *d'albertisii*)	X *D. johannis*	= *D.* Red Wings
D. antennatum (as *d'albertisii*)	X *D.* Judy Leroy	= *D.* Lowana Tiki
D. antennatum (as *d'albertisii*)	X *D. mirbelianum*	= *D.* Moses Mark
D. Lowana Nioka	X *D. johannis*	= *D.* Allyn Sands
D. Lowana Nioka	X *D. canaliculatum*	= *D.* Glenys King
D. Lowana Nioka	X *D. speciosum*	= *D.* Justin Wall
D. Lowana Nioka	X *D.* Star of Gold	= *D.* Natalie Wall
D. Allyn Sands	X *D.* Rachelle Simpson	= *D.* Allyn Supreme
D. Allyn Sands	X *D. discolor* (as *D. undulatum*)	= *D.* Allyn King
D. Verninha	X *D.* Yellow Curls	= *D.* Wee Cheow Lay

Dendrobium baileyi F. Muell.

Fragm. Phyt. Aust. 8:173 (1874)

Figs 13 and 14, Plates 13 and 14

SYNONYMS

Dendrobium keffordii F. M. Bail., *Proc. Roy. Soc. Qld.* 1:2 (1884)
Callista baileyi (F. Muell.) Kuntze, *Rev. Gen. Pl.* 2:654 (1891)
Grastidium baileyi (F. Muell.) Rauschert, *Feddes Repert.* 94:446 (1983)

DESCRIPTION

Named after F. M. Bailey, Queensland Government Botanist. Found by F. M. Bailey in the mountains behind Rockingham Bay, north Queensland.

Very slender wiry stems up to about 120 cm long. In cultivation in NSW the pseudobulbous stems seem to average about 40 to 60 cm. Stem bears numerous grass-like leaves for the greater part of its length.

Fragrant flowers, mostly in pairs and often facing each other, produced on short lateral racemes. Yellowish-green in colour and densely speckled and streaked with red-purple on the inside. Filiform ends to the sepals and petals are red-purple. Flowers short lived; often after they have been widely open for about 2 hours, they curl round and become intertwined (as seen in Fig. 13)—no doubt an act to assist pollination. Generally stay intertwined for a few days before withering completely. Pollinia translucent cream.

FLOWERING PERIOD

Irregularly throughout the year but most commonly in January or February (mid to late summer).

DISTRIBUTION

Lowland north-east Queensland, occasionally up to 450 m, from Burdekin River (20° 00′S) to at least as far north as the McIlwraith Range (13° 40′S).

HABITAT

Mostly on trees but occasionally on rocks overhanging creeks, or in mangrove swamps. Mainly in the coastal areas.

CULTURE

Slab culture best; if in pots a very open mix necessary as species requires plenty of moisture. Even through the winter it expects more water than the average *Dendrobium*. Minimum winter temperature of 15°C.

HYBRIDS

Nil at present. I attempted a cross with *D. macropus* subsp. *gracilicaule* in 1976 and the seed appeared good and plentiful, but did not germinate. In view of the short-lived nature of the flowers I have not tried again.

Dendrobium beckleri F. Muell.

See *D. schoeninum* Lindley

Figs 13 & 14 *Dendrobium baileyi* F. Muell. Sent to author from the Cairns area of north Queensland

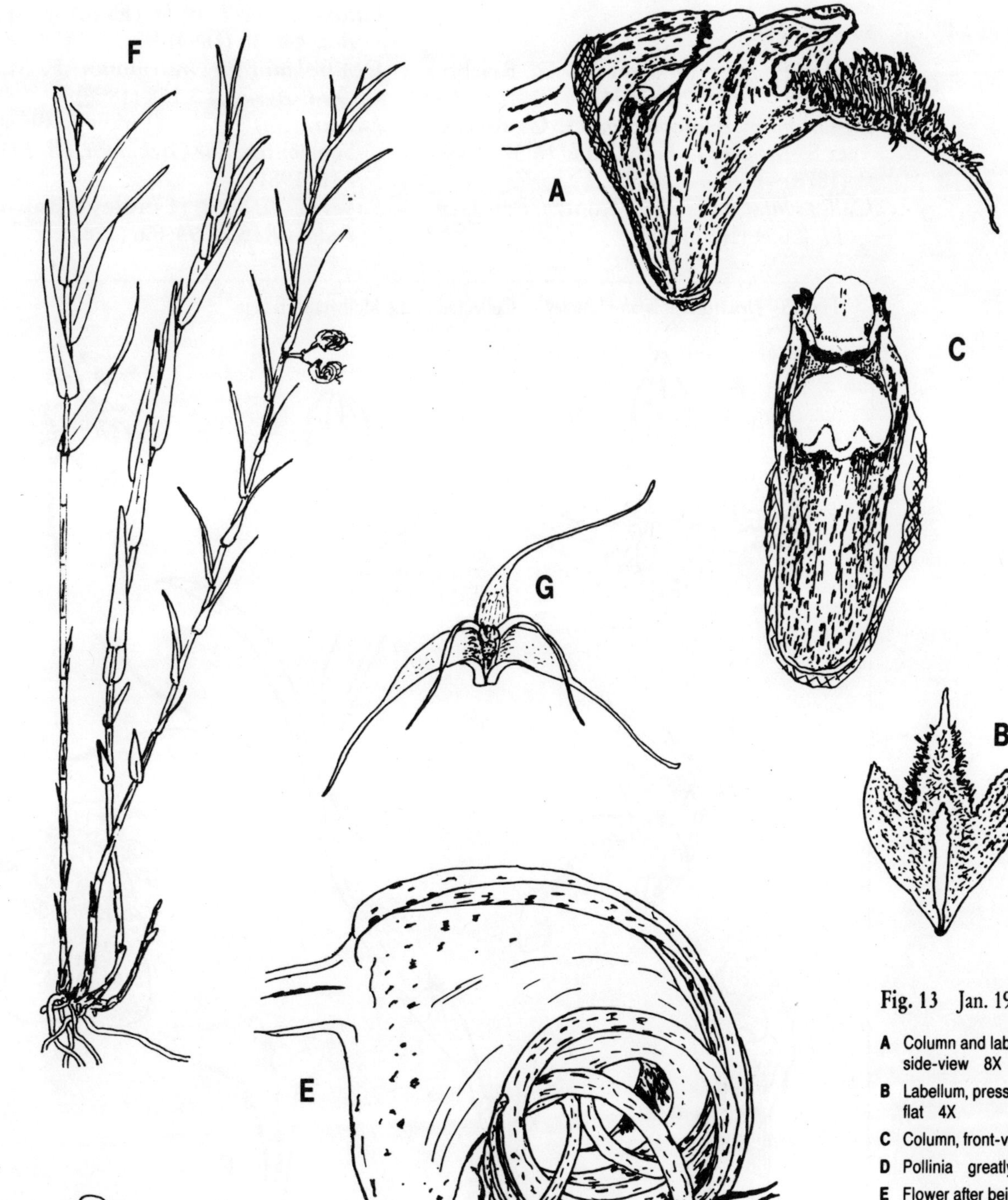

Fig. 13 Jan. 1972

A Column and labellum, side-view 8X
B Labellum, pressed flat 4X
C Column, front-view 12X
D Pollinia greatly enlarged
E Flower after being open all day 10X

Fig. 14 Feb. 1972

F Plant $^{1}/_{3}$X
G Flower, freshly opened 2X

Dendrobium bifalce Lindley

J. Bot. Lond. 2:237 (1843)
Fig. 15, Plate 15

SYNONYMS

Doritis bifalcis (Lindley) H. G. Reichb., *Xenia Orchid.* 2:7 (1874)

Dendrobium chloropterum H. G. Reichb. et S. Moore, *J. Bot. Lond.* 16:37, t.196 (1878)

Callista bifalcis (Lindley) Kuntze, *Rev. Gen. Pl.* 2:654 (1891)

Phyllorchis chloroptera (H. G. Reichb.) Kuntze, *Rev. Gen. Pl.* 2:677 (1891)

Bulbophyllum oncidiochilum Krzl., *Bot. Jahrb. Syst.* 17:485 (1893)

Latourea oncidiochila (Krzl.) Krzl., *Öst. Bot.* 44:336 (1894)

Dendrobium breviracemosum F. M. Bail., *Q/ld. Agric. J.* 3:158 (1898)

Latourorchis bifalce (Lindley) Brieger, in Schlechter, *Die Orchideen*, ed. 3 (11/12): 731 (1981)

Sayeria bifalcis (Lindley) Rauschert, *Feddes Repert.* 94:466 (1983)

Fig. 15 *Dendrobium bifalce* Lindley Collected in the McIlwraith Range

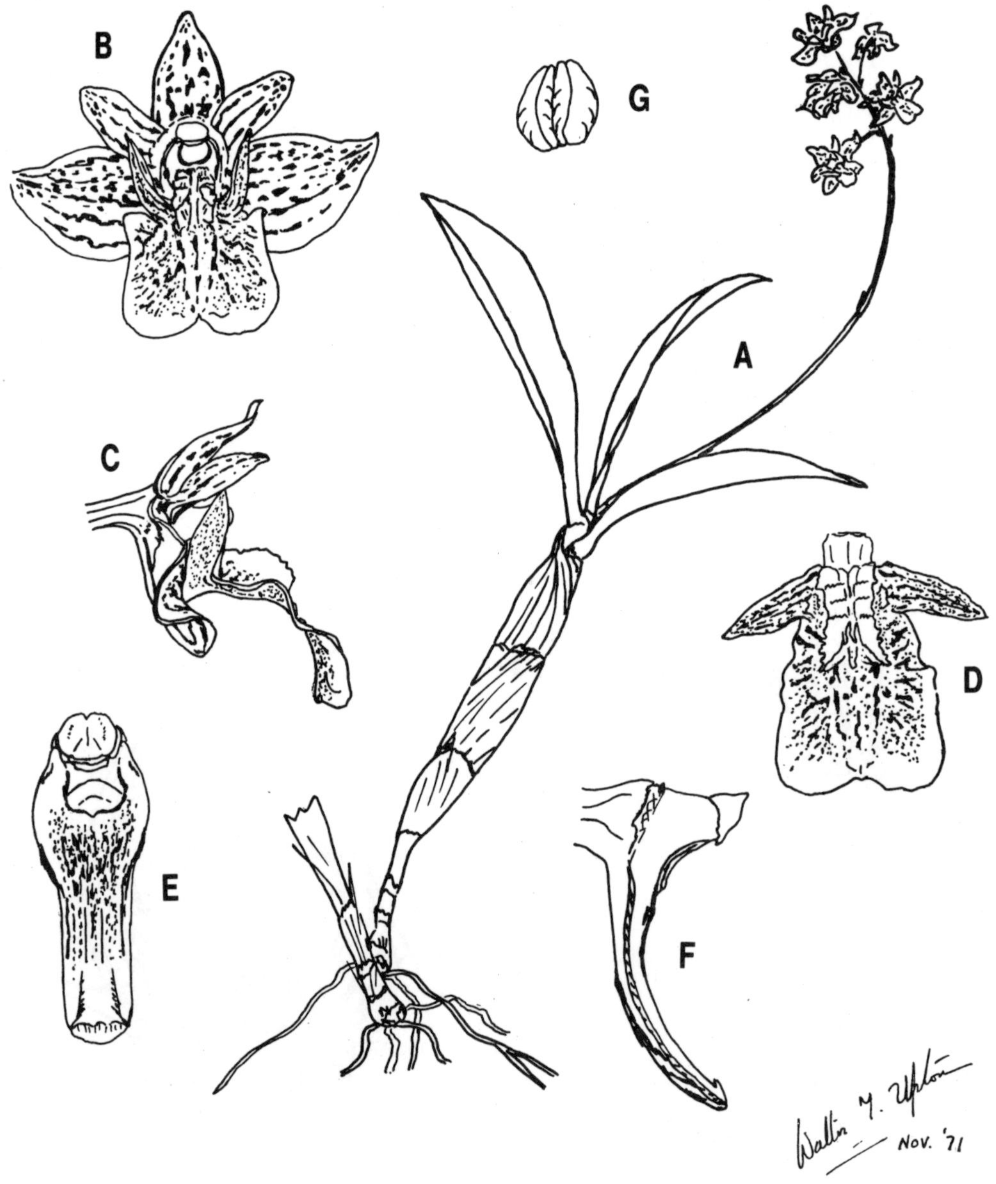

A Plant $^1/_3$X
B Flower, front-view 2X
C Flower, side-view 2X
D Labellum, pressed flat 2X
E Column, front-view 4X
F Column, side-view 4X
G Pollinia greatly enlarged

DESCRIPTION

It is understood to be described from specimens collected in Fiji about 1843, later recorded from the island of Saibai, geographically part of Papua though politically part of Queensland. It was first named in Australia by F. M. Bailey as *Dendrobium breviracemosum* in 1898. In June 1945, R. L. Hunter collected specimens in the Portland Roads area of north Queensland.

Pseudobulbous stems, mostly fusiforme in shape, up to 60 cm long with two to three fairly broad, ovate leaves at the top. Eight to 12 flowers borne on racemes, up to 20 cm long, and from the top of the pseudobulbous stems. Sepals and petals pale greenish to yellowish with numerous dark purplish-brown spots and markings. Labellum fawn to green with reddish-brown markings, disc white. Flowers long lasting.

FLOWERING PERIOD

May to June (late autumn to early winter).

DISTRIBUTION

From the Daintree River, north Queensland (16° 20′S) (where it is very rare), north to Cape York Peninsula and then through the islands on to New Guinea (where it is quite common from sea level up to 1000 m or more), also in Timor, New Britain and the Solomon Islands.

HABITAT

Lowland rainforests up to 550 m, occasionally on rocks. Frequently seen high up in trees or in less-dense areas where the light penetrates well.

CULTURE

Grows equally well on slabs or in pots. In pots the mix must be very open. Needs a warm sunny position as it must have high light. Abundant moisture when growing, but care must be taken not to damp off the new growth when it first appears. Needs a decided rest after flowering, until the new growth appears. Minimum winter temperature 15°C.

Schlechter (1982b), referring to species in the section *Latouria* as being adverse to continuous high temperatures, quoted *D. bifalce* as an exception. Writing about *D. bifalce* in New Guinea he said '. . . is found in the hot coastal zone, growing there usually on tall trees overhanging the sea, where it receives a certain amount of coolness. That it is capable also of bearing severe heat and much sun is proved by specimens which I observed for two years on some dead trees, where they flowered regularly and seemed to thrive.'

HYBRID

D. bifalce	X *D. discolor* (as *D. undulatum*)	= *D.* Gold Stripe

Dendrobium bigibbum Lindley

Pax. Flower Gdn. 3:25 f.245 (1853)

Plates 5, 202, 203 and 209

DESCRIPTION

Name derived from *bi*, meaning two, and *gibbus*, meaning hump—referring to the two-humped or two-chinned spur. Common name is Cooktown orchid.

Much has been written on this very widespread and varied species. Perhaps the two most interesting and informative articles are by Blake (1966) and S. C. Clemesha (1978). These two articles contain all the information on the complex background of this magnificent orchid and its numerous named varieties.

D. bigibbum was named in 1853 from a specimen stated to be collected on Mount Adolphus Island in the Torres Strait. It was in cultivation at Kew in 1824 (H. G. Reichb. in *Gard. Chron.* ser. 2, 10:748 (1878)).

In November 1959, the Parliament of Queensland passed an act declaring that 'The Cooktown Orchid (*Dendrobium bigibbum* var. *phalaenopsis*) shall be the Floral Emblem of the State of Queensland'. A postage stamp depicting the Cooktown orchid was issued on 10 July 1968 as one of a set of six stamps of the State Floral Emblems; its value was 25 cents.

A variable species of two subspecies, only one of which occurs in Australia and is comprised of three varieties. Synonyms and descriptions are given for each variety. *D. bigibbum* Lindley subsp. *laratensis* Clemesha (syn. *D. phalaenopsis* Fitzg. var. *schroederianum* Hort. ex Masters) is found in the Tanimbar Islands formerly Timor Laut (Plate 19). Although not an Australian native, this subspecies has been regularly confused with *D. bigibbum* Lindley subsp. *bigibbum*.

D. bigibbum Lindley subsp. *bigibbum* var. *bigibbum*

Fig. 16, Plate 8

SYNONYMS

Dendrobium sumneri F. Muell., *Fragm. Phyt. Aust.* 6:94 (1867)

Dendrobium bigibbum Lindley var. *candidum* H. G. Reichb., *Gard. Chron.* ser. II, 9:168 (1878)

Dendrobium bigibbum Lindley var. *sumneri* (F. Muell.) F. M. Bail., *Syn. Q/ld. Fl.* 509 (1883)

Dendrobium bigibbum Lindley subvar. *candidum* (H. G. Reichb.) Veitch in Veitch & Sons, *Man. Orchid. Pl.* 1, pt 3:23 (1888)

Dendrobium bigibbum Lindley var. *albomarginatum* Linden, *Lindenia* 7:317 (1891), non F. M. Bail. (1891)

Callista bigibba (Lindley) Kuntze, *Rev. Gen. Pl.* 2:654 (1891)

Callista sumneri (F. Muell.) Kuntze, *Rev. Gen. Pl.* 2:655 (1891)

Dendrobium bigibbum Lindley var. *albopurpuratum* Hort., *Gard. Chron.* ser. 3, 18:396 (1895)

Fig. 16 *Dendrobium bigibbum* Lindley subsp. *bigibbum* var. *bigibbum* Collected just north of Nichol Creek, north Queensland

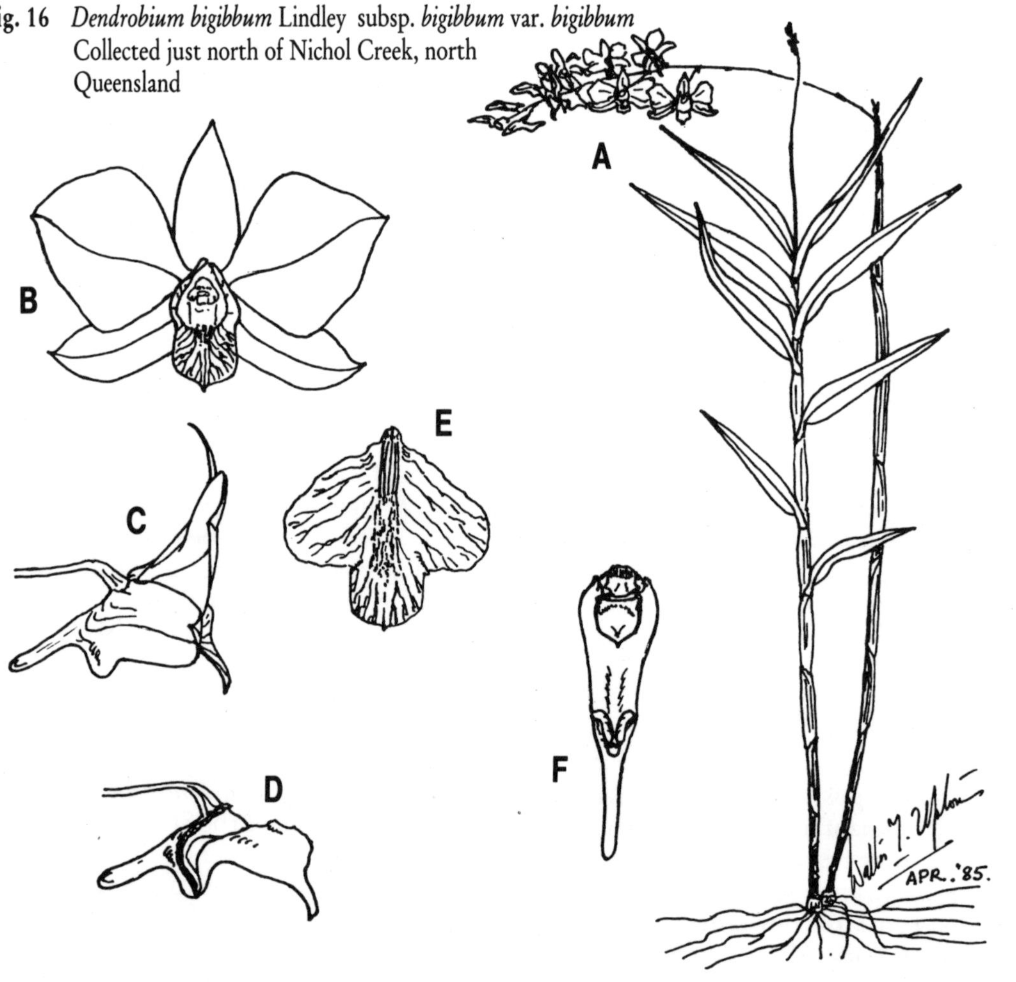

A Plant $^1/_5$X
B Flower, front-view 1X
C Flower, side-view 1X
D Column and labellum, side-view 1X
E Labellum, pressed flat 1X
F Column, front-view 2X

Dendrobium bigibbum Lindley var. *album*, F. M. Bail., *Q/ld. Fl.* 5:1524 (1902)

DESCRIPTION

Slender pseudobulbous stems, often quite purplish, up to 120 cm long (averaging about 40 to 60 cm). Few to about 12 narrow, lanceolate leaves in the upper part of the pseudobulbous stem.

Racemes generally about 20 to 40 cm long and bear up to 20 rosy, mauve to purple flowers; very occasionally white and bicolour flowers. Petals and sepals usually curved backwards, mid-lobe of lip blunt with its end sometimes slightly notched. Callus frequently white; flower generally paler than var. *superbum*.

FLOWERING PERIOD

Main period is March to July (autumn to winter), but sometimes flowers seen as early as January and as late as October.

DISTRIBUTION

Cape York Peninsula, from the Archer River (13° 20′S), north through some of the Torres Strait islands into Papua New Guinea, where very few collections have been made to date.

HABITAT

In semi-arid regions mostly at low altitudes, on small trees and rocks. Grows in strong light on narrow branches, the light being only slightly filtered by the trees; often in very exposed places on rocks. Near the coast frequently seen on trees overhanging salt water. From January to April during the wet season, receives plenty of rain, creating a hot humid atmosphere, and then experiences a long dry season. By the time the dry has started, their growth is completed and the plants become dormant and begin their period of blooming. In some harsh habitats plants are leafless, no doubt to help survival in the harsh conditions.

I saw some small plants on a low tree in the open forest by the side of the road just after Nichol Creek, approximately 13° 00′S, in July—a very dry area at that time of the year. They were not in flower.

D. bigibbum Lindley subsp. *bigibbum* var. *superbum* Hort. ex H. G. Reichb. *Gard. Chron.* ser. II, 10:748 (1878)

Fig. 17, Plates 16, 17 and 205

SYNONYMS

Dendrobium phalaenopsis Fitzg., *Gard. Chron.* 14:341 (1880)
Dendrobium bigibbum Lindley var. *macranthum* F. M. Bail., *Bot. Bull. Dept. Agric. Q/ld.* 10:24 (1895)
Dendrobium bigibbum Lindley var. *phalaenopsis* (Fitzg.) F. M. Bail., *Syn. Q/ld. Fl.* 509 (1883)

DESCRIPTION

Name is from *superbus*, meaning excellent, splendid, superb. Discovered by J. Gould Veitch in 1865. This variety is also extremely variable. It is distinguished from the variety *compactum* mainly by the size of pseudobulbous stems.

Plants same as type variety. Flowers, while different from the type, have characteristics difficult to define clearly. The most typical differences are listed in the table.

Character	var. *bigibbum*	var. *superbum*
Flower size	3 to 5 cm	3·5 to 7 cm
Sepals and petals	Generally recurved	Petals only slightly recurved
Colour	Typically mid-lilac	Typically deep-lilac
Labellum mid-lobe	More or less rounded	Usually pointed
Keels on disc of labellum	Five low keels extend to about halfway along the mid-lobe and converge towards each other at front. Front half of keels beset with short hairs, usually white but occasionally pale-lilac. These become more or less interwoven and appear to form a mat of generally white hairs	Three to seven, but usually five keels, fringed, or merely rugose. Usually mauve in front half
Column foot	Generally at a 90° angle to column	Frequently at a greater angle

Fig. 17 *Dendrobium bigibbum* Lindley subsp. *bigibbum* var. *superbum* Hort. ex H. G. Reichb.
From Cooktown area, about 1965

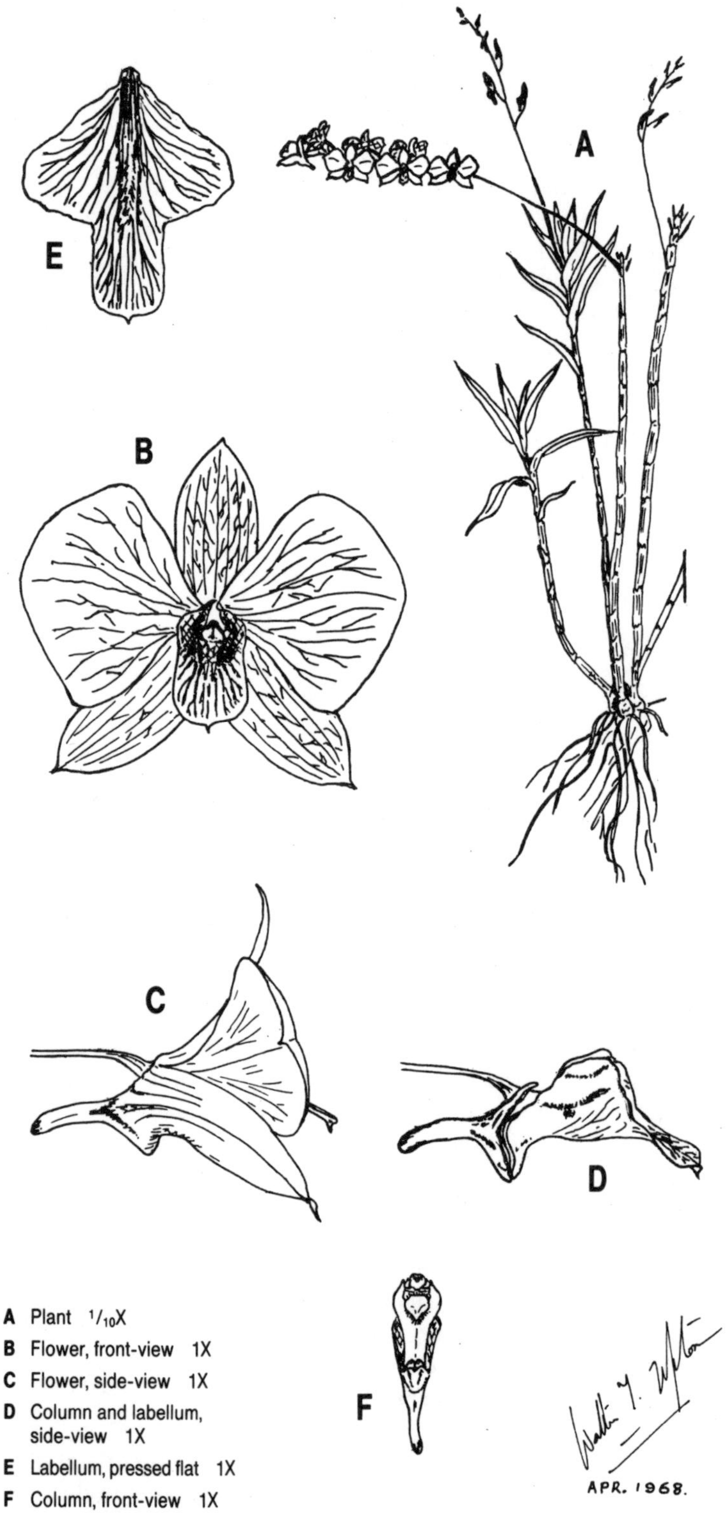

A Plant $^{1}/_{10}$X
B Flower, front-view 1X
C Flower, side-view 1X
D Column and labellum, side-view 1X
E Labellum, pressed flat 1X
F Column, front-view 1X

FLOWERING PERIOD

It would appear that in many instances this variety flowers slightly earlier than the type variety.

DISTRIBUTION

Cape York Peninsula. From Mt Molloy (145° 20′E, 16° 40′S) to the Gilbert River (141° 20′E, 16° 40′S), then north to the Archer River (13° 20′S).

HABITAT

Similar habitat to type variety. I saw about a dozen large plants of this variety in full flower a short way down a dry sandy creek off the Stewart River (14° 10′S) in mid-July—a pretty sight. Plate 5 shows the habitat of the bed of the Stewart River looking towards the creek; plants were seen about 30 m up the creek. Fairly intense light available. Height approximately 200 m. Plants were on, I believe, a *Eugenia* sp. The bank of the creek was about 4 m high and the plants were approximately 4 m to 8 m up the tree (near the top). The tree was approximately 20 cm in diameter. The forest was fairly open.

D. bigibbum Lindley subsp. *bigibbum* var. *compactum* C. T. White

Aust. Orchid Review 6:53 (1941)
(based on *D. phalaenopsis* Fitzg. var. *compactum*)
Fig. 18, Plate 18

SYNONYMS

Dendrobium bigibbum Lindley var. *phalaenopsis* (Fitzg.) F. M. Bail. forma *compactum* (C. T. White) G. L. Piper, *Aust. Orchid Rev.* 15:40 (1950)
Dendrobium bigibbum Lindley subvar. *compactum* (C. T. White) St. Cloud, *Nth. Q'ld. Nat.* 24, no. 115:18 (1956)
Dendrobium bigibbum Lindley forma *compactum* (C. T. White) St. Cloud, *Nth. Q'ld. Nat.* 24, no. 115:18 (1956)
Dendrobium bigibbum var. *superbum* Hort. ex H. G. Reichb., subvar. *compactum* (C. T. White) Dockr., *Aust. Indg. Orchids* 1:442 (1969)

DESCRIPTION

The original description was from a plant cultivated by Mr A. Hilsden of Brisbane and obtained from Hartley's Creek, Qld. The plants illustrated were obtained from the Mossman River.

Pseudobulbous stems short and stout, 5 to 25 cm long, but averaging about 10 to 12 cm. In full sun, pseudobulbous stems show a lot of purple pigment, frequently leaves are purplish-bronze.

Racemes usually carry three to ten flowers which are very variable, as they are in var. *superbum.* Quite similar to var. *superbum*, except petals tend to droop downwards a little more, and fewer flowers on the raceme. Occasionally white or white with a purple-lip flowers are found.

FLOWERING PERIOD

From January to October but the main period is from March to July (autumn to winter).

DISTRIBUTION

From the Macalister Range, the coastal range between the Barron River (16° 50′S) and the Mossman River (16° 30′S). This variety is coming close to extinction in the wild; only isolated colonies in almost inaccessible areas are now seen. What a great

Fig. 18 *Dendrobium bigibbum* Lindley subsp. *bigibbum* var. *compactum* C. T. White
Purchased from Limberlost Nursery, Cairns

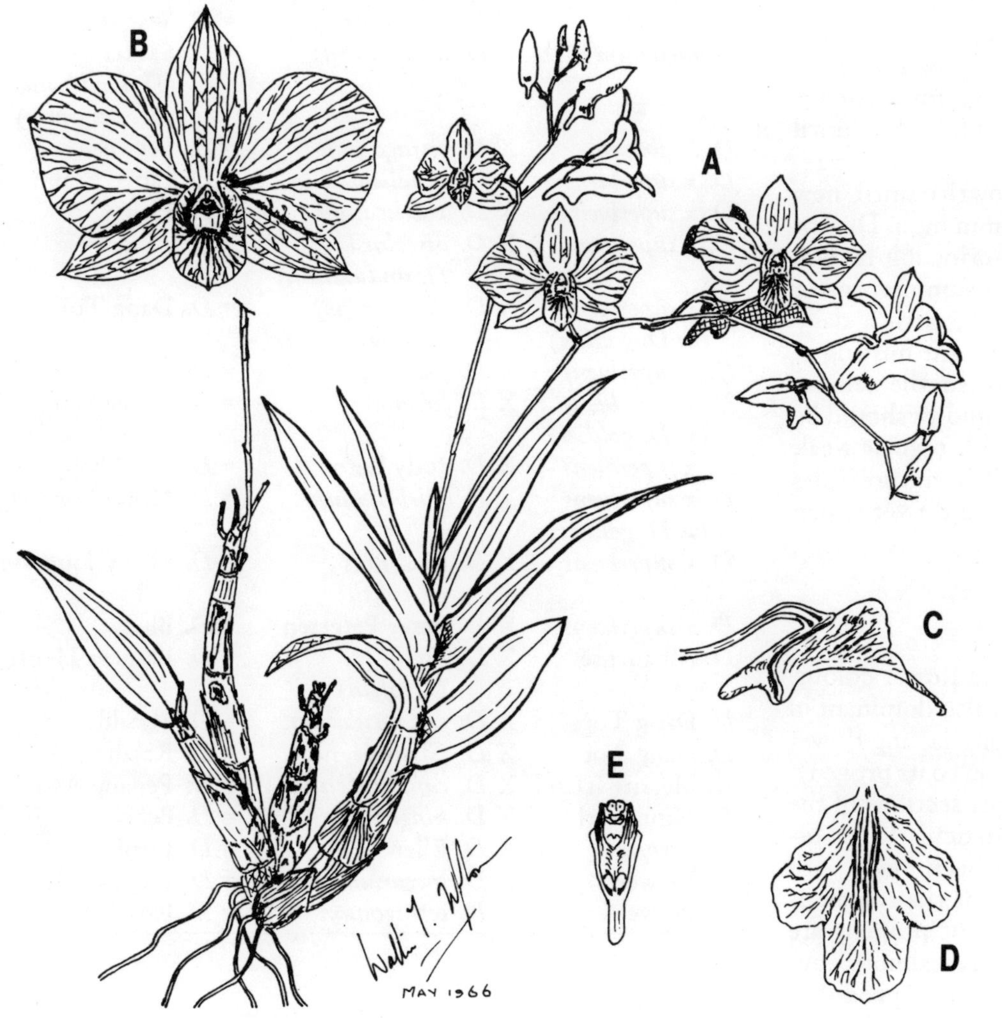

A Plant $^1/_2$X
B Flower, front-view 1X
C Column and labellum, side-view 1X
D Labellum, pressed flat 1X
E Column, front-view 1X

loss, once whole cliff faces were covered with plants—a splendid sight when in flower.

HABITAT

On rocks, frequently in fully-exposed situations and often devoid of leaf-mould, between 300 and 600 m. Occasionally on trees, but usually low down (up to 3 m).

CULTURE

Essentials for the successful growing of *D. bigibbum* and its varieties are excellent drainage, high light intensity, plenty of water while in full growth, very little water in winter, good air movement and a good rest after flowering. Minimum winter temperature 12°C.

D. bigibbum starts its growth cycle in late winter to early spring, so complete potting early, to ensure full benefit of growing season. After potting, mist foliage regularly (not during cold spells) until growth activity has started, then water carefully (not over new growth) until new growth is about 50 to 75 mm high. Do not water in the early days of spring if it is cold, rainy or overcast. In mid-summer, water two to three times a week, with a daily misting in the morning. A light misting at night can be beneficial when the temperatures are over 33°C. Humidity should be kept high in the summer. A regular weak fertilizing is beneficial. Slab culture suits well. If in pots, a small pot and a very open mix is essential.

IN HYBRIDIZATION

D. bigibbum is dominant in flower colour, and the var. *compactum* is also dominant in plant size. The species' long-lasting flower quality is generally passed on to its progeny. It is freely fertile with most sections of the subgenus *Dendrobium*, particularly *Spatulata.* A few hybrids have been made with the section *Dendrocoryne* of the subgenus *Athecebium* and the resultant progeny are reasonably cold tolerant. An almost constant mauve to purple colour is produced by this species with an occasional white.

HYBRIDS

D. bigibbum (as *D. compactum*)	X *D. affine* (as *D. dicuphum*)	= *D.* Merritt Island
D. bigibbum	X *D. agrostophyllum*	= *D.* Pink Ballerina
D. bigibbum	X *D. antennatum* (as *D. d'albertisii*)	= *D.* Albertine
D. bigibbum	X *D.* Aussie Ira	= *D.* Aussie Angel
D. bigibbum (as *D. compactum*)	X *D. canaliculatum*	= *D.* Mini Pearl
D. bigibbum	X *D. discolor*	= *D. x superbiens*
D. bigibbum	X *D. fleckeri*	= *D.* Big Fleck
D. bigibbum	X *D.* Gloucester Sands	= *D.* Gloucester Gem
D. bigibbum	X *D.* Hilda Poxon	= *D.* Brinawa
D. bigibbum	X *D.* Judy Leroy	= *D.* Eleebana
D. bigibbum	X *D. kingianum*	= *D.* Esme Poulton
D. bigibbum	X *D.* Mini Pearl	= *D.* Cherry Dance
D. bigibbum	X *D. mirbelianum*	= *D.* Princess Stephanie
D. bigibbum	X *D. nindii* (as *D. tofftii*)	= *D.* Australia
D. bigibbum	X *D.* Rachelle Simpson	= *D.* Donie
D. bigibbum	X *D. speciosum*	= *D.* Warringah
D. bigibbum	X *D. x superbiens*	= *D.* Saskia (*D. bigibbum* var. *venosum*)
D. bigibbum	X *D. tetragonum*	= *D.* Peewee
D. x superbiens	X *D. canaliculatum*	= *D.* Lloyd Stainton
D. x superbiens	X *D. x delicatum*	= *D.* Elsie Cox
D. x superbiens	X *D. discolor* (as *D. undulatum*)	= *D.* Albanense
D. x superbiens (*as D. goldiei*)	X *D. discolor* (as *D. undulatum*)	= *D.* Dang Toi
D. x superbiens	X *D. johannis*	= *D.* D'Bush Chic
D. x superbiens (as *D. goldiei*)	X *D. johannis*	= *D.* Three Star
D. x superbiens	X *D.* Judy Leroy	= *D.* Gold Baby
D. x superbiens (as *D. goldiei*)	X *D. mirbelianum*	= *D.* Moira Stewart
D. x superbiens	X *D. nindii* (as *D. tofftii*)	= *D.* Blue Gloucester
D. x superbiens	X *D.* Peter Petersen	= *D.* Bluey
D. Albanense	X *D. nindii* (as *D. tofftii*)	= *D.* Frances Henton
D. Dang Toi	X *D. canaliculatum*	= *D.* Matilda
D. Dang Toi	X *D.* Champagne	= *D.* Rajah
D. Merritt Island	X *D. canaliculatum*	= *D.* Penang Angel
D. Mini Pearl	X *D. kingianum*	= *D.* Berry
D. Peewee	X *D.* Ellen	= *D.* Lorikeet
D. Peewee	X *D. speciosum*	= *D.* Elegant Heart
D. Peewee	X *D. tetragonum*	= *D.* Rosella

ETHNOBOTANY

Dendrobium bigibbum Lindley. 'The aboriginal people of northern Cape York Peninsula roasted the bark of the stems to produce yellow strips and used them to decorate weapons and domestic implements. The women used the strips to make chainwork belts for the men and cross shoulder bands for themselves' (Lawler, 1984).

Dendrobium bowmanii Benth.

Fl. Aust. 6:286 (1873)
Fig. 19, Plate 20

SYNONYMS

Dendrobium mortii sensu F. M. Bail., *Syn. Q/ld. Fl.* 512 (1883) and *Q/ld. Fl.* 5:1534 (1902)

Dendrobium mortii (attributed erroneously to Mueller by later authors), as listed in Dockrill *Aust. Indg. Orchids* 1:360 (1969).

DESCRIPTION

Clements, while recently at Kew, sighted the type sheet for this species and determined it to be the first naming of this taxon, and therefore the earliest correct name for *D. mortii* sensu Dockrill non F. Muell. Confusion about the correct identity of *D. mortii* and *D. bowmanii* seems to have originated from F. M. Bailey (1883 and 1902) (Pers. Comm. M. Clements, 1986).

Named in honour of E. Bowman who, according to Bentham (1873), collected material of the species from the 'Bersaker Range and Port Cooper'. Probably first collected by A. Cunningham, collection cited by Bentham (1873) as 'On trees and rocks, Rodds Bay'. Common name is small pencil orchid.

Similar to *D. teretifolium* in plant habit but smaller. Slender stems become pendulous as they grow due to their weight. Plants generally about 40 to 60 cm long. Leaves cylindrical or terete, up to 15 cm long, mostly around 6 to 9 cm.

Fragrant, lemon-scented flowers generally produced singly or in pairs, but occasionally up to four flowers produced.

Fig. 19 *Dendrobium bowmanii* Benth.
From Cedar Creek, north-west of Brisbane

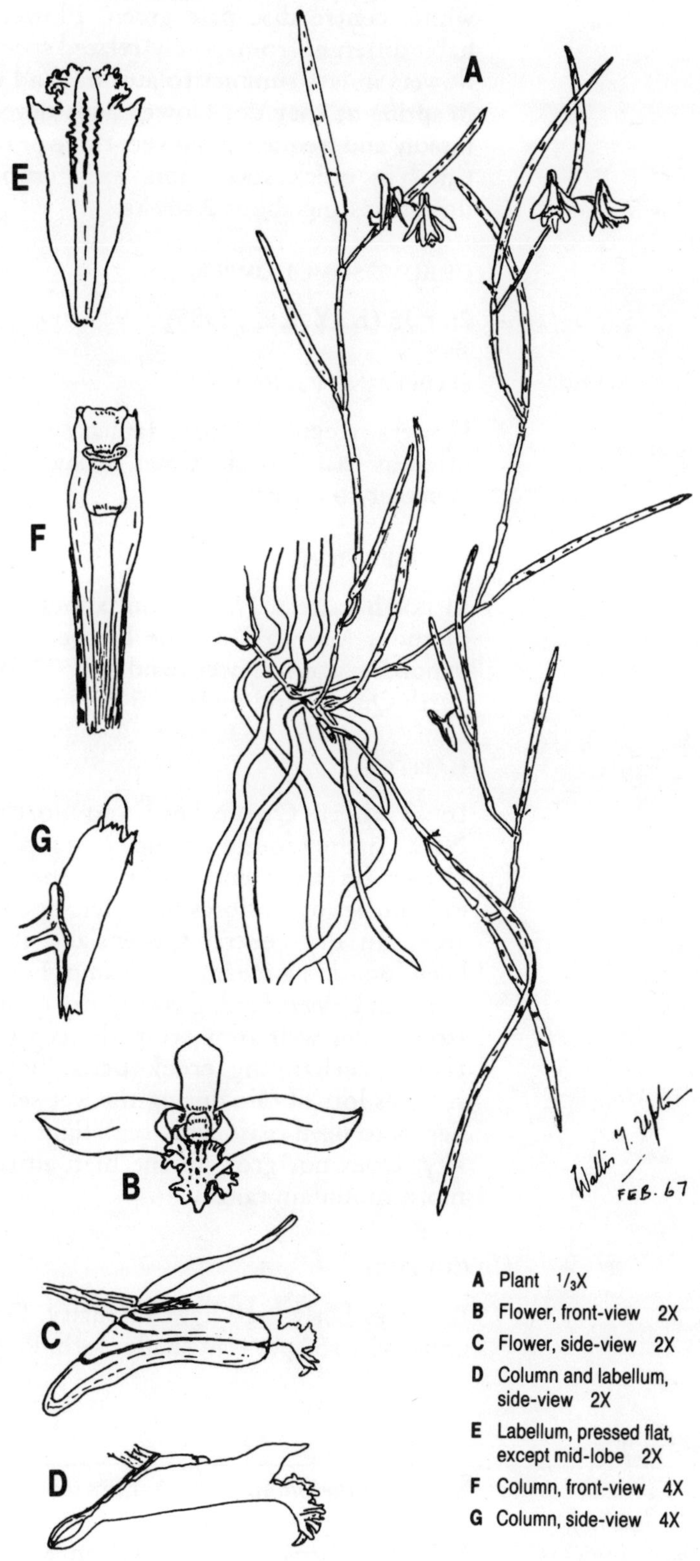

Very pale creamy-green to yellowish-green, mostly with purple-brown striate markings on the rear of the segments. Labellum white, centre disc pale green. Flowering habit different from closely-related species; flowers in late summer to autumn and not in spring as they do. Flower buds develop rapidly and frequently flowers three or four times in quick succession, each crop of flowers lasting about 2 weeks.

CHROMOSOME NUMBER

2n = 38 (K. Y. Lim, 1985).

FLOWERING PERIOD

Flowers irregularly from January to June often as many as six times in succession (summer to early winter).

DISTRIBUTION

In north-eastern NSW from about Coffs Harbour (30° 15′S) to the Herbert River in north-eastern Queensland (18° 30′S). In north Queensland up to 170 km inland.

HABITAT

In southern Queensland and northern NSW, mainly found on the coastal plains. Grows on trees and rocks in the drier extremities of rainforests. Prefers open dry areas. In dry central Queensland it has been found in a mangrove swamp. Further north in Queensland, grows in dry country (sometimes well away from the coast) on trees overhanging creek beds; here it receives lots of rain during the wet season. Seems to need reasonably high light intensity. Does not grow in the high-altitude, moist mountain ranges.

CULTURE

Requires fairly high light intensity. Grow under 40 to 50 per cent shade cloth. Prefers slab culture, pieces of hardwood or cork. Necessary to remove straggly pieces from plant to stop it becoming untidy. Needs lots of water in hot summer months. Minimum temperature of 2°C.

IN HYBRIDIZATION

Only two hybrids have been registered to date and in both, progeny are approximately midway between the two parents. Both hybrids to date were registered under the species name *D. mortii*.

HYBRIDS

D. bowmanii (as *D. mortii*)	X *D. cucumerinum*	= *D.* Limestone
D. bowmanii (as *D. mortii*)	X *D. linguiforme*	= *D.* Glenn Wall

Dendrobium calamiforme Lodd.

Edwards's Bot. Reg. 27:misc 9 (1841)

Fig. 20, Plate 12

SYNONYMS

Dendrobium teretifolium R.Br. var. *fasciculatum* Rupp., *Proc. Linn. Soc. N.S.W.* 60:157 t.4 (1935)
Dendrobium teretifolium R.Br. var. *album* C. T. White, *Proc. Roy. Soc. Q/ld.* 4:33 (1936)
Dendrobium baseyanum St. Cloud, *Nth. Q/ld. Nat.* 23 (110):1 (1955)
Dockrillia baseyana (St. Cloud) Rauschert, *Feddes Repert.* 94:446 (1983)

DESCRIPTION

In the original description it states 'a pretty and graceful epiphyte from New Holland, imported by Messrs. Loddiges'. Name comes from *calamus*, meaning stalk, and *forma*, meaning shape—referring to the shape of the leaves. In recent research Clements determined the change in status for this taxon and the earlier specific name is again recognized (Pers. comm. M. Clements, 1986).

I have not seen *D. baseyanum* St. Cloud, and further research may confirm that it is a distinct taxon. For the present it has been placed as a synonym of *D. calamiforme*, which is consistent with Dockrill's interpretation of the taxon. A drawing of the flower is shown in Dockrill (1969).

Plant similar to *D. teretifolium* but leaves frequently more robust and longer. Normally bears two racemes from the top of the stem, each with up to 15 flowers. Very showy plant in full flower. Flowers white, cream or light yellow with a few short, light-red to purple striations near the base inside and on the side lobes of the white labellum. Mid-lobe of the labellum frequently less accuminate than other closely allied species (Fig. 73).

FLOWERING PERIOD

August to November (late winter to spring). Requires lots of sun to flower.

DISTRIBUTION

North Queensland species from at least as far south as the Burdekin River (20° 00′S), northward to at least the Iron Range and Portland Roads (12° 35′S). Species very plentiful throughout most of its range.

HABITAT

Growing on trees and rocks from sea level to 1200 m. In fairly open forests to rainforests, swamps and even in some areas totally exposed to the sun. I have seen medium-sized plants in rainforest in the Iron Range in quite heavy shade. In the high ranges, rocks are frequently the main host.

CULTURE

Similar culture to *D. teretifolium*, except that it requires a minimum winter temperature of approximately 8°C (46°F).

Fig. 20 *Dendrobium calamiforme* Lodd. From Daintree River area, north Queensland

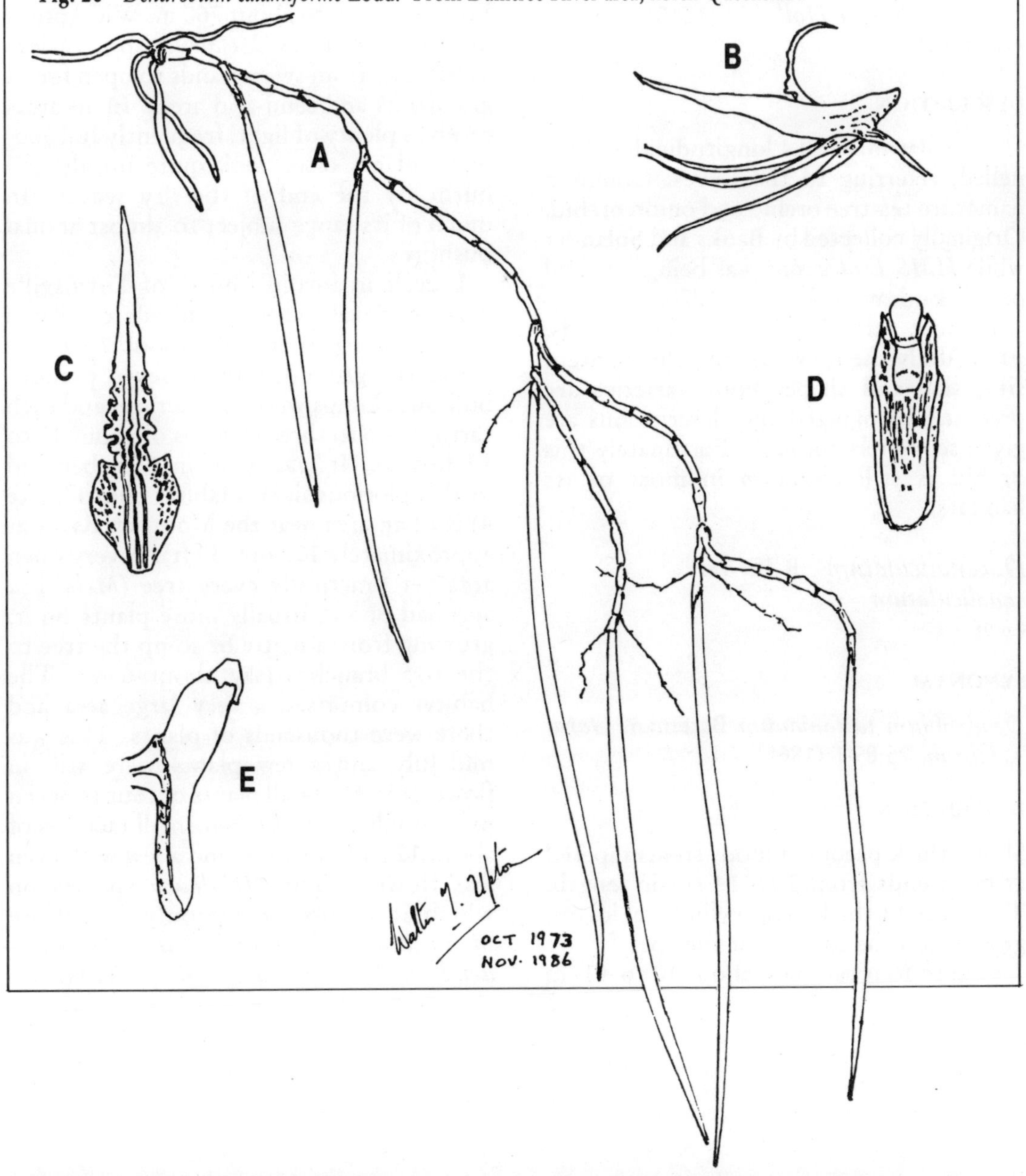

A Plant 1/3X
B Flower, side-view 1X
C Labellum, pressed flat 2Y
D Column, front-view 4X
E Column, side-view 4X

IN HYBRIDIZATION

Two natural hybrids have been reported:

HYBRIDS

D. calamiforme (as *D. teretifolium* var. *fasciculatum*)	X *D. linguiforme*	= *D.* x *grimesii*
D. calamiforme (as *D. teretifolium* var. *fasciculatum*)	X *D. rigidum*	= *D.* x *foederatum*

This species has been grown and used under the name of *D. teretifolium* var. *fasciculatum* for many years, so any hybrids using this taxon would have been registered under *D. teretifolium*.

Dendrobium canaliculatum R.Br.

Prodr. Fl. Novae Holl. 333 (1810)
Plates 4 and 22

DESCRIPTION

Canaliculatum means longitudinally channelled, referring to the leaves. Common names are tea-tree orchid and onion orchid. Originally collected by Banks and Solander while *HMS Endeavour* was being repaired on the Endeavour River in 1770.

A most variable species in all regards, particularly the flowers—in colour, intensity, size and shape. Four varieties are presently recognized and descriptions are given separately for each. Fortunately this orchid is still common in most of its habitats.

D. canaliculatum R.Br. var. *canaliculatum*

Fig. 21

SYNONYM

Dendrobium tattonianum Bateman, *Gard. Chron.* 25:890 (1865)

DESCRIPTION

Short, thick pseudobulbous stems tapered at both ends from 3 to 14 cm in length. Two to eight thick, semi-cylindrical leaves, grooved on the top-side at the apex.

One to four racemes, about 10 to 40 cm long, bearing up to 60 flowers, commonly about 25 flowers. Striking, tri-coloured, fragrant flowers with slightly twisted sepals and petals. White to pale green in the basal half and then yellow to the tips. Labellum white with rich mauve to purple markings. Flowers last approximately 5 weeks.

FLOWERING PERIOD

Mid-July to October (late winter to spring).

DISTRIBUTION

From the Fitzroy River (23° 40′S) south of Rockhampton, northward to Cape York and Papua New Guinea and westward into the northern part of the Northern Territory and adjacent islands.

HABITAT

From sea level to about 760 m. Widespread on paperbark trees (*Melaleuca* sp.). Variety of habitats from swamplands to open forest grasslands and semi-arid areas. In all areas receives plenty of light, frequently full sunlight; plants often look quite untidy and burnt by the end of the dry season. In much of its range subject to almost annual bushfires.

I recall an area just north of Townsville near sea level, where hundreds of plants were growing in trees at about 2 to 4 m above the ground, four to seven pseudobulbous clumps were the average, and each carried two to three racemes of about 10 to 12 flowers. It was early in October and made a glorious sight. Habitat photo (Plate 4) is of an area near the Moorhen River at approximately 100 m (330 ft). A very open area yet practically every tree (*Melaleuca* sp.) had one or usually more plants on it, growing from a metre or so up the tree to the top branches (say about 6 m). The habitat comprised a very large area and there were thousands of plants. This was mid-July and a few plants were still in flower, mostly small plants of four to seven pseudobulbous stems and small racemes of about 12 to 15 flowers, and a few with even less flowers. The *Melaleuca* species on which it is most commonly found are *M. viridiflora*, *M. symphiocarpa*, *M. leucadendron*, *M. argentea* and *M. dealbata*.

Fig. 21 *Dendrobium canaliculatum* R.Br. Collected at Tully, north Queensland, February 1959 by R. Miller

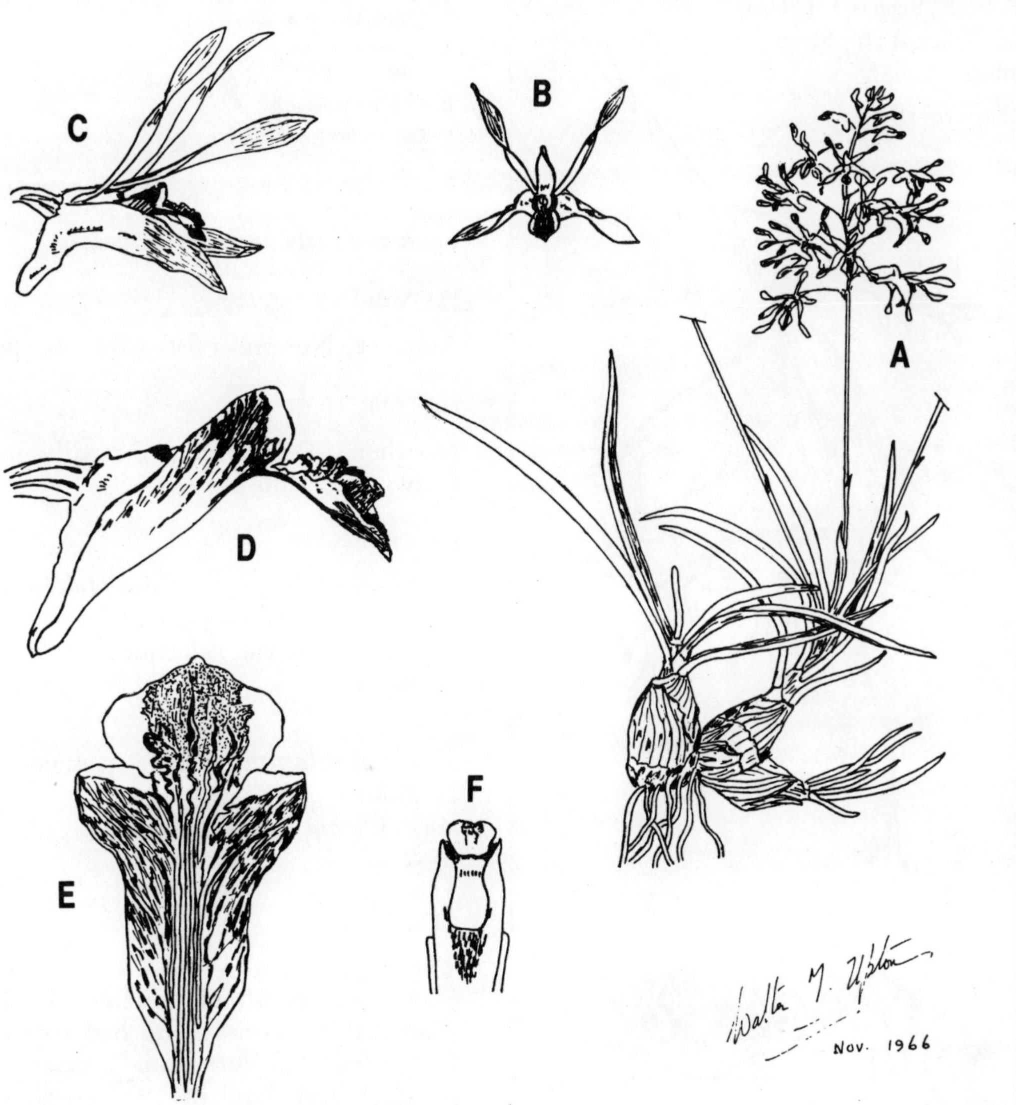

A Plant $^1/_3$X
B Flower, front-view 1X
C Flower, side-view 2X
D Column and labellum, side-view 4X
E Labellum, pressed flat 4X
F Column, front-view 4X

D. canaliculatum R.Br. var. *foelschei* (F. Muell.) Rupp et T. E. Hunt

Proc. Linn. Soc. N.S.W. 72:240 (1948)

Fig. 22, Plate 21

BASIONYM

Dendrobium foelschei F. Muell., *Sth. Sci. Rec.* 2:230 (1882)

SYNONYM

Callista foelschei (F. Muell.) Kuntze, *Rev. Gen. Pl.* 2:654 (1891)

DESCRIPTION

Named after Paul Foelsche. Plant collected near Port Darwin.

Pseudobulbous stems very similar and in many cases indistinguishable from the type variety. The pseudobulbous stems are, if anything, less swollen than Queensland varieties, and leaves more terete.

Flowers white to yellow with variously-coloured tips to the petals, i.e. green, yellow, grey, purple or brown. Flowers generally smaller and the floral segments narrower than the type variety. Generally

Fig. 22 *Dendrobium canaliculatum* R.Br. var. *foelschei* (F. Muell.) Rupp et T. E. Hunt Collected by Ben Wallace at Daly River, Northern Territory

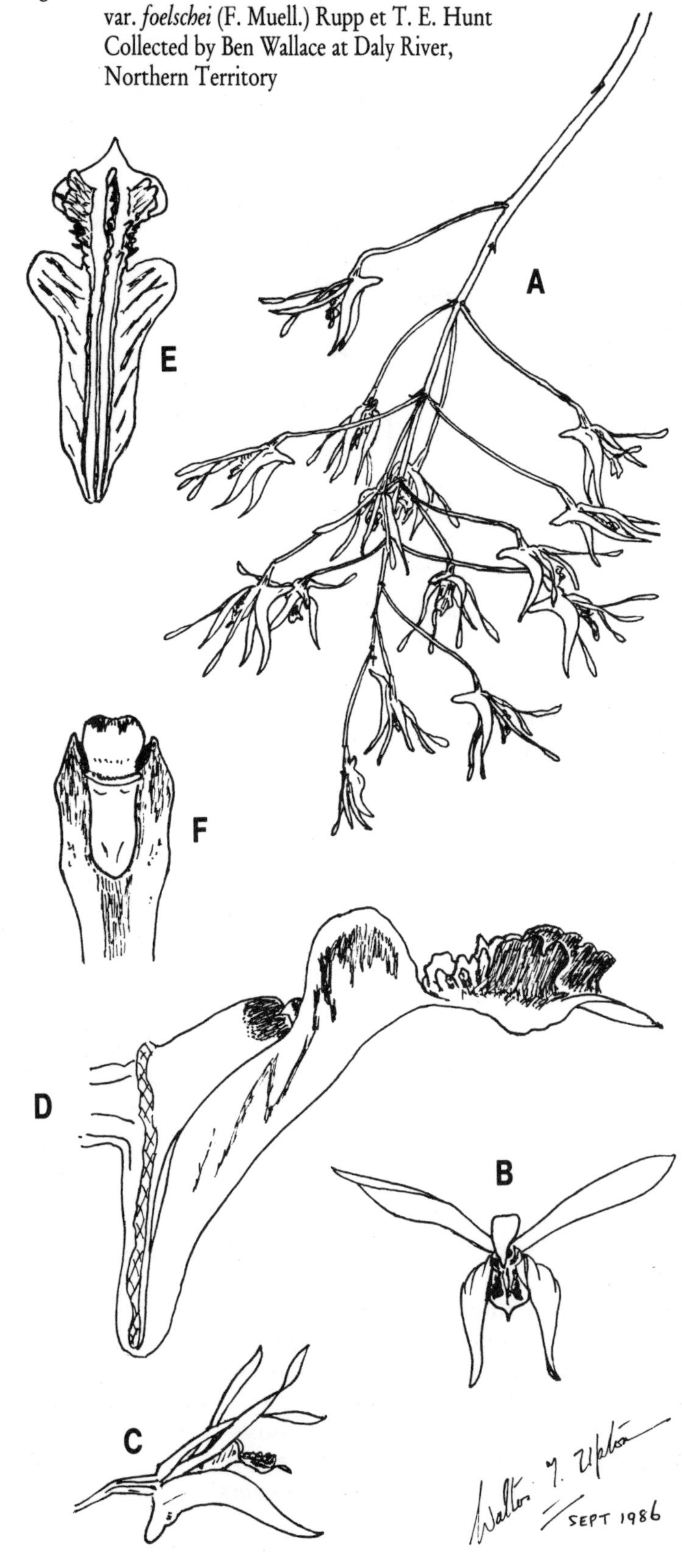

A Raceme 1X
B Flower, front-view 2X
C Flower, side-view 2X
D Column and labellum, side-view 8X
E Labellum, pressed flat 4X
F Column, front-view 8X

inflorescence pendulous and flowers do not open as widely as the type variety.

FLOWERING PERIOD

August to November (late winter to spring).

DISTRIBUTION

Northern Territory, principally in the Darwin and Gulf District.

HABITAT

Widespread in swamps on *Melaleuca* sp. and along the flood plains of rivers in the north, such as the Adelaide and Alligator Rivers.

D. canaliculatum R.Br. var. *nigrescens* Nicholls

Aust. Orchid Rev. 7:40 (1942)
Fig. 23

DESCRIPTION

The original description was taken from a plant sent to W. H. Nicholls by F. A. Weinthal of Sydney, who had received it from a friend at Port Douglas near Cairns, north Queensland in 1932, with other plants collected in that district.

Plants similar to type variety but racemes frequently shorter. Shape of flowers mostly like type variety although some have either wider or shorter segments or both wider and shorter. The main difference is the colour of the segments. They are generally varying sepia shades—lighter towards the base and sometimes greenish towards the base. The labellum has various markings from blood-red to mauve to purple on a yellow to cream or white background.

DISTRIBUTION

Throughout Cape York Peninsula north of the Mitchell River (16° 30′S), extending to some of the islands of Torres Strait.

13 *D. baileyi* segments widely spread on opening

14 *D. baileyi* segments spiralled and twisted after first two hours

15 *D. bifalce*

16 *D. bigibbum* var. *superbum*

17 *D. bigibbum* var. *superbum* white form

18 *D. bigibbum* var. *compactum*

19 *D. bigibbum* subsp. *laratensis*, Tanimbar Islands

20 *D. bowmanii*

21 *D. canaliculatum* var. *foelschei*

22 *D. canaliculatum*

23 *D. cancroides*

24 *D. carrii*

25 *D. carronii*

26 *D. cucumerinum*

27 *D. discolor*

28 *D. falcorostrum* 'Anembo'

Fig. 23 *Dendrobium canaliculatum* R. Br. var. *nigrescens* Nicholls

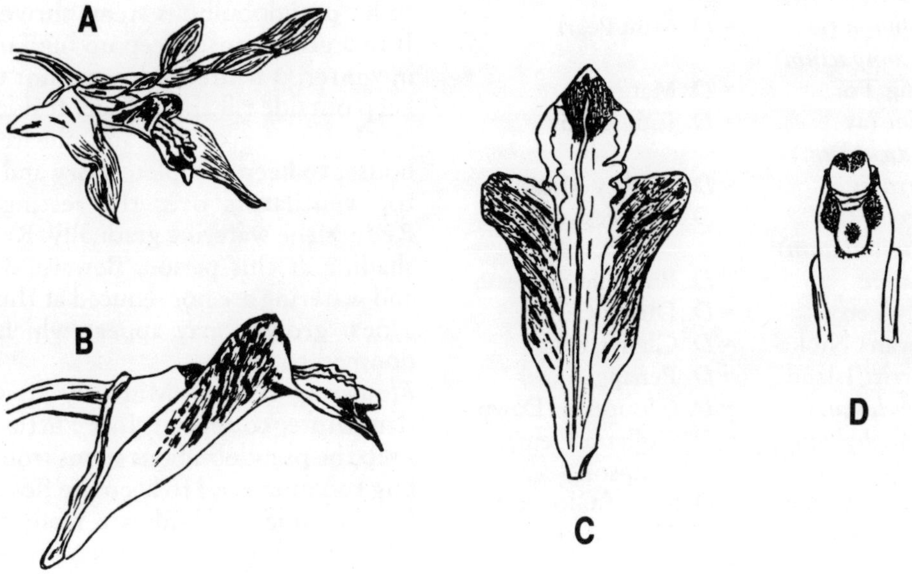

A Flower, side-view 2X
B Column and labellum, side-view 4X
C Labellum, pressed flat 4X
D Column, front-view 4X

HABITAT

The same as for the type variety.

D. canaliculatum R.Br. var. *pallidum* Dockr.
Nth. Q/ld. Nat. 24, No. 115:21 (1956)

DESCRIPTION

Name comes from *pallidum*, meaning pale. Originally collected by C. J. Leroy at the Starcke River, Cape York Peninsula in 1954.

Plants similar to the type variety.

Flowers white with pale-yellowish to green tips to the segments. Sepals and petals not usually twisted. Labellum white, mid-lobe short and broad, and keels very tall.

FLOWERING PERIOD

August to October (late winter to spring).

DISTRIBUTION

Eastern Cape York Peninsula (14° 50′S).

HABITAT

Open forest on the river flats.

CULTURE

These plants do not grow best if kept permanently on the piece of tree on which they were collected or purchased, nor do they like to be in a pot. They do best on a piece of cork (from Portugal). They have three quite distinct seasons: growing, resting and flowering.

Growing season November to March (late spring to early autumn). Copious watering at the roots, plenty of fresh air, high temperatures, but not high humidity. In Sydney they can be grown outside during this period. In their natural habitat this is the 'wet season'; rain falls nearly every day, mostly late afternoon, so watering at this time of day is beneficial. Frequently, during the hottest days, water in the mornings as well. By the end of this season the pseudobulbous stems will have reached full size. In a well-grown plant, they will be swollen and occasionally split, but this does not seem to hurt them. The new leads will appear from the base of the pseudobulbous stems; do not start heavy watering until these are well up, say about 4 to 5 cm. Before really heavy watering starts, ensure the plant is growing; it is no use watering a plant if the roots are not in active growth.

HYBRIDS

D. canaliculatum	X *D. antennatum* (as *D. d'albertisii*)	= *D.* Lowana Nioka
D. canaliculatum	X *D. bigibbum* (as *D. compactum*)	= *D.* Mini Pearl
D. canaliculatum	X *D.* Dang Toi	= *D.* Matilda
D. canaliculatum	X *D. affine* (as *D. dicuphum*)	= *D.* Judy Leroy
D. canaliculatum	X *D. carronii*	= *D.* Mighty Mite
D. canaliculatum	X *D. discolor* (as *D. undulatum*)	= *D.* Gloucester Sands
D. canaliculatum	X *D. johannis*	= *D.* Rachelle Simpson
D. canaliculatum	X *D.* Judy Leroy	= *D.* Dusky Girl
D. canaliculatum	X *D.* Lowana Nioka	= *D.* Glenys King
D. canaliculatum	X *D.* Merritt Island	= *D.* Penang Angel
D. canaliculatum	X *D. mirbelianum*	= *D.* Gloucester Dawn
D. canaliculatum	X *D. monophyllum*	= *D.* Collette Fleur Simpson
D. canaliculatum	X *D.* Peter Petersen	= *D.* Son of Gloucester
D. canaliculatum	X *D.* Rachelle Simpson	= *D.* Black Mickey Mouse
D. canaliculatum	X *D. speciosum*	= *D.* Memoria Lloyd Bradford
D. canaliculatum	X *D. x superbiens*	= *D.* Lloyd Stainton
D. canaliculatum	X *D. teretifolium*	= *D.* Ayr Queensland
D. Allyn Sands	X *D. discolor* (as *D. undulatum*)	= *D.* Allyn King
D. Judy Leroy	X *D. antennatum* (as *D. d'albertisii*)	= *D.* Lowana Tiki
D. Judy Leroy	X *D. bigibbum*	= *D.* Eleebana
D. Judy Leroy	X *D.* Colin	= *D.* Jombock
D. Judy Leroy	X *D. johannis*	= *D.* Blue Tiki
D. Judy Leroy	X *D.* Star of Gold	= *D.* Andrew Upton
D. Judy Leroy	X *D. x superbiens*	= *D.* Gold Baby
D. Judy Leroy	X *D. tetragonum*	= *D.* Duke
D. Gloucester Sands	X *D. bigibbum*	= *D.* Gloucester Gem
D. Gloucester Sands	X *D. discolor* (as *D. undulatum*)	= *D.* Dryander Mountain
D. Lowana Nioka	X *D. johannis*	= *D.* Allyn Sands
D. Lowana Nioka	X *D. speciosum*	= *D.* Justin Wall
D. Lowana Nioka	X *D.* Star of Gold	= *D.* Natalie Wall
D. Mini Pearl	X *D. bigibbum*	= *D.* Cherry Dance
D. Mini Pearl	X *D. kingianum*	= *D.* Berry
D. Rachelle Simpson	X *D.* Allyn Sands	= *D.* Allyn Supreme
D. Rachelle Simpson	X *D. bigibbum*	= *D.* Donie
D. Rachelle Simpson	X *D. discolor* (as *D. undulatum*)	= *D.* Allyn Prince

Season of rest April to mid-July (autumn to early winter). Keep drier than most dendrobes while resting. It does not hurt to let pseudobulbous stems shrivel a little. If in a greenhouse, keep up high near glass in winter, if winter is dry and not too cold, keep outside till about June. It does not do any harm, if grown in an unheated greenhouse, to keep completely dry and near the top ventilators over the resting period. Reduce the watering gradually. Reduce the shading at this period. Beware, if shading and watering are not reduced at this period, a new growth may appear which will be doomed to failure.

Flowering season Mid-July to October (late winter to spring). Just a little water to keep the pseudobulbous stems from shrivelling too much and to keep the flowers firm. By the time the buds are ready to open, watering should have been increased to about every fourth day. Minimum water temperature 13°C (55°F). Slab culture suits well, and is the best. But, if grown in a pot, a very open mix is necessary and as small a pot as possible is essential.

IN HYBRIDIZATION

When crossed with large pseudobulbous plants it reduces the plant size considerably. Generally produces good upright racemes with numerous flowers.

ETHNOBOTANY

Aboriginal people of northern Queensland used this orchid for food. The pseudobulbs are edible when stripped of the old leaves and baked (Lawler, 1984).

Dendrobium cancroides T. E. Hunt

Nth. Q/ld. Nat. 14, No. 83:30 (1947)

Fig. 24, Plate 23

SYNONYM

Grastidium cancroides (T. E. Hunt) Rauschert, *Feddes Repert.* 94:448 (1983)

DESCRIPTION

Cancroides means crab-like, and refers to the shape of the flower. Originally found

Fig. 24 *Dendrobium cancroides* T. E. Hunt
Plant grown by Peter Dunstan, Sydney

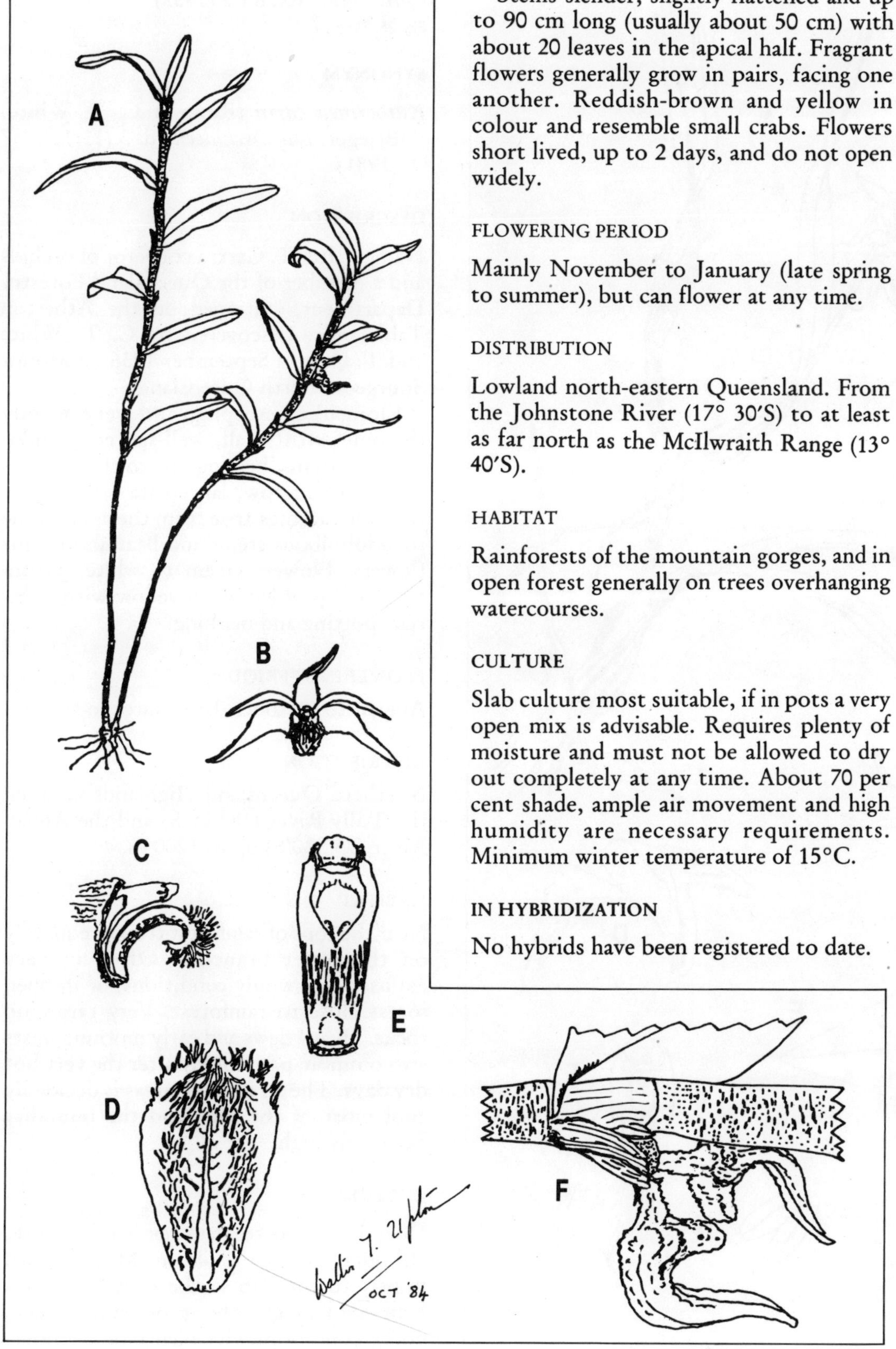

A Plant $^1/_5$X
B Flower front, open about 3 hours 1X
C Column and labellum, side-view 2X
D Labellum, pressed flat 4X
E Column, front-view 4X
F Part of stem, showing leaf base, flower bracts and flowers just before opening 2X

by J. H. Wilkie from near Babinda, north Queensland.

Stems slender, slightly flattened and up to 90 cm long (usually about 50 cm) with about 20 leaves in the apical half. Fragrant flowers generally grow in pairs, facing one another. Reddish-brown and yellow in colour and resemble small crabs. Flowers short lived, up to 2 days, and do not open widely.

FLOWERING PERIOD

Mainly November to January (late spring to summer), but can flower at any time.

DISTRIBUTION

Lowland north-eastern Queensland. From the Johnstone River (17° 30′S) to at least as far north as the McIlwraith Range (13° 40′S).

HABITAT

Rainforests of the mountain gorges, and in open forest generally on trees overhanging watercourses.

CULTURE

Slab culture most suitable, if in pots a very open mix is advisable. Requires plenty of moisture and must not be allowed to dry out completely at any time. About 70 per cent shade, ample air movement and high humidity are necessary requirements. Minimum winter temperature of 15°C.

IN HYBRIDIZATION

No hybrids have been registered to date.

Fig. 25 *Dendrobium carrii* Rupp et C. T. White
Collected by S. Clemesha, Mt. Lewis, north Queensland, 1964

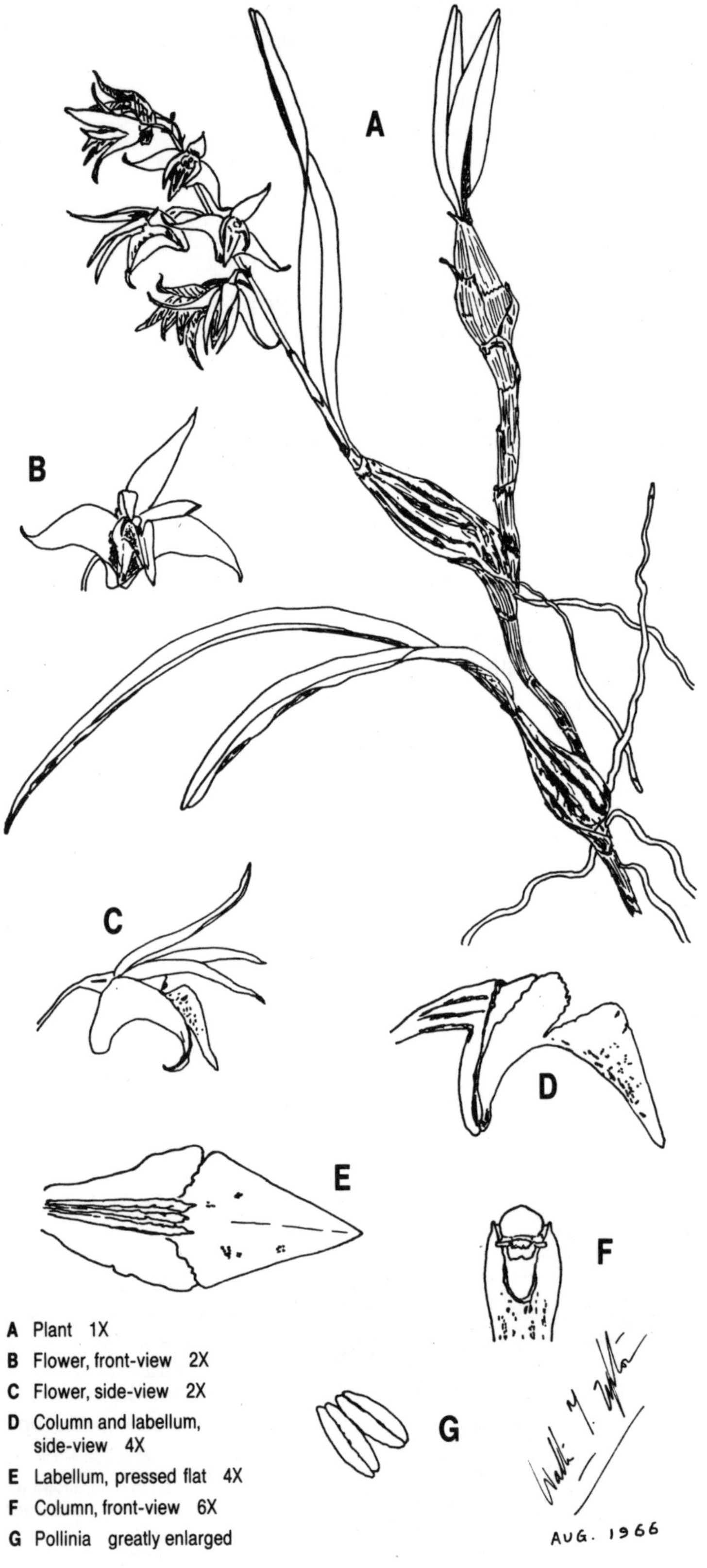

- **A** Plant 1X
- **B** Flower, front-view 2X
- **C** Flower, side-view 2X
- **D** Column and labellum, side-view 4X
- **E** Labellum, pressed flat 4X
- **F** Column, front-view 6X
- **G** Pollinia greatly enlarged

Dendrobium carrii Rupp et C. T. White

Qld., Nat. 10:26 t.2 (1937)
Fig. 25, Plate 24

SYNONYM

Katherinea carrii (Rupp et C. T. White) Brieger, *Die Orchideen* ed. 3 (11/12):737 (1981)

DESCRIPTION

Named after T. Carr, a collector of orchids and a member of the Queensland Forestry Department, stationed on the Atherton Tablelands. Discovered by C. T. White and T. Carr in September 1936 on Mount Spurgeon, north Queensland.

Elongated, creeping and very woody rhizome with small, well-spaced pseudobulbous stems, bearing one to three (generally two) narrow, lanceolate leaves. One to three racemes arise from the apex of the pseudobulbous stems and bear about nine flowers. Flowers cream to white, a contrasting lip of lemon to yellow with light-red spotting and marking.

FLOWERING PERIOD

August to October (late winter to spring).

DISTRIBUTION

Northern Queensland Highlands between the Tully River (18° 00′S) and the Annan River (15° 30′S) up to 1200 m.

HABITAT

Near the tops of rainforest trees, frequently on the outer branches. Often in very exposed and windy conditions, or in open forest close to rainforest. Very rarely on rocks. Heavy dews and early morning mists are common, particularly after the very hot dry days. The air in these areas is decidedly cool most of the time and the humidity extremely high.

CULTURE

Does best tied to slabs, preferably cork, although slow to establish. Must be tied tightly to the slab so the roots can get a firm hold as quickly as possible. Shade-house conditions with plenty of air movement suit it well in the Sydney area of NSW, although a little bit of warmth in the mid-winter seems to be beneficial. Minimum winter temperature of 6°C. Water heavily while growing; responds well to a misting at night in the very hot months.

IN HYBRIDIZATION

No hybrids have been registered to date.

Dendrobium carronii Lavarack et Cribb

Austrobaileya 1(5):497–501 (1982)

Fig. 26, Plate 25

DESCRIPTION

Named in honour of William Carron (1823–1876), botanist to the Kennedy expedition in 1848.

Plant almost identical to *D. canaliculatum* R.Br. except generally smaller in all its vegetative parts and possibly more purple. Pseudobulbous stems short, tapered at both ends, up to 5 cm long. Up to four thick, semi-cylindrical leaves.

Racemes arching, one to three per pseudobulbous stem, up to 20 cm long with two to 12 flowers. Fragrant flowers differ quite considerably from *D. canaliculatum*; sepals white to greenish with few purple suffusions, petals dark-brown to purple with a white base, twisted once or not at all. Lip green to green-yellow. Flowers last approximately 3 weeks.

FLOWERING PERIOD

August to October (late winter to spring), although plants given to me by Kevin McFarlane in 1978 were in flower on the 3 May.

Fig. 26 *Dendrobium carronii* Lavarack et Cribb
Collected at Tozer's Gap, north Queensland, 1981

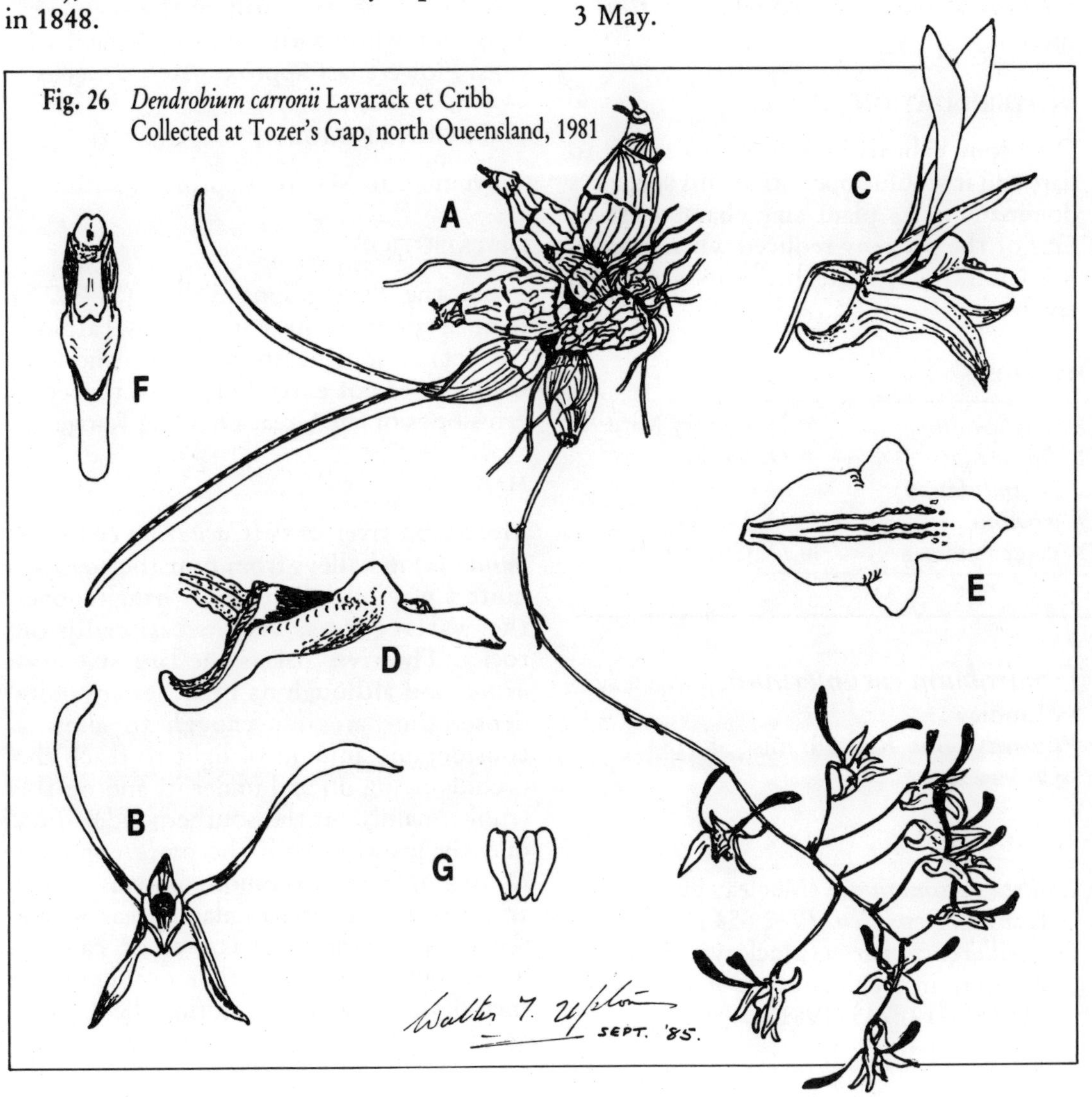

A Plant ²/₃X
B Flower, front-view 2X
C Flower, side-view 2X
D Column and labellum, side-view 3X
E Labellum, pressed flat 3X
F Column, front-view 4X
G Pollinia greatly enlarged

DISTRIBUTION

So far found in a few localities near Coen (13° 57′S), then north to the Jardine River and New Guinea.

HABITAT

From sea level to 500 m. In open forests and swamps. Common hosts are various *Melaleuca* species. Always areas of high humidity. I have seen it at Tozer's Gap growing on *Melaleuca viridiflora* in open forest; the plants were generally about 2 to 4 m from the ground. Frequent swampy areas were present, the trees were spaced well apart and the ground was covered with low grasses.

CULTURE

Refer to the culture notes under *D. canaliculatum*, as they apply equally well to this species.

IN HYBRIDIZATION

Only four hybrids have been registered to date and it would appear that this species is dominant in its plant size characteristics. Size of the progeny reduced when crossed with a large pseudobulbous-stemmed species.

HYBRIDS

D. carronii	X *D. canaliculatum*	= *D.* Mighty Mite
D. carronii	X *D. discolor* (as *D. undulatum*)	= *D.* Valda
D. carronii	X *D. johannis*	= *D.* Minnie
D. carronii	X *D.* Peter Petersen	= *D.* Cherub

Dendrobium cucumerinum Macleay ex Lindley

Edwards's Bot. Reg. 28: misc 58 (1842)

Fig. 27, Plate 26

SYNONYMS

Callista cucumerinum (Macleay ex Lindley) Kuntze, *Rev. Gen. Pl.* 2:654 (1891)

Dockrillia cucumerina (Macleay ex Lindley) Brieger, in Schlechter *Die Orchideen* ed. 3 (11/12):745 (1981)

DESCRIPTION

Cucumerinum means like a cucumber—referring to the leaves. Common name is cucumber orchid. Found by William Sharp Macleay on *Casuarina* trees at Hunter Creek, a tributary of the Nepean River, at Brownlow Hill near Camden, NSW. Macleay sent it to Messrs Loddiges in England and Lindley described it.

Creeping rhizomes, sometimes in dense masses, but most frequently in long, drawn-out branching rhizomes up to 2 to 5 cm long. Dark-green, tuberculate leaves up to 4 cm long, resembling small gherkins.

Slightly perfumed flowers, approximately 2·5 cm in diameter, on a short raceme with a very short peduncle. Usually about four or five flowers, occasionally up to 12 in its native habitat. Flowers green-white through cream to pale-yellow with reddish-purple streaks in the basal half of the segments. Labellum white with reddish-purple markings. Flowers last approximately 2 weeks.

FLOWERING PERIOD

November to May (late spring to autumn).

DISTRIBUTION

From the Burragorang Valley (34° 00′S) near Sydney (where it is now almost extinct) to south-eastern Queensland (26° 00′S). Does not extend inland to the western slopes of the Great Dividing Range.

HABITAT

Mostly on river oaks (*Casuarina cunninghamiana*) in valleys from near the coast to quite a way inland, generally hanging over the watercourses, very occasionally on rocks. The river oak is the largest *Casuarina*, and although its branches are quite dense, they are fine enough to allow a considerable amount of light to reach the orchids living on and under it, and on the trunk (mainly on the southern side). Frequently growing with the more common *D. linguiforme.* Common only in a few areas, seems to prefer inland areas where the temperature range is greater, causing heavy night dews and early morning mists, then drying out quickly during the day.

Fig. 27 *Dendrobium cucumerinum* Macleay ex Lindley Collected near Gloucester, NSW, 1963.

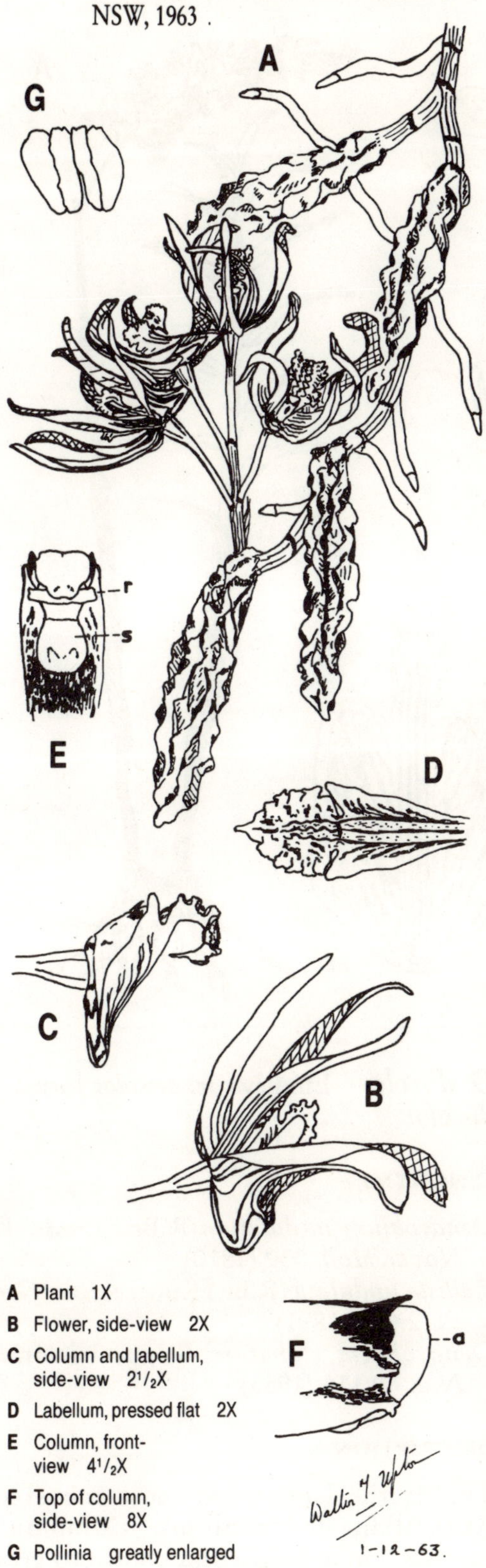

A Plant 1X
B Flower, side-view 2X
C Column and labellum, side-view 2½X
D Labellum, pressed flat 2X
E Column, front-view 4½X
F Top of column, side-view 8X
G Pollinia greatly enlarged

CULTURE

I find this species very difficult to grow well. Tied tightly to rough, hardwood blocks seems the best treatment to get it going. Enjoys a very weak fertilizer, say about a quarter the strength shown on the packets of proprietary brands, N:P:K as near 30:10:10 as possible. Water heavily when it is growing well in the hot months; a misting late evening on hot nights is beneficial. Prefers quite a bit of shade. Good air movement is essential. Minimum winter temperature of 0°C.

IN HYBRIDIZATION

No strikingly obvious dominant characteristics at this stage, although the leaves of the few hybrids I have seen are quite rough, almost bordering on tuberculate. Flowers generally midway between the parents. A hybrid made with *D. bigibbum*, very slow growing and leaves almost tubular in shape; it has not flowered yet, so has not been registered.

HYBRIDS

D. cucumerinum	X *D. bowmanii* (as *D. mortii*)	= *D.* Limestone
D. cucumerinum	X *D. schoeninum* (as *D. beckleri*	= *D.* Burie
D. cucumerinum	X *D. teretifolium*	= *D.* Goose Bumps

Dendrobium discolor Lindley

Edwards's Bot. Reg. 27: misc. 21, t.52 (1841)

Fig. 28, Plate 27

DESCRIPTION

Discolor means of different colours. Common name is golden orchid. Discovered by Robert Brown, naturalist on *HMS Investigator*, when Captain Flinders was mapping the Queensland Coast during the first circumnavigation of Australia.

This species was originally named by Robert Brown as *D. undulatum* in 1810, and was known by this name until 1963 when A. D. Hawkes reported on the correct status for this taxon in the September issue of the *Aust. Orchid Rev.* He stated

A Plant 1/8X
B Flower, front-view 1X
C Flower, side-view 1X
D Labellum, pressed flat 2X
E Column, front-view 2X
F Column, side-view, column foot in cross-section 2X

Fig. 28 *Dendrobium discolor* Lindley
A From near Wenlock River, Cape York, 1981 **B to F** From north of Cairns area, north Queensland

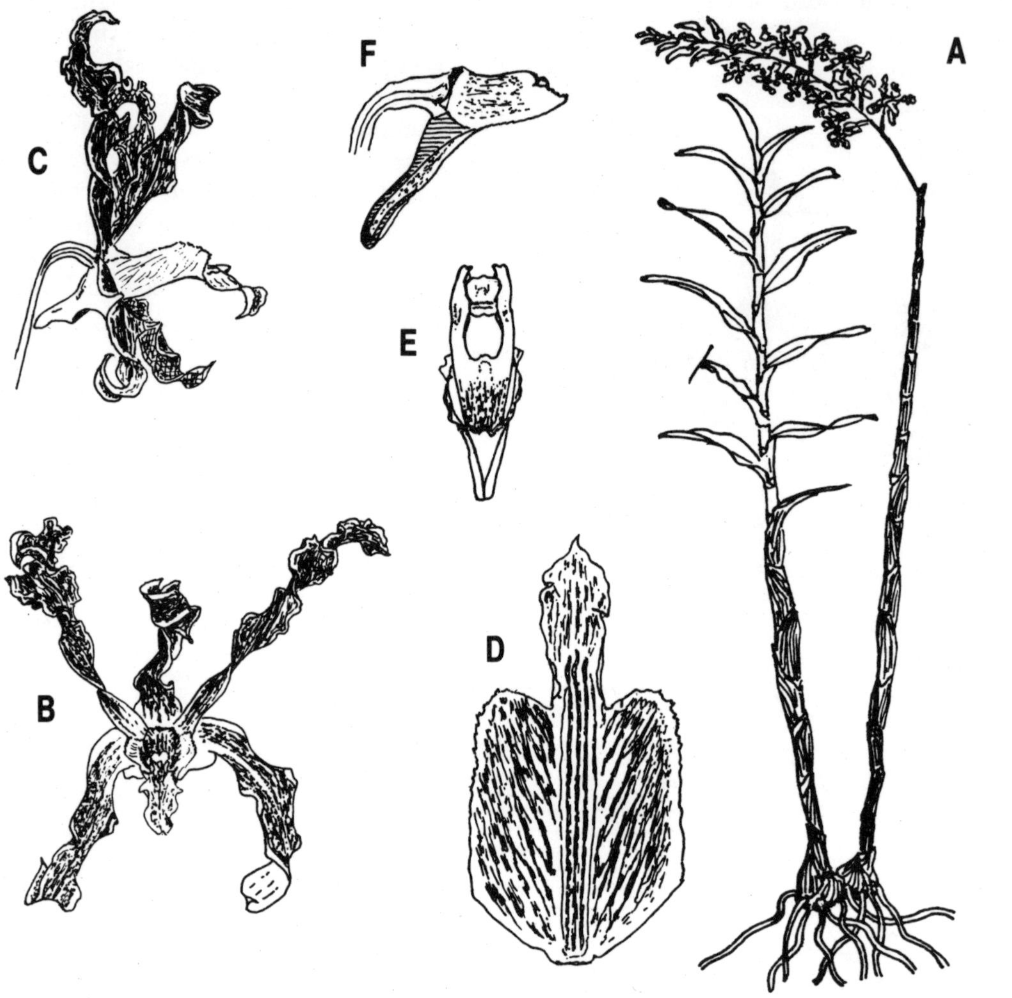

'. . . *Dendrobium undulatum* of Robert Brown was predated by three years by another *Dendrobium undulatum*, this of Persoon (*Synopsis Plantarum* 2:524, 1807). Even though this is now referred, with some degree of doubt, to a species of the American genus *Xylobium*, its existence makes the Australian *Dendrobium undulatum* a homonymic name, and hence unusable under contemporary rules of botanical nomenclature.' The next available name was *D. discolor*.

It is a complex species. In Australia, at present, there are three named varieties and one of these, *D. discolor* var. *discolor* (an extremely variable variety), has two named forms. Each variety and form are described separately.

D. discolor Lindley var. *discolor* forma *discolor*

SYNONYMS

Dendrobium undulatum R.Br., *Prodr. Fl. Novae Holl.* 332 (1810)
Callista undulata (R.Br.) Kuntze, *Rev. Gen. Pl.* 2:655 (1891)
Dendrobium elobatum Rupp, *Victorian Nat.* 69:116 (1953)

DESCRIPTION

This is the largest and most robust of Australian dendrobiums. Commonly exceeds 2 m in height and known to grow to 5 m. Frequently grows into large clumps of considerable dimensions.

Pseudobulbous stems with a small basal swelling, then more or less tapered to each end except for a swelling in the middle or somewhere between the middle and the base. Numerous leaves extremely coriaceous, produced on the upper two-thirds of the pseudobulbous stem. Up to six arching racemes, up to 60 cm long, produced from the upper part of the pseudobulbous stem. (In highland forms, pseudobulbous stems greenish-gold in colour and up to 5 m. In lowland forms, pseudobulbous stems shorter and stouter and also greater in number and lack the greenish tinge, being subject to more direct sunlight.)

The open forest country seems to promote longer racemes of larger flowers. Widely-expanding, fragrant, long-lasting flowers, occasionally up to 80 per raceme, usually about 30 to 40. Extremely variable in size (4 to 8 cm), twisting of segments, size of individual floral parts, colour, etc. The colour is from dark-brown through to golden-yellow often turning reddish as flowers wither. A dingy brown bordered with yellow is quite common. Lip generally whitish, veined to varying degrees with violet.

Kitchen (1970) reports two unusual colour forms from the tip of Cape York: (1) 'the flowers are white on the back and faintly striated with blue on the front. The labellum is yellow and white'; (2) '. . . collected . . . about 200′ from the first, but the sepals and petals are suffused with pale brown instead of blue'.

FLOWERING PERIOD

Main season July to October (winter to spring), but can often occur as early as April and as late as December.

DISTRIBUTION

Eastern part of north Queensland from approximately Rockhampton (23° 30′S), northward to the tip of Cape York Peninsula, to many of the islands of Torres Strait and into Papua New Guinea where it is also very common.

HABITAT

Common species, frequently swamp dwelling, mainly a lowland plant but also encountered on the high ranges. Grows on rocks, cliffs and trees, mostly in exposed or lightly-shaded areas. Not uncommonly seen right on the coast, sitting on the tops of rocks fully exposed to salt spray as well as sun. In scrub behind the sand hills, *D. discolor* can be frequently found on fallen trees; and plants torn from the trees by strong winds continue to grow where they fall, with their roots completely buried in the sand. In the Iron Range area, I have seen it in full flower high up in larger trees in the rainforest, a majestic sight. In some of its habitats, the offshore islands for instance, it receives no water, except dew at night, for approximately 5 months of the year.

D. discolor Lindley var. *discolor* forma *broomfieldii* (Fitzg.) Dockr.

Aust. Pl. 3:28–30 (1964)

BASIONYM

Dendrobium undulatum R.Br. var. *broomfieldii* Fitzg., *Aust. Orch.* 2(3):(t.7) (1888)

SYNONYM

Dendrobium undulatum R.Br. var. *carterae*. F. M. Bail., *Q/ld. Agric. J.* 27:306 (1911).

DESCRIPTION

Fitzgerald (1888) stated 'I have named it after my friend Captain Broomfield, who obtained it from Northern Australia, flowered it in his hot-houses, and kindly afforded me the opportunity of depicting it and many other Orchids'.

Plant similar to type form, with flowers that are generally a clear golden yellow to canary yellow, with white keels on the labellum.

DISTRIBUTION

Uncommon throughout the range of the type species.

D. discolor Lindley var. *fimbrilabium* (H. G. Reichb.) Dockr.

Aust. Pl. 3:28–30 (1964)

DESCRIPTION

Name comes from *fimbriae*, meaning fringe, and *labia*, meaning lip—referring to the fringed labellum. I have no knowledge of this variety except the original description which reads: 'Plant similar to that of var. *discolor* but flowers with a labellum which has fimbriate lateral lobes'.

DISTRIBUTION

Could be looked for in areas in which var. *discolor* occurs in quantity.

D. discolor Lindley var. *fuscum* (Fitzg.) Dockr.

Aust. Pl. 3:28–30 (1964)

BASIONYM

Dendrobium fuscum Fitzg., *Sydney Morning Herald* 3 (24/9/1879)

DESCRIPTION

Name comes from *fuscus*, meaning dark—referring to the brown flowers.

Fitzgerald's letter to the Editor of the *Sydney Morning Herald* with his original description states: 'In the Queensland Court of the Exhibition, among specimens of *Dendrobium undulatum*, and confounded with it, is a new species closely allied to *D. undulatum*, but inferior to it except in foliage. It can never, owing to the dinginess of the flowers, be a favourite with florists, but it is interesting to the botanist as another addition to the largest genus of our Australian epiphytal orchids. I am indebted to Mr J. S. Edgar, of the Botanic Gardens, Rockhampton—who kindly gave me one of the plants—for the opportunity of figuring and describing this addition to our flora. It has been sent to the Exhibition from Port Denison. It flowers in September.'

The original description of the flowers only is given here as the plant is similar to *D. discolor:* '. . . sepals broadly lanceolate acute, the lower ones falcate three-quarters of an inch, reddish brown, lighter on the edges (not undulate); petals linear oblong, undulate towards the top, 1 inch darker than the sepals, margins light, labellum half an inch, linear, two lines broad, lateral lobes incurved, central lobe about 1 line long, acute; plate of the disc, 5, much raised towards the centre lobe; spur, about two lines, obtuse, slightly recurved; column short, broad, the wings higher than the anther, emarginate; anther very flat.'

CULTURE

D. discolor seems to be continuously in growth, but does benefit from a short rest in winter when kept slightly dry.

Although mainly a north Queensland lowland plant, it is reasonably tolerant of winter temperatures down to about 6°C (43°F), or even lower if its roots are dry on the coldest nights. Needs very high humidity in the hottest months; does not like the hot westerly winds experienced in late summer in Sydney, NSW.

High light intensity essential if it is to flower well. Grows well in a pot, in a very open mix, and equally well on a large piece of cork, provided it is watered frequently in summer. The large pseudobulbous stems take many months to mature, and the plant flowers off the older pseudobulbous stems as the new growths develop. Frequent weak fertilizing is a great benefit.

HYBRIDS

D. discolor (as (*D. undulatum*)	X *D. affine* (as *D. dicuphum*)	= *D.* Chieno
D. discolor (as (*D. undulatum*)	X *D.* Allyn Sands	= *D.* Allyn King
D. discolor (as *D. undulatum*)	X *D. antennatum* (as *D. d'albertisii*)	= *D.* Verninha
D. discolor (as (*D. undulatum*)	X *D. bifalce*	= *D.* Gold Stripe
D. discolor (as (*D. undulatum*)	X *D. bigibbum*	= *D. x superbiens*

HYBRIDS (con't)

D. discolor (as (*D. undulatum*)	X *D. canaliculatum*	= *D.* Gloucester Sands
D. discolor (as (*D. undulatum*)	X *D. carronii*	= *D.* Valda
D. discolor (as (*D. undulatum*)	X *D.* Gloucester Sands	= *D.* Dryander Mountain
D. discolor (as (*D. undulatum*)	X *D. johannis*	= *D.* Johulatum
D. discolor (as (*D. undulatum*)	X *D. mirbelianum*	= *D.* Champagne
D. discolor (as (*D. undulatum*)	X *D. nindii* (as *D. tofftii*)	= *D.* Peter Petersen
D. discolor (as (*D. undulatum*)	X *D.* Peter Petersen	= *D.* Camp Island
D. discolor (as (*D. undulatum*)	X *D.* Rachelle Simpson	= *D.* Allyn Prince
D. discolor (as (*D. undulatum*)	X *D. x superbiens*	= *D.* Albanense
D. discolor (as (*D. undulatum*)	X *D. x superbiens* (as *D. goldiei*)	= *D.* Dang Toi
D. discolor (as (*D. undulatum*)	X *D. tetragonum*	= *D.* Willowie Gold
D. discolor (as (*D. undulatum*)	X *D. trilamellatum* (as *D. semi-fuscum*)	= *D.* John's Charm
D. Champagne	X *D. discolor* (as (*D. undulatum*)	= *D.* Yellow Curls
D. Champagne	X *D. johannis*	= *D.* Sauterne
D. Dang Toi	X *D. canaliculatum*	= *D.* Matilda
D. Dang Toi	X *D.* Champagne	= *D.* Rajah
D. Dang Toi	X *D. nindii* (as *D. tofftii*)	= *D.* Frances Henton
D. Gloucester Sands	X *D. bigibbum*	= *D.* Gloucester Gem
D. Peter Petersen	X *D. canaliculatum*	= *D.* Son of Gloucester
D. Peter Petersen	X *D. carronii*	= *D.* Cherub
D. x superbiens	X *D. bigibbum*	= *D.* Saskia (or *D. bigibbum* var. *venosum*)
D. x superbiens	X *D. canaliculatum*	= *D.* Lloyd Stainton
D. x superbiens	X *D. x delicatum*	= *D.* Elsie Cox
D. x superbiens	X *D. johannis*	= *D.* D'Bush Chic
D. x superbiens (as *D. goldiei*)	X *D. johannis*	= *D.* Three Star
D. x superbiens	X *D.* Judy Leroy	= *D.* Gold Baby
D. x superbiens (as *D. goldiei*)	X *D. mirbelianum*	= *D.* Moira Stewart
D. x superbiens	X *D. nindii* (as *D. tofftii*)	= *D.* Blue Gloucester
D. x superbiens	X *D.* Peter Petersen	= *D.* Bluey
D. Verninha	X *D.* Yellow Curls	= *D.* Wee Cheow Lay
D. Willowie Gold	X *D. speciosum*	= *D.* Manning

IN HYBRIDIZATION

Most of the progeny have strong, stout pseudobulbous stems. If crossed with a small pseudobulbous-stemmed species, stem length is fortunately not dominant. When crossed outside its own section, the twisting of the segments is not dominant. Not particularly colour dominant. Smallish mid-lobe to the labellum does not generally come through, although the good side lobes do. When crossed with *D. nindii*, the mid-lobe of the labellum is, of course, small because both the species have small mid-lobes. Many delightful hybrids have been made with this species, particularly within its own section *Spatulata*, producing long-ish racemes of many flowers.

ETHNOBOTANY

Lawler (1984) states, 'In 1976 I witnessed aboriginal Australians at the Lockhart River Community in northern Queensland crafting weapons using traditional methods and materials; strips of the outer skin of this orchid are used for binding,' and 'In Queensland a paste from young canes was used by Chinese herbalists to draw boils, and a spirit extract of the crushed mature canes was used as a liniment cum cure-all, especially for ringworm (E. Brennan, personal communication)'.

Dendrobium fairfaxii F. Muell. et Fitzg.

Sydney Mail 360 (21/9/1872)

Fig. 29, Plate 29

SYNONYMS

Dendrobium teretifolium R.Br. var. *fairfaxii* (F. Muell. et Fitzg.) F. M. Bail., *Syn. Qld. Fl.* 1:512 (1883)

Callista fairfaxii (F. Muell. et Fitzg.) Kuntze, *Rev. Gen. Pl.* 2:654 (1891)

Dockrillia fairfaxii (F. Muell. et Fitzg.) Rauschert, *Feddes Repert.* 94:446 (1983)

DESCRIPTION

Named after the editor of the *Sydney Mail*, Mr Fairfax. Type was collected at Mount Tomah in August 1872 by R. D. Fitzgerald.

Fig. 29 *Dendrobium fairfaxii* F. Muell. et Fitzg.
From near Blackheath, NSW

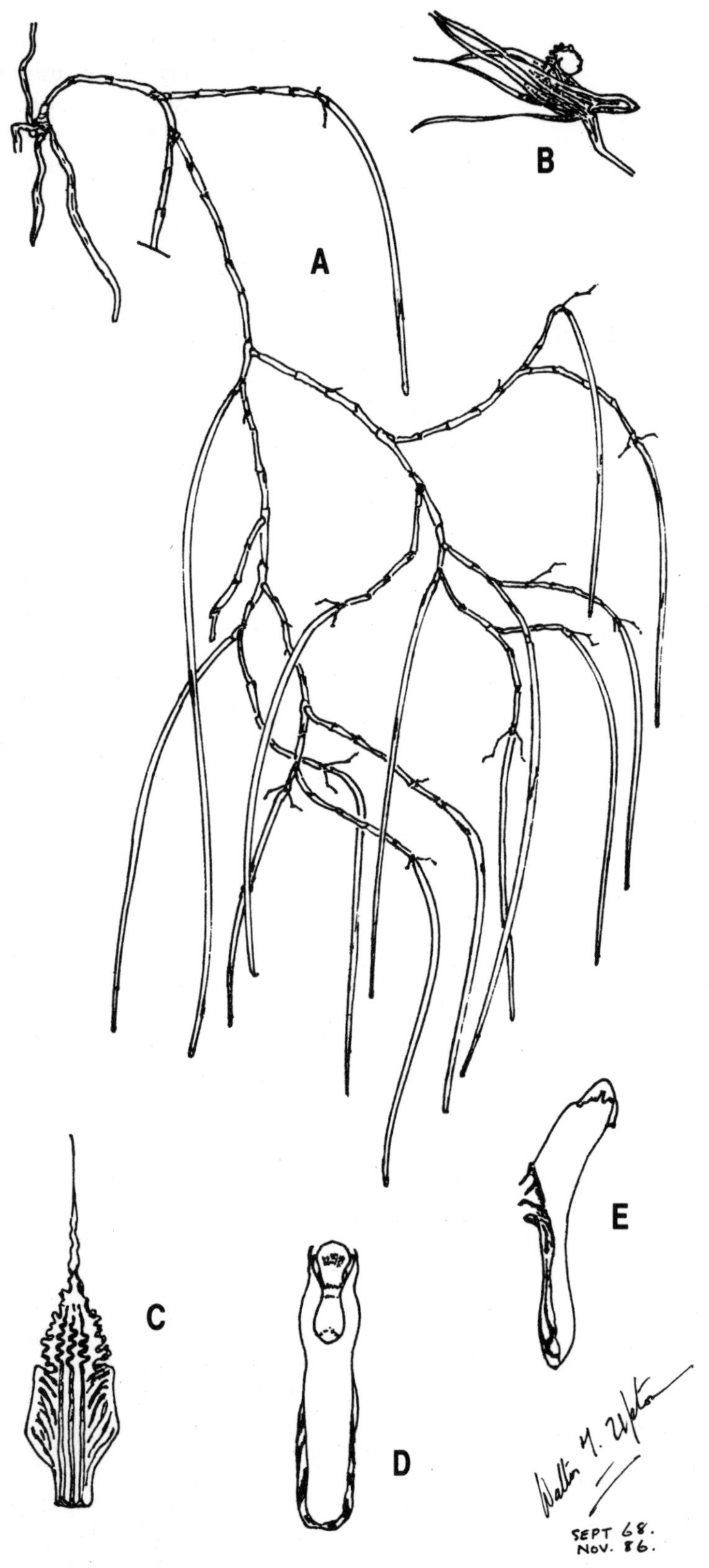

In recent research Clements has determined the change in status for this taxon and the earlier specific name is again recognized (Pers. Comm. M. Clements, 1986).

Grows in rainforests where it receives less sun. Leaves more slender and generally longer than those of *D. teretifolium.*

Racemes bear one to four or more flowers, but normally two. Not as floriferous as *D. teretifolium*, and segments of the flowers usually less recurved, and prominently striate on the basal half with reddish purple (Fig. 73). Keels of the labellum also reddish purple. Labellum devoid of spots and these are replaced by striation. Has an almost waxy appearance.

FLOWERING PERIOD

August to October (late winter to spring).

DISTRIBUTION

From the Blue Mountains (34° 00′S) of central NSW to at least the upper reaches of the Clarence River in NSW (28° 30′S) and possibly into the McPherson Ranges— in this area it overlaps with a closely allied species, referred to here as *Dendrobium* sp. aff. *fairfaxii.*

HABITAT

On rainforest trees mostly in the ranges up to at least 900 m.

CULTURE

Similar culture to *D. teretifolium.* This species is quite slow to establish, sometimes taking up to 3 years. Therefore necessary to tie it on to its host very, very tightly.

IN HYBRIDIZATION

This species has been grown and used under the name *D. teretifolium* var. *fairfaxii* for many years, so any hybrids using this taxon would have been registered under *D. teretifolium.*

A Plant $^1/_3$X
B Flower, side-view 1X
C Labellum, pressed flat 2X
D Column, front-view 4X
E Column, side-view 4X

Dendrobium sp. aff. *fairfaxii*

Fig. 30, Plate 30

SYNONYMS

Dendrobium teretifolium R.Br. var. *aureum* F. M. Bail., *Q/ld. Fl.* 5:1534 (1902)
Dendrobium teretifolium R.Br. var. *fairfaxii* (F. Muell. et Fitzg.) F. M. Bail. forma *aureum* (F. M. Bail.) Clemesha, *Orchadian* 2:11 (1965)

DESCRIPTION

Clements is researching this taxon, and it is still unnamed at specific rank, something that will be attended to in the near future. (Pers. Comm., M. Clements, 1986).

Collected by J. F. Bailey at Killarney, south-east Queensland about 1902.

Plant form is frequently of a straggling, rambling habit in appearance. Appears to be two vegetative forms, one has leaves that are quite slender and frequently about 50 cm or more long, the other is much more robust — no doubt the habitat conditions could cause this.

Very similar to *Dendrobium fairfaxii* except that it has larger flowers; racemes bear up to seven or more flowers; segments often slender and more recurved; colour from greenish-yellow to a deep golden-yellow; and filiform apex of the mid-lobe of the labellum is longer (Fig. 73).

Flowers are heavily marked with reddish-purple striations.

Fig. 30 *Dendrobium* sp. aff. *fairfaxii*
A From the Tweed Range, south-east Queensland, 1976
B–E From Mt Gannon, south-east Queensland, 1960

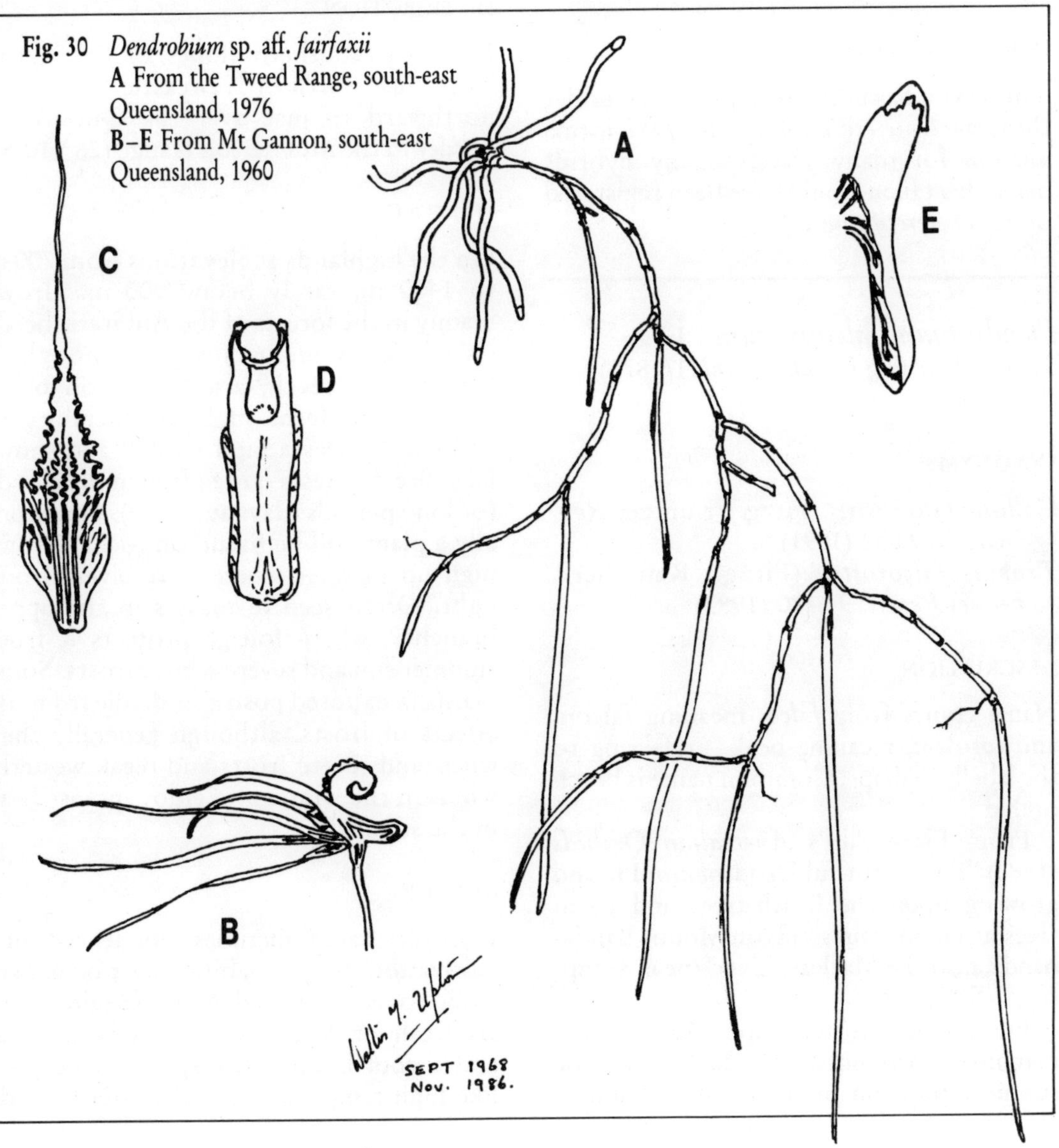

A Plant 1/3X
B Flower, side-view 1X
C Labellum, pressed flat 2X
D Column, front-view 4X
E Column, side-view 4X

FLOWERING PERIOD

July to September (winter to spring).

DISTRIBUTION

Possibly from the Richmond River (28° 50′S) in northern NSW, northward to south-eastern Queensland, near Nambour (26° 40′S).

HABITAT

Mainly in rainforests from sea level to 1000 m, more tolerant of brighter light conditions than *D. fairfaxii.*

CULTURE

Similar culture to *D. teretifolium* except, like *D. fairfaxii*, it is slow to establish.

IN HYBRIDIZATION

This taxon has been grown and used under the name *D. teretifolium* var. *fairfaxii* forma *aureum* for many years, so any hybrids using this taxon would have been registered under *D. teretifolium.*

Dendrobium falcorostrum Fitzg.

Sydney Morning Herald 7 (18/11/1876)
Fig. 31, Plate 28

SYNONYMS

Callista falcorostris (Fitzg.) Kuntze, *Rev. Gen. Pl.* 2:654 (1891)
Tropilis falcorostra (Fitzg.) Rauschert, *Feddes Repert.* 94:470 (1983)

DESCRIPTION

Name comes from *falco*, meaning falcon, and *rostrum*, meaning beak—referring to the labellum shape. Common name is beech orchid.

From Fitzgerald's *Australian Orchids* (1882) 'This beautiful *Dendrobium* I found growing upon the Brush-trees and Fern-trees in a mountain scrub on Mount Banda-banda, near the Macleay River' (near Kempsey, NSW).

Pseudobulbous stems up to 50 cm, most commonly around 15 to 20 cm. In the sunniest position in their native habitat, they become yellow in colour as they mature. Two to five leaves are produced at the top of the pseudobulbous stems.

The near-terminal racemes are quite short, to 18 cm long, and generally carrying about 10 to 15 flowers (occasionally up to 20), which are intensely fragrant during the warmer hours of the day. The strong perfume is sweetish and not like any other *Dendrobium.* Sepals and petals usually a snowy glistening white and labellum white with purple and orange markings. Occasionally sepals and petals have a few purple spots at the base. Flowers last 2 to 3 weeks.

FLOWERING PERIOD

August to October (spring).

DISTRIBUTION

From the southern slopes of Barrington Tops, just north of Newcastle (32° 15′S), northward to just over the Queensland border in the McPherson Range (28° 10′S).

HABITAT

On the highlands at elevations from 700 m to 1400 m, rarely below 900 m. Grows mainly in the forests of the Antarctic beech (*Nothofagus moorei*) and, although it favours the beech, also grows on other trees and tree ferns. Habitat generally cool and experiences a high rainfall and snowfalls. Beech forests quite often in the clouds for long periods (days at a time). Occasionally a plant will be found on rocks. Grows high up in trees where it receives strong light. Often seen in masses in the upper branches, where foliage protects it from summer sun and severe winter frosts. Some plants in exposed positions do die from the effects of frosts, although generally they withstand severe frosts and bleak westerly winds in the winter and enjoy intense heat in the summer.

CULTURE

Difficulties are sometimes experienced with cultivation, but its habitat must be understood to really succeed. Major requirements are fresh air, lots of water, medium shade and a cool humid atmosphere. Does not like high temperatures and westerly winds when grown at low elevations. Prefers to be hung in an exposed, but shady, position (it needs room to breathe).

Needs a slightly richer potting mix than most Australian dendrobiums or, if grown on a slab, frequent weak fertilizing. There is no need to withhold water during the winter months as is common for most members of the genus. New growth generally starts in late spring, early summer and matures quickly if grown well. Minimum winter temperature of approximately 0°C.

IN HYBRIDIZATION

Although not a very dominant species for shape, it has had much influence on hybridizing within the section *Dendrocoryne.* Eighty-eight of the 210 hybrids made between species within this section have *D. falcorostrum* at least once in their parentage. It has also been used twice in hybrids where two of the species were from the subgenus *Dendrobium*; *D.* Andrew Upton (*D.* Star of Gold X *D.* Judy Leroy) and *D.* Natalie Wall (*D.* Star of Gold X *D.* Lowana Nioka). Most dominant characteristics are its lip and habit of raceme. Vegetative form and habit also come through in many progeny.

Best when used as a pollen parent, indeed seldom if ever is a plant seen in its natural habitat with seed capsules. Their absence is very noticeable.

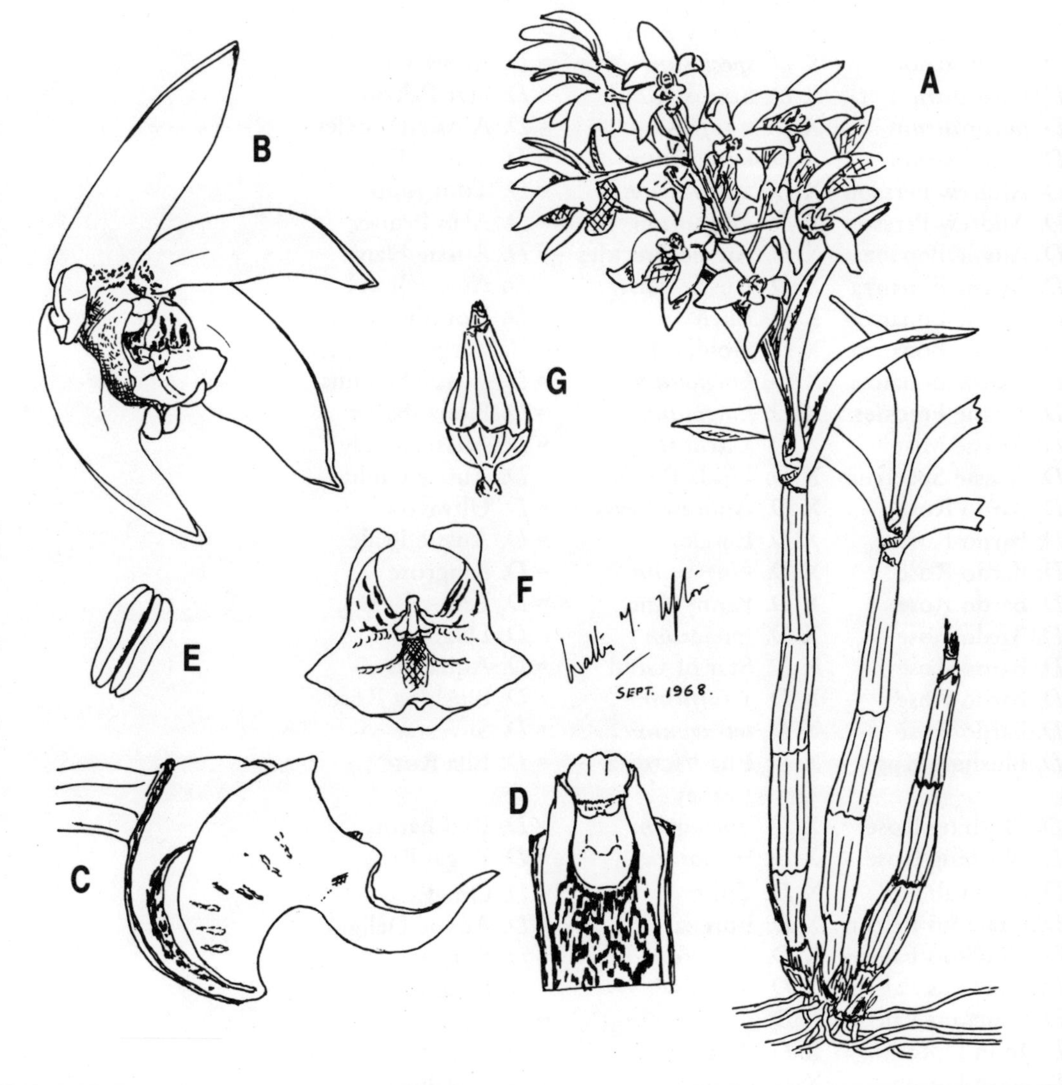

Fig. 31 *Dendrobium falcorostrum* Fitzg. Collected near Dorrigo, NSW

A Plant $^{2}/_{5}$X
B Flower 2X
C Column and labellum, side-view 4X
D Column, front-view 4X
E Pollinia greatly enlarged
F Labellum, front-view 4X
G Immature pseudobulbous stem $^{2}/_{5}$X

HYBRIDS

D. falcorostrum	X *D. adae*	= *D.* Leonie Jane
D. falcorostrum	X *D.* Aussie Freckles	= *D.* Aussie Majestic
D. falcorostrum	X *D.* Aussie Ira	= *D.* Aussie Bonanza
D. falcorostrum	X *D.* Bardo Rose	= *D.* Aussie Bounty
D. falcorostrum	X *D. x delicatum*	= *D.* Delicate Falcon
D. falcorostrum	X *D.* Ellen	= *D.* Warrambool
D. falcorostrum	X *D.* Eureka	= *D.* Nooro
D. falcorostrum	X *D. fleckeri*	= *D.* Peter
D. falcorostrum	X *D. macropus* subsp. *gracilicaule* (as *D. gracilicaule*)	= *D.* Susan
D. falcorostrum	X *D. x gracillimum*	= *D.* Gracious Falcon
D. falcorostrum	X *D.* Hastings	= *D.* Aussie Spice
D. falcorostrum	X *D.* Hilda Poxon	= *D.* Miara
D. falcorostrum	X *D. kingianum*	= *D.* Bardo Rose
D. falcorostrum	X *D.* Kingrose	= *D.* King Falcon
D. falcorostrum	X *D. mortii* (as *D. tenuissimum*)	= *D.* Amaroo
D. falcorostrum	X *D.* Peter	= *D.* Aussie Joy
D. falcorostrum	X *D.* Penny Ann	= *D.* Twinkle
D. falcorostrum	X *D. speciosum*	= *D.* Andrew Persson
D. falcorostrum	X *D.* Star of Gold	= *D.* Star Falcon
D. falcorostrum	X *D. x suffusum*	= *D.* Aussie Freckles
D. falcorostrum	X *D. tetragonum*	= *D.* Star of Gold
D. Andrew Persson	X *D. x delicatum*	= *D.* Tom Jupp
D. Andrew Persson	X *D. kingianum*	= *D.* Alan Printer
D. Aussie Bonanza	X *D.* Aussie Freckles	= *D.* Aussie Flame
D. Aussie Bonanza	X *D.* Aussie Mist	= *D.* Aussie Jewel
D. Aussie Bonanza	X *D.* Ellen	= *D.* Dot Sheen
D. Aussie Bonanza	X *D.* Golden Fleck	= *D.* Margery Angeleri
D. Aussie Bonanza	X *D. kingianum*	= *D.* Aussie Sequins
D. Aussie Freckles	X *D. kingianum*	= *D.* Fiona Banks
D. Aussie Mist	X *D. x delicatum*	= *D.* Aussie Sandy
D. Aussie Spectrum	X *D.* Hilda Poxon	= *D.* Aussie Child
D. Bardo Rose	X *D.* Andrew Persson	= *D.* Ultravox
D. Bardo Rose	X *D.* Eureka	= *D.* Aussie Pride
D. Bardo Rose	X *D. kingianum*	= *D.* Kingrose
D. Bardo Rose	X *D.* Penny Ann	= *D.* Penny Rose
D. Bardo Rose	X *D. speciosum*	= *D.* Dilgry
D. Bardo Rose	X *D.* Star of Gold	= *D.* Aquarius
D. Bardo Rose	X *D. x suffusum*	= *D.* Blushing Rose
D. Bardo Rose	X *D. tetragonum*	= *D.* Silvester
D. Blushing Rose	X *D.* Ella Victoria Leaney	= *D.* Ella Rose
D. Blushing Rose	X *D. kingianum*	= *D.* Red Baron
D. Blushing Rose	X *D. speciosum*	= *D.* Regal Rose
D. Carawah	X *D.* Tully	= *D.* Omega
D. Gracious Falcon	X *D.* Eureka	= *D.* Aussie Delight
D. Gracious Falcon	X *D. speciosum*	= *D.* Karloo
D. Gracious Falcon	X *D.* Sunglow	= *D.* Kalang
D. Gracious Falcon	X *D. tetragonum*	= *D.* Gracious Star
D. John Upton	X *D.* Bardo Rose	= *D.* John Rose
D. John Upton	X *D.* Colin	= *D.* Coljohn

29 *D. fairfaxii*

30 *D.* sp. aff. *fairfaxii*

31 *D. fellowsii*

32 *D. fleckeri*

33 *D. insigne*

34 *D. jonesii*

35 *D.* sp. aff. *jonesii* 'Jill'

36 *D. johannis*

37 *D. kingianum* from Gloucester area NSW

38 *D. kingianum* 'No. Ten'

39 *D. kingianum* white form

40 *D. kingianum* 'White Mid-Lobe'

41 *D. linguiforme* var. *linguiforme*

42 *D. linguiforme* var. *huntianum*

HYBRIDS (con't)

D. John Upton	X *D.* Jamie Upton	= *D.* Colsboys
D. John Upton	X *D.* Kungara	= *D.* Wakehurst
D. John Upton	X *D.* Ku-Ring-Gai	= *D.* Ring of Fire
D. John Upton	X *D. speciosum*	= *D.* Ingleside
D. John Upton	X *D. tetragonum*	= *D.* Bright Spark
D. Kim	X *D.* Colin	= *D.* Fiesta
D. Kim	X *D.* Ellen	= *D.* John Upton
D. Kingrose	X *D.* Aussie Ira	= *D.* Aussie Lust
D. Kingrose	X *D.* Aussie Starlight	= *D.* Aussie Utmost
D. Kingrose	X *D.* Bardo Rose	= *D.* Tawhai
D. Kingrose	X *D.* Eureka	= *D.* Aussie Hero
D. Kingrose	X *D. kingianum*	= *D.* Joy Wray
D. Kingrose	X *D. specio-kingianum*	= *D.* Debra Wray
D. Kingrose	X *D.* Yondi	= *D.* Jenni
D. Peter	X *D.* Aussie Ira	= *D.* Aussie Ember
D. Peter	X *D.* Bardo Rose	= *D.* Aussie Mist
D. Peter	X *D. x delicatum*	= *D.* Memoria Eric Gordon
D. Peter	X *D.* Ellen	= *D.* Aussie Treat
D. Peter	X *D.* Eureka	= *D.* Aussie Sunshine
D. Peter	X *D.* Gillian Leaney	= *D.* Gilpet
D. Peter	X *D.* Golden Fleck	= *D.* Whispering Jack
D. Peter	X *D.* Hilda Poxon	= *D.* Berserker
D. Peter	X *D. kingianum*	= *D.* Kim
D. Peter	X *D.* speciosum	= *D.* Aussie Spectrum
D. Peter	X *D.* Star of Gold	= *D.* Aussie Aurora
D. Peter	X *D.* Sunglow	= *D.* Tweed
D. Peter	X *D. tetragonum*	= *D.* Amber
D. Star of Gold	X *D.* Aussie Ira	= *D.* Aussie Parade
D. Star of Gold	X *D.* Aussie Mist	= *D.* Aussie Luck
D. Star of Gold	X *D. x delicatum*	= *D.* Star One
D. Star of Gold	X *D.* Ellen	= *D.* Kookaburra
D. Star of Gold	X *D.* Hilda Poxon	= *D.* Tarean
D. Star of Gold	X *D.* Judy Leroy	= *D.* Andrew Upton
D. Star of Gold	X *D. kingianum*	= *D.* Star of Dawn
D. Star of Gold	X *D.* Ku-Ring-Gai	= *D.* Star Child
D. Star of Gold	X *D.* Lowana Nioka	= *D.* Natalie Wall
D. Star of Gold	X *D. jonesii* (as *D. ruppianum*)	= *D.* Star Delight
D. Star of Gold	X *D. speciosum*	= *D.* Star of Riverdene
D. Star of Gold	X *D.* Sunglow	= *D.* Carawah
D. Susan	X *D. adae*	= *D.* Green Goddess
D. Susan	X *D.* Ellen	= *D.* Sweet Ellen
D. Susan	X *D. kingianum*	= *D.* Kit Murdoch

Dendrobium fellowsii F. Muell.

Fragm. Phyt. Aust. 7:63 (1870)
Fig. 32, Plate 31

SYNONYMS

Dendrobium bairdianum F. M. Bail., *Syn. Q/ld. Fl.* suppl. 1:53 (1886)
Callista bairdiana (F. M. Bail.) Kuntze, *Rev. Gen. Pl.* 2:654 (1891)
Dendrobium giddinsii T. E. Hunt, *Nth. Q/ld. Nat.* 15, No. 87:25 (1948)
Sayeria bairdiana (F. M. Bail.) Rauschert, *Feddes Repert.* 94:466 (1983)

DESCRIPTION

D. fellowsii was collected by John Dallachy at Rockingham Bay, prior to 1870. Type sheet sighted by Clements and recognized as an earlier name for *D. bairdianum* (Pers. Comm. M. Clements, 1986). *D. bairdianum* was found by J. W. R. Stuart between Herberton and Mourilyan Harbour, north Queensland and named after Mr J. C. Baird. (See *Orchadian* Vol. 7, No. 6, Dec. 1982, p. 140, for true affinities of *D. fellowsii* (as *D. bairdianum*) to section *Eleutheroglossum* Krzl.)

Pseudobulbous stems commonly about 6 to 12 cm, but can grow to 25 cm. The two to five narrow, lanceolate leaves are short lived.

Near-terminal, shortish racemes produce two to seven long-lasting flowers of heavy substance. The petals and sepals are pale-green to yellowish-green, fading (as the flowers become old), to a buff colour. The very dark purple markings on the labellum, together with the raised white platform-like callus on it, form a beautiful contrast to the rest of the flower.

CHROMOSOME NUMBER

2n = 36 (K. Y. Lim, 1985).

FLOWERING PERIOD

Mainly from November to January (spring to summer), occasionally earlier. Often flowers more than once over this period.

DISTRIBUTION

From Burdekin River (20° 00′S) to Bloomfield River (16° 00′S), from 450 to 1000 m (1480 to 3280 ft).

Fig. 32 *Dendrobium fellowsii* F. Muell.
Collected Mt. Lewis, north Queensland, 1964

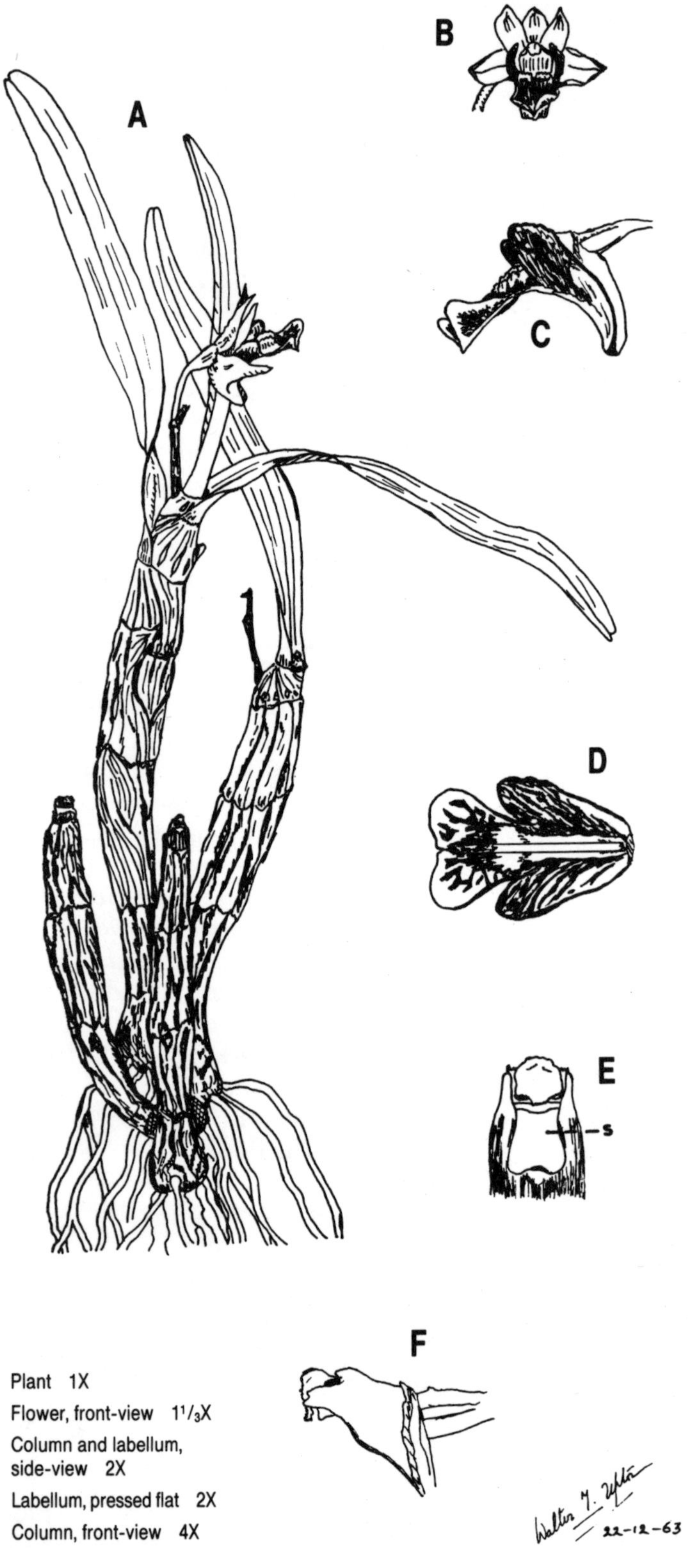

A Plant 1X
B Flower, front-view 1¹/₃X
C Column and labellum, side-view 2X
D Labellum, pressed flat 2X
E Column, front-view 4X
F Column, side-view 4X

HABITAT

In open *Eucalyptus* and *Casuarina* forests mostly below the level of the highland rainforests, but where they receive a considerable amount of moisture at night. These areas, though hot and dry in the day, are much cooler and moister at night than lower altitude areas, causing heavy dews and morning mists. This species is never abundant and usually grows high on its host where it is out of reach of the frequent grass fires. Less frequently it grows on rocks. Favourite hosts are the bloodwoods (*Eucalyptus* spp.) and the forest oak (*Allocasuarina torulosa*).

CULTURE

Very difficult to grow in my opinion, although it will grow well in an open bushhouse in Sydney, NSW, in the summer. Requires a minimum of 13°C (55°F) in the winter. Plenty of air movement at all times. Needs cool, moist, sheltered conditions and must never be allowed to dry out completely. After flowering a short rest is required until new growth appears.

IN HYBRIDIZATION

Only one hybrid has been made and flowered to my knowledge. It is *D. fellowsii* X *D. bigibbum*—unnamed. Flower colour, size and plant form seem to favour *D. bigibbum*, but the flower shape resembles *D. fellowsii.*

Dendrobium fleckeri Rupp et C. T. White

Q/ld. Nat. 10:25, t.1 (1937)
Fig. 33, Plate 32

SYNONYM

Tropilis fleckeri (Rupp et C. T. White) Rauschert, *Feddes Repert.* 94:470 (1983)

DESCRIPTION

Originally collected by Dr Hugo Flecker in the Upper Mossman River rainforest country, near Mt Spurgeon, in 1935. Named after Flecker, a radiologist who was also a zoologist and botanist, and the

Founder and Foundation President of the North Queensland Naturalists Club.

Pseudobulbous stems very slender and up to 35 cm long, generally about 20 cm, with two to four lanceolate leaves at the top. Two vegetative forms, one with very slender stems, and the other with shorter, more robust stems.

Short racemes from near the top of the stem produce one to three flowers, averaging about 25 mm in diameter. Flowers have a musky fragrance and are from greenish-yellow, through yellow to apricot and occasionally almost brown, though most commonly an apricot-yellow. Labellum whitish at the base, becoming pale yellow with brick red to purple markings; mid-lobe margins densely ciliate. Callus on labellum very prominent. Flowers last about 10 to 14 days.

CHROMOSOME NUMBER

2n = 38 (K. Y. Lim, 1985).

FLOWERING PERIOD

August to January (spring to summer). Frequently flowers continuously for about 2 months with the greatest number out in October to November (spring).

DISTRIBUTION

South-eastern Cape York Peninsula from the Johnstone River (17° 30′S) to the Annan River (15° 30′S).

HABITAT

Generally confined to forests near the top of mountains from 800 m to 1550 m, seldom less than 900 m. Here it grows on trees and rocks. When growing on exposed rocks its stem is usually quite short. It is very common on Mt Bartle Frere, often covering large rock faces. Heavy dews and fogs are a common occurrence in the greater part of its habitat.

CULTURE

An easy-to-grow orchid, with the result that it is quite common in collections in Australia. Its major requirements are good air movement, high humidity, a fair amount of light and plenty of water in the growing season.

In pots needs a very open mix and should almost dry out between waterings, although plants should never become completely dry, not even in the winter. If grown on slabs, requires at least daily watering throughout the year. In the very hot summer days frequent mistings are most beneficial. Winter temperatures down to as low as 0°C for very short periods do no harm, but it is preferable to keep the minimum winter temperature to about 6°C.

IN HYBRIDIZATION

Not dominant generally, but does show some dominance in habit of raceme, number of flowers and labellum shape. Most dominant vegetative characteristic is the slender pseudobulbous stem. Colour mostly comes through when crossed with another non-dominant species.

HYBRIDS

D. fleckeri	X *D.* Aussie Ira	= *D.* Aussie Glow
D. fleckeri	X *D. affine* (as *D. dicuphum*)	= *D.* Reg Burns
D. fleckeri	X *D. bigibbum*	= *D.* Big Fleck
D. fleckeri	X *D.* Ella Victoria Leaney	= *D.* Wyuna
D. fleckeri	X *D.* Ellen	= *D.* Aussie Dream
D. fleckeri	X *D. falcorostrum*	= *D.* Peter
D. fleckeri	X *D. macropus* subsp. *gracilicaule* (as *D. gracilicaule*)	= *D.* Golden Fleck
D. fleckeri	X *D.* Hilda Poxon	= *D.* Wirruna
D. fleckeri	X *D. kingianum*	= *D.* Hastings
D. fleckeri	X *D. jonesii* (as *D. ruppianum*)	= *D.* Eureka
D. fleckeri	X *D. speciosum*	= *D.* Sunglow
D. fleckeri	X *D. tetragonum*	= *D.* Golden Glory
D. Aussie Gypsy	X *D. speciosum*	= *D.* Aussie Frontier
D. Aussie Mist	X *D.* Aussie Bonanza	= *D.* Aussie Jewel
D. Aussie Mist	X *D. x delicatum*	= *D.* Aussie Sandy
D. Aussie Mist	X *D.* Star of Gold	= *D.* Aussie Luck
D. Aussie Spectrum	X *D.* Hilda Poxon	= *D.* Aussie Child
D. Eureka	X *D. adae*	= *D.* Warooa
D. Eureka	X *D.* Aussie Starlight	= *D.* Aussie Zest
D. Eureka	X *D.* Bardo Rose	= *D.* Aussie Pride
D. Eureka	X *D.* Colin	= *D.* Colreka
D. Eureka	X *D. x delicatum*	= *D.* Aussie Love
D. Eureka	X *D.* Ella Victoria Leaney	= *D.* Aussie Treasure

Fig. 33 *Dendrobium fleckeri* Rupp et C. T. White Collected by S. Clemesha, Mt. Lewis, north Queensland

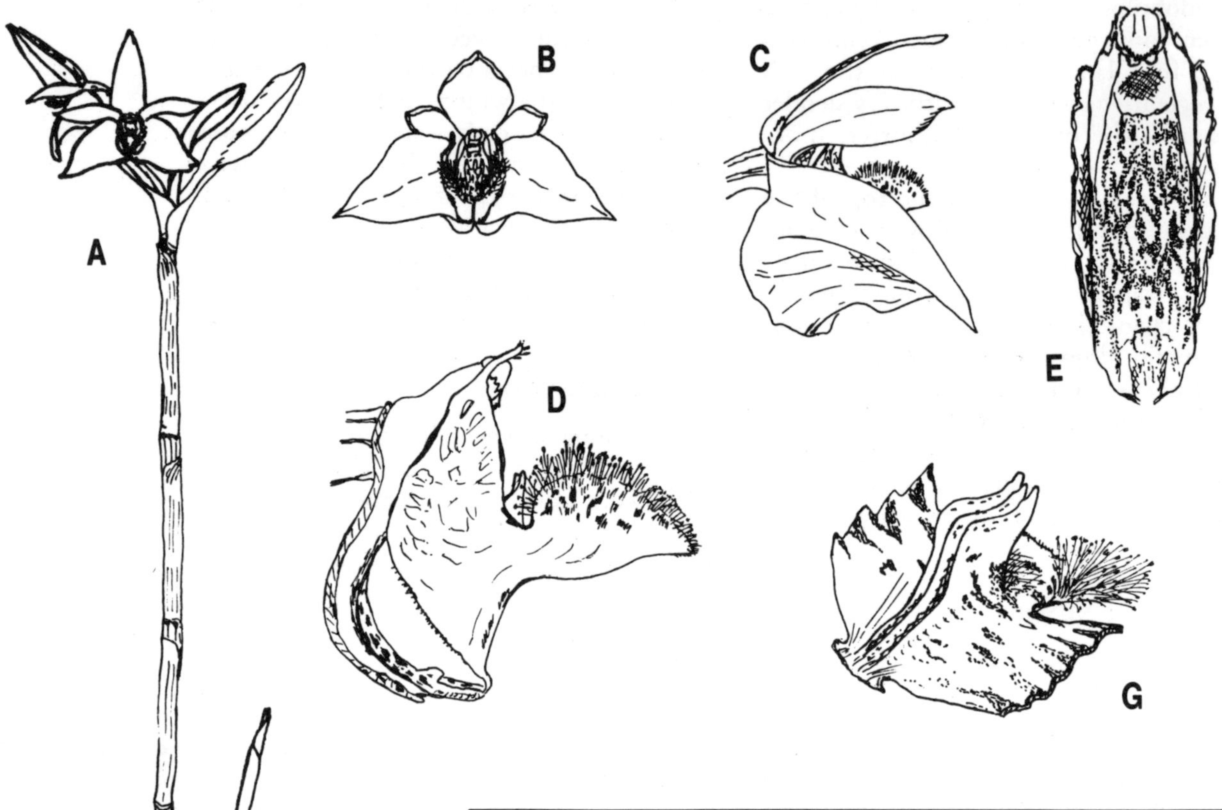

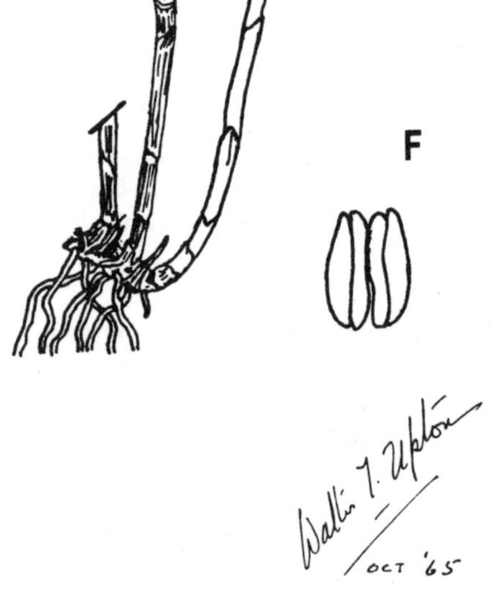

A Plant 2/3X
B Flower, front-view 1 1/3X
C Flower, side-view 2X
D Column and labellum, side-view 4X
E Column, front-view 4X
F Pollinia greatly enlarged
G Callus on labellum about 6X

HYBRIDS (con't)

D. Eureka	X *D.* Ellen	= *D.* Ocker's Dream
D. Eureka	X *D.* *falcorostrum*	= *D.* Nooro
D. Eureka	X *D.* Gracious Falcon	= *D.* Aussie Delight
D. Eureka	X *D.* Hastings	= *D.* Arcadia
D. Eureka	X *D.* Hilda Poxon	= *D.* Aussie Magic
D. Eureka	X *D.* *kingianum*	= *D.* Yondi
D. Eureka	X *D.* Kingrose	= *D.* Aussie Hero
D. Eureka	X *D.* *jonesii* (as *D. ruppianum*)	= *D.* Aussie Ecru
D. Eureka	X *D.* *macropus* (as *D. gracilicaule*)	= *D.* Nugget
D. Eureka	X *D.* *speciosum*	= *D.* Lynette Banks
D. Eureka	X *D.* *tetragonum*	= *D.* Nunkumbil
D. Eureka	X *D.* Wyuna	= *D.* Barnard
D. Golden Fleck	X *D.* Aussie Bonanza	= *D.* Margery Angeleri
D. Golden Fleck	X *D.* Aussie Glow	= *D.* Tambar
D. Golden Fleck	X *D.* Hastings	= *D.* Peach Glow
D. Golden Fleck	X *D.* Hilda Poxon	= *D.* Hot Shot
D. Golden Fleck	X *D.* Jamie Upton	= *D.* Green Star
D. Golden Fleck	X *D.* Peter	= *D.* Whispering Jack
D. Golden Fleck	X *D.* *tetragonum*	= *D.* Jamie Upton

HYBRIDS (con't)

D. Gulginni	X *D. tetragonum*	= *D.* Coral Sun
D. Hastings	X *D.* Aussie Ira	= *D.* Aussie Gift
D. Hastings	X *D. falcorostrum*	= *D.* Aussie Spice
D. Hastings	X *D.* Hilda Poxon	= *D.* Peter Hewitt
D. Hastings	X *D. kingianum*	= *D.* Rutherford Surprise
D. Hastings	X *D. jonesii* (as *D. ruppianum*)	= *D.* Aussie Gypsy
D. Hastings	X *D. speciosum*	= *D.* Wonga
D. Hastings	X *D. tetragonum*	= *D.* Peach Star
D. Hastings	X *D.* Wonga	= *D.* Rutherford Supreme
D. Jamie Upton	X *D. speciosum*	= *D.* Lemon Glow
D. Jamie Upton	X *D.* Wonga	= *D.* Karen Nair
D. John Upton	X *D.* Bardo Rose	= *D.* John Rose
D. John Upton	X *D.* Colin	= *D.* Coljohn
D. John Upton	X *D.* Jamie Upton	= *D.* Colsboys
D. John Upton	X *D.* Kungara	= *D.* Wakehurst
D. John Upton	X *D.* Ku-Ring-Gai	= *D.* Ring of Fire
D. John Upton	X *D. speciosum*	= *D.* Ingleside
D. John Upton	X *D. tetragonum*	= *D.* Bright Spark
D. Kim	X *D.* Colin	= *D.* Fiesta
D. Kim	X *D.* Ellen	= *D.* John Upton
D. Nunkumbil	X *D. adae*	= *D.* Whitehart
D. Peter	X *D.* Aussie Ira	= *D.* Aussie Ember
D. Peter	X *D.* Bardo Rose	= *D.* Aussie Mist
D. Peter	X *D. x delicatum*	= *D.* Memoria Eric Gordon
D. Peter	X *D.* Ellen	= *D.* Aussie Treat
D. Peter	X *D.* Eureka	= *D.* Aussie Sunshine
D. Peter	X *D. falcorostrum*	= *D.* Aussie Joy
D. Peter	X *D.* Gillian Leaney	= *D.* Gilpet
D. Peter	X *D.* Hilda Poxon	= *D.* Berserker
D. Peter	X *D. kingianum*	= *D.* Kim
D. Peter	X *D. speciosum*	= *D.* Aussie Spectrum
D. Peter	X *D.* Star of Gold	= *D.* Aussie Aurora
D. Peter	X *D.* Sunglow	= *D.* Tweed
D. Peter	X *D. tetragonum*	= *D.* Amber
D. Sunglow	X *D.* Blushing Star	= *D.* Blushing Sun
D. Sunglow	X *D.* Gracious Falcon	= *D.* Kalang
D. Sunglow	X *D. kingianum*	= *D.* Gulginni
D. Sunglow	X *D.* Hilda Poxon	= *D.* Graeme Banks
D. Sunglow	X *D.* Star of Gold	= *D.* Carawah
D. Sunglow	X *D. tetragonum*	= *D.* Sun Star
D. Sunglow	X *D.* Wyuna	= *D.* Tully
D. Tully	X *D.* Carawah	= *D.* Omega
D. Wonga	X *D.* Hilda Poxon	= *D.* Vida Banks
D. Wonga	X *D. kingianum*	= *D.* King Wong
D. Wonga	X *D. speciosum*	= *D.* Woko
D. Wyuna	X *D. speciosum*	= *D.* Nerang
D. Yondi	X *D.* Kingrose	= *D.* Jenni
D. Yondi	X *D. speciosum*	= *D.* Jesmond Gem

Dendrobium insigne (Blume) H. G. Reichb. ex Miq.

Fl. Ind. Bat. 3:640 (1855)

Fig. 34, Plate 33

BASIONYM

Dichopus insignis Blume, *Mus. Bot. Lug.— Bat.* 2:176 (1856)

SYNONYMS

Dendrobium gazellae Krzl., *Bot. Jahrb. Syst.* 7:436 (1886)
Dendrobium lyperanthiflorum Krzl., *Öst. Bot.* 334 (1894)
Dendrobium obcuneatum F. M. Bail., *Q/ld. Agric. J.* 17:231 (1906)
Dendrobium pentactis Krzl., *Engl. Pflanzenreich* 45:200 (1910)

DESCRIPTION

Name is from *insignis*, meaning distinguished. First collected in Australian Territory in Torres Strait on the Island of Saibai by Geoff Stocker in 1978. It appeared on a 30 cent stamp of Papua New Guinea in 1975.

Wiry stems up to 60 cm long, normally 25 to 40 cm in collections, frequently growing into large clumps. Leaves thick and fleshy.

Very short racemes of two flowers. Flowers dull to bright yellow with reddish-brown markings. Labellum cream to white, sometimes with a few yellow-brown spots. Depth of colour and density of markings vary greatly. The flowers generally last up to 2 days and are very fragrant. On one plant I have, the flowers remained in excellent condition for 3 days. Seeds very large for genus.

Schlechter (1982c) places this species in the section *Dichopus* although he stated 'in habit it agrees exactly with *Grastidium*'. He regarded it as a separate section 'on account of the peculiar moveable appendage on the lower margin of the stigma'. Fig. 34 shows this appendage and also how the structure of the column differs from *Grastidium* (see *D. baileyi*). The stigma is very low down and the side-lobes of the clinandrium are greatly reduced.

CHROMOSOME NUMBER

2n = 36 plus two additional fragment-like chromosomes in their karyotypes (K. Y. Lim, 1985).

FLOWERING PERIOD

Throughout the year.

DISTRIBUTION

Torres Strait Islands, also in Papua New Guinea and Bougainville.

HABITAT

Epiphyte on small branches in open forests and on mangroves in various areas. It seems quite a common plant in Papua New Guinea where it is found on mangroves and forest trees.

Schlechter (1982d) remarks on this species 'I saw it in company with *D. bifalce* Lindley on dead trees at the beach in especially hot localities where it was exposed all day to the strongest rays of the sun' . . . 'It grows on isolated trees, very often in the

Fig. 34 *Dendrobium insigne* (Blume) H. G. Reichb. ex Miq. Collected on Saibai Island, Torres Strait

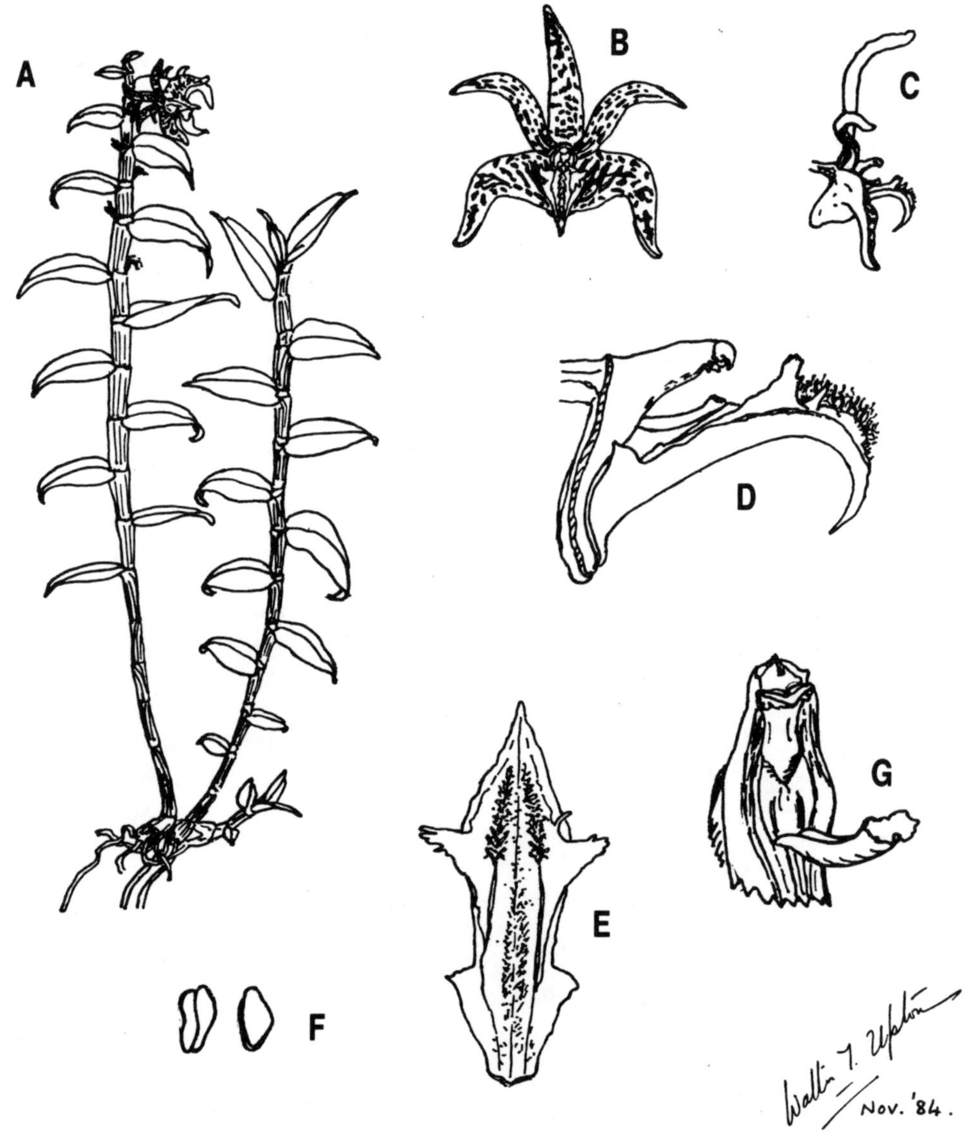

A Plant $^{1}/_{3}$X
B Flower, front-view 1X
C Flower, side-view 1X
D Column and labellum, side-view 3X
E Labellum, pressed flat 3X
F Pollinia greatly enlarged
G Column, front-view, showing appendage and large rostellum, anther removed 6X

most glaring sun and is able to resist long dry periods apparently without danger'.

CULTURE

Preferable to grow in a pot, the smallest pot possible that will allow for 2 year's growth, in very open medium. Does not like to be overpotted. I have seen it growing quite happily on a slab. Requires minimum winter temperature of at least 12°C, high light intensity, high humidity and plenty of water during its growing period. Keep just moist the rest of the year.

IN HYBRIDIZATION

No hybrids have been registered using this species.

Dendrobium johannis H. G. Reichb.

Gard. Chron. 25:890 (1865)

Fig. 35, Plate 36

SYNONYMS

Dendrobium undulatum R.Br. var. *johannis* (H. G. Reichb.) F. M. Bail., *Syn. Q/ld. Fl.* 509 (1883)

Callista johannis (H. G. Reichb.) Kuntze, *Rev. Gen. Pl.* 2:655 (1891)

DESCRIPTION

In 1865 Reichenbach described *D. johannis* from a plant collected in 'north Australia' by John G. Veitch in 1863. Plant is named after John Gould Veitch.

Pseudobulbous stems up to 30 cm. Leaves up to eight, generally about four, at the top of the stem.

Racemes to 25 cm long, carrying up to 20 very fragrant flowers (generally about 10). Flowers approximately 3·5 cm in diameter, opening widely. Sepals and petals from yellow brown to a dark purple-brown, frequently lighter towards the base. Most frequently occurring colour is dark purple-brown. Labellum white to yellow with purple striations, mid-lobe generally sulphur yellow. The flowers are long lasting and emit a strong and not very pleasant odour. The notch in the anther cap is an aid to identification of this species.

Fig. 35 *Dendrobium johannis* H. G. Reichb.
Collected near Falls Creek, Cape York, north Queensland, 1981

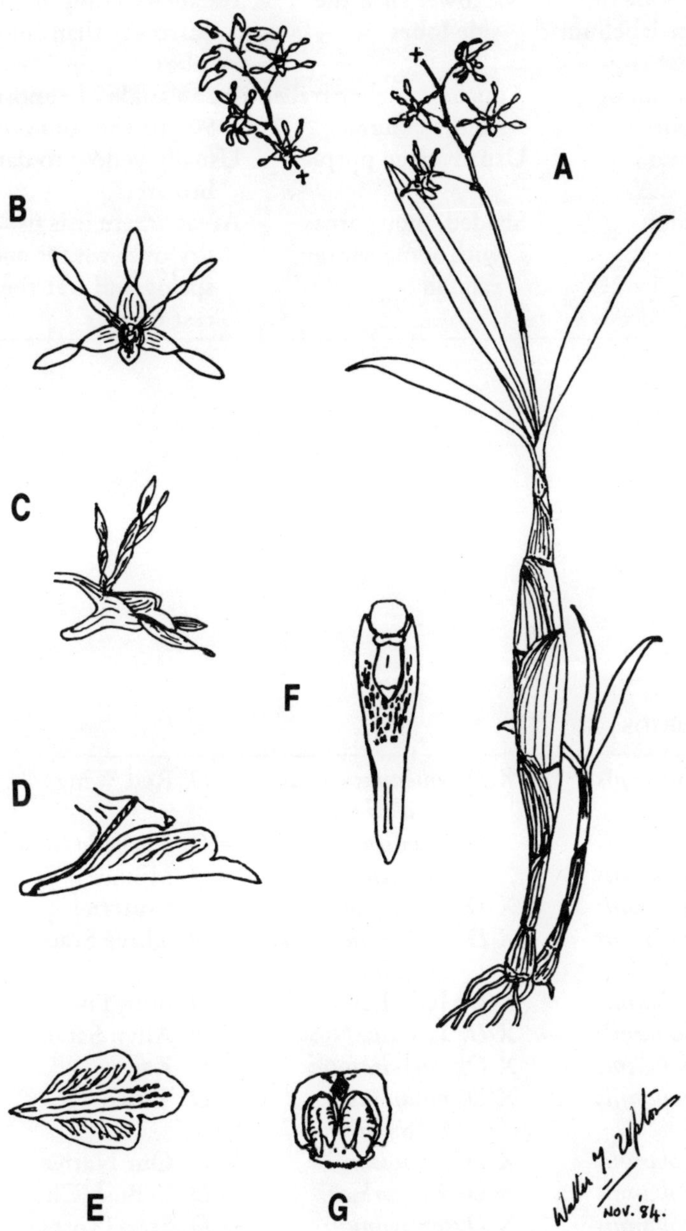

A Plant $^1/_3$X
B Flower, front-view 1X
C Flower, side-view 1X
D Column and labellum, side-view 2X
E Labellum, pressed flat 2X
F Column, front-view 4X
G Looking into anther approx. 10X

FLOWERING PERIOD

March to June (autumn to early winter).

DISTRIBUTION

From near Coen (14° 00'S) eastern Cape York Peninsula, northward to the Torres Strait Islands and to Papua New Guinea.

	D. johannis	*D. trilamellatum*
Flowering time	Autumn	Spring
Anther cap	Deeply cleft	Only a small notch
Mid-lobe of the labellum	Narrower than the side-lobes	Broader to only slightly narrower than side-lobes
Mentum or spur	Approximately parallel to the column	At an angle of almost 90° to the column
Colour	Usually deep purple	Usually yellow to dark brown
Habitat	Shaded, moist areas with some winter rainfall	Areas where it is usually dry over winter and spring and wet the rest of year

HYBRIDS

D. johannis	X *D. antennatum* (as *D. d'albertisii*)	= *D.* Red Wings
D. johannis	X *D. canaliculatum*	= *D.* Rachelle Simpson
D. johannis	X *D. carronii*	= *D.* Minnie
D. johannis	X *D.* Champagne	= *D.* Sauterne
D. johannis	X *D. x superbiens* (as *D. goldiei*)	= *D.* Three Star
D. johannis	X *D.* Judy Leroy	= *D.* Blue Tiki
D. johannis	X *D.* Lowana Nioka	= *D.* Allyn Sands
D. johannis	X *D. mirbelianum*	= *D.* Kaipu
D. johannis	X *D. nindii* (as *D. tofftii*)	= *D.* Tusco
D. johannis	X *D. speciosum*	= *D.* Our Native
D. johannis	X *D. x superbiens*	= *D.* D'Bush Chic
D. johannis	X *D. tetragonum*	= *D.* Star Twist
D. johannis	X *D. discolor* (as *D. undulatum*)	= *D.* Johulatum
D. Allyn Sands	X *D.* Rachelle Simpson	= *D.* Allyn Supreme
D. Allyn Sands	X *D. discolor* (as *D. undulatum*)	= *D.* Allyn King
D. Rachelle Simpson	X *D. bigibbum*	= *D.* Donie
D. Rachelle Simpson	X *D. canaliculatum*	= *D.* Black Mickey Mouse
D. Rachelle Simpson	X *D. discolor* (as *D. undulatum*)	= *D.* Allyn Prince

HABITAT

On trees in fairly open forest areas, the margins of rainforests and in swampy places. Areas where there is some winter rainfall and good humidity. Cambourne (1936) states that 'A few *D. johannis* were noticed growing around the salt water swamps'. Host trees include *Lophostemon suaveolens* and *Dillenia alata.*

Lavarack (1984) gives a full description of *D. johannis* and gives in great detail the difference between *D. johannis* and *D. trilamellatum* (as *D. semifuscum*). These differences are listed briefly in the table.

CULTURE

This is a species that requires a decided rest in the winter, after it has flowered. In Sydney, NSW, I hang it up high in the greenhouse during this period and almost forget about it for 2 months. Repot as soon as new growth is about 5 to 7 cm high and new root movement commenced.

Requires ample air movement at all times, high light intensity, heavy watering while in full growth and medium humidity. Minimum winter temperature of about 15°C (60°F).

Either pot or slab culture is satisfactory. If potted, requires a very open medium.

IN HYBRIDIZATION

Does not appear to be any absolutely dominant characteristics obvious in progeny. Perhaps the size of the flowers, the twisting of the segments and the mid-lobe of the labellum, together with plant form, are the most dominant. Most of its progeny are delightful.

ETHNOBOTANY

'The aboriginal people of western Cape York Peninsula heat the skin over a fire to accentuate the yellow color and then weave or plait it into a string base. Larger strands are worn by women as girdles or breastlets; the shorter lengths are worn by men as arm-, leg-, or headbands. Strips of the orchid skin are also used to decorate weapons and ceremonial objects. When a new born child is presented to its father, the umbilical cord is decorated with orchid skin and hung around the baby's neck' (Lawler, 1984).

Dendrobium johnsoniae F. Muell.

Wing's South. Science Rec. 2:95 (1882)
Fig. 36

SYNONYMS

Dendrobium macfarlanei H. G. Reichb., *Gard. Chron.* 520 (1882) non. F. Muell.
Dendrobium niveum Rolfe, *Gard. Chron.* 104 (1891)
Dendrobium monodon Krzl., *Engl. Pflanzenreich* Heft 45:254 (1910)
Callista macfarlanei (H. G. Reichb.) Kuntze, *Rev. Gen. Pl.* 2:655 (1891)
Latourorchis johnsoniae (F. Muell.) Brieger, in Schlechter *Die Orchideen* ed. 3 (11/12): 731 (1981)
Sayeria johnsoniae (F. Muell) Rauschert, *Feddes Repert.* 94:467 (1983)

DESCRIPTION

Much has been written about *D. johnsoniae* in Australia and, although there is great doubt if it has ever been found growing in Australia, I have included it in this work. In the Melbourne National Herbarium there is a specimen in Baron Ferdinand von Mueller's collections, on which Mueller has written 'Persieh, Endeavour River, 1884'.

Named after the daughter of the Rev. Johnson of Surrey Hills, NSW. Mueller received the plant from Rev. J. Chalmers who, it is understood, collected it in 1882 on the east Peninsula of New Guinea. Fig. 36 is of a plant said to come from the collection at Lae, New Guinea.

Stems up to 40 cm long carrying three to four thin but leathery leaves about 8 to

Fig. 36 *Dendrobium johnsoniae* F. Muell. Drawing of a New Guinea plant

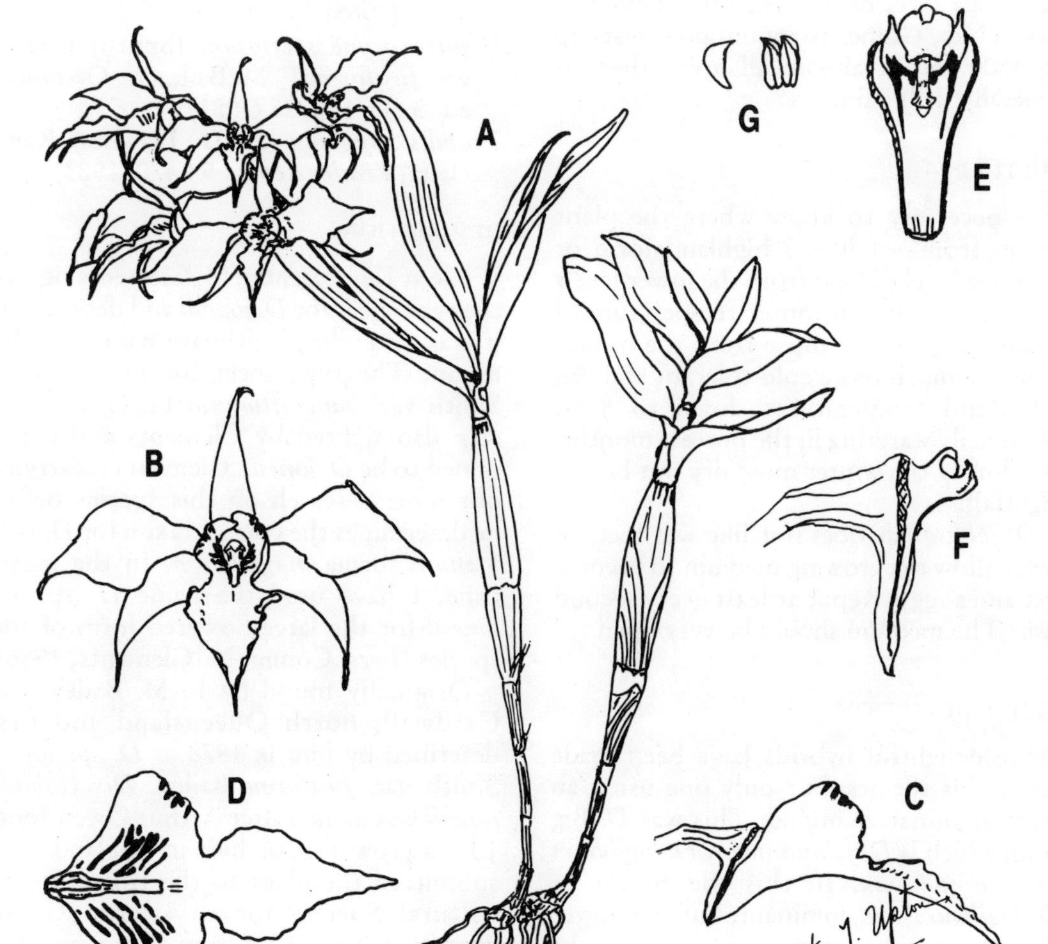

A Plant $^1/_3$X
B Flower, front-view $^2/_3$X
C Column and labellum, side-view 1X
D Labellum, pressed flat 1X
E Column, front-view 2X
F Column, side-view 2X
G Pollinia, side and front view greatly enlarged

10 cm long. The pure glistening white flowers are from 65 mm to 125 mm across their large petals. Generally about six flowers are carried, although up to 12 have been seen. Flowers frequently last up to 2 months in perfect condition.

CHROMOSOME NUMBER

2n = 38 (K. Y. Lim, 1985).

FLOWERING PERIOD

Main flowering period April to July (autumn to winter).

DISTRIBUTION

Doubtful recording from Australia. New Guinea and Solomon Islands.

HABITAT

From sea level on the islands offshore in Papua New Guinea to mountain forests up to 1500 m. In almost all areas there is generally daily rain.

CULTURE

It is necessary to know where the plant comes from—if it is a highland form or from sea level. Those from the lower levels require a minimum winter temperature of about 15°C, while those from the mountains are much more cold tolerant and can withstand temperatures down to 8°C. Need daily watering in the hottest months, but during the winter must dry out before nightfall.

D. johnsoniae does not like wet feet, so never allow the growing medium to become wet and soggy. Repot at least every second year. The medium should be very open.

IN HYBRIDIZATION

Some delightful hybrids have been made using this species, but only one using an Australian native orchid. This was *D.* Big John which is *D. johnsoniae* X *D. bigibbum* var. *compactum*. In this the colour of *D. bigibbum* was dominant, but the flower shape favoured *D. johnsoniae*, including the labellum. By using the short-stemmed form of *D. bigibbum*, the plant was compact and about 20 cm tall.

Dendrobium jonesii Rendle

J. Bot. Lond. 39:197 (1901)

Fig. 37, Plate 34

SYNONYMS

Dendrobium speciosum Smith var. *fusiforme* F. M. Bail., *Proc. Linn. Soc. N.S.W.* 2:277 (1878)

Dendrobium fusiforme (F. M. Bail.) F. M. Bail., *Q/ld. Fl.* 5:1527 (1902), non Thouars (1822)

Dendrobium speciosum Smith var. *bancroftianum* H. G. Reichb., *Gard. Chron.* (new series) 15:782 (1881)

Dendrobium fusiforme (F. M. Bail.) F. M. Bail. var. *blackburnii* Nicholls, *Victorian Nat.* 66:255 (1950)

Dendrobium ruppianum A. D. Hawkes, *Aust. Orchid Rev.* 29:40 (1964)

Dendrobium ruppianum A. D. Hawkes var. *blackburnii* (Nicholls) Dockr., *Aust. Pl.* 3:224 (1965)

Dendrocoryne speciosum (Smith) Brieger var. *fusiforme* F. M. Bail., *Die Orchideen* ed. 3 (11/12):725 (1981)

Tropilis ruppiana (A. D. Hawkes) Rauschert, *Feddes Repert.* 94:470 (1983)

DESCRIPTION

While at Kew recently, M. Clements sighted the type sheet for *D. jonesii* and determined it was an earlier legitimate name for this taxon. The type sheet for *D. speciosum* Smith var. *bancroftianum* H. G. Reichb., was also sighted by Clements and determined to be *D. jonesii*. Clements is carrying out more research on this species before he determines the correct taxon for *D. ruppianum* forma *magnificum*. In the meantime, I have used the name *D.* sp. aff. *jonesii* for the large-flowered form of this species (Pers. Comm. M. Clements, 1986).

Originally found by F. M. Bailey near Cardwell, north Queensland and first described by him in 1878 as *D. speciosum* Smith var. *fusiforme* Bailey. *Dendrobium jonesii* was named after Arthur Owen Jones J.P., a grower of orchids in England, who submitted the plant to the Royal Horticultural Society for an award. Rendle described the species from this plant. The plant was sent to Jones by J. Sparks of north Queensland. Common name is oak orchid.

Robust plant frequently growing into large clumps with an extensive, strong root system. Pseudobulbous stems up to 50 cm tall, most commonly about 20 to 30 cm. Up to seven leaves near the apex of the stem.

Erect to arching racemes are from 15 to 35 cm long carrying up to 30 or more white to cream (rarely pale-yellow) fragrant flowers that release a heavy spicy perfume in the mornings of warm sunny days. Flowers turn to pale yellow-cream with age or when pollinated. Seldom open very wide and are about 2 cm in diameter. Labellum marked with purple. Flowers last about 10 days.

You will notice from the synonyms of *D. jonesii* that the var. *blackburnii* is given as a synonym of this species, but for the horticulturists I give the original description of it. 'This compact variety differs from the typical form chiefly in its much more robust pseudobulbs which are 12–22 cm long x 2·5–3·0 cm diam. The pendulous racemes have shortly segmented pale primrose-yellow flowers, whereas in the typical form the creamy-white flowers are born erect with somewhat filiform "spidery" segments. Fragrance of the flowers persists for some time.'

The locality given was 'Font Hill Station, west of Port Douglas, Qld.'

Fig. 37 *Dendrobium jonesii* Rendle Collected in Tully Falls area

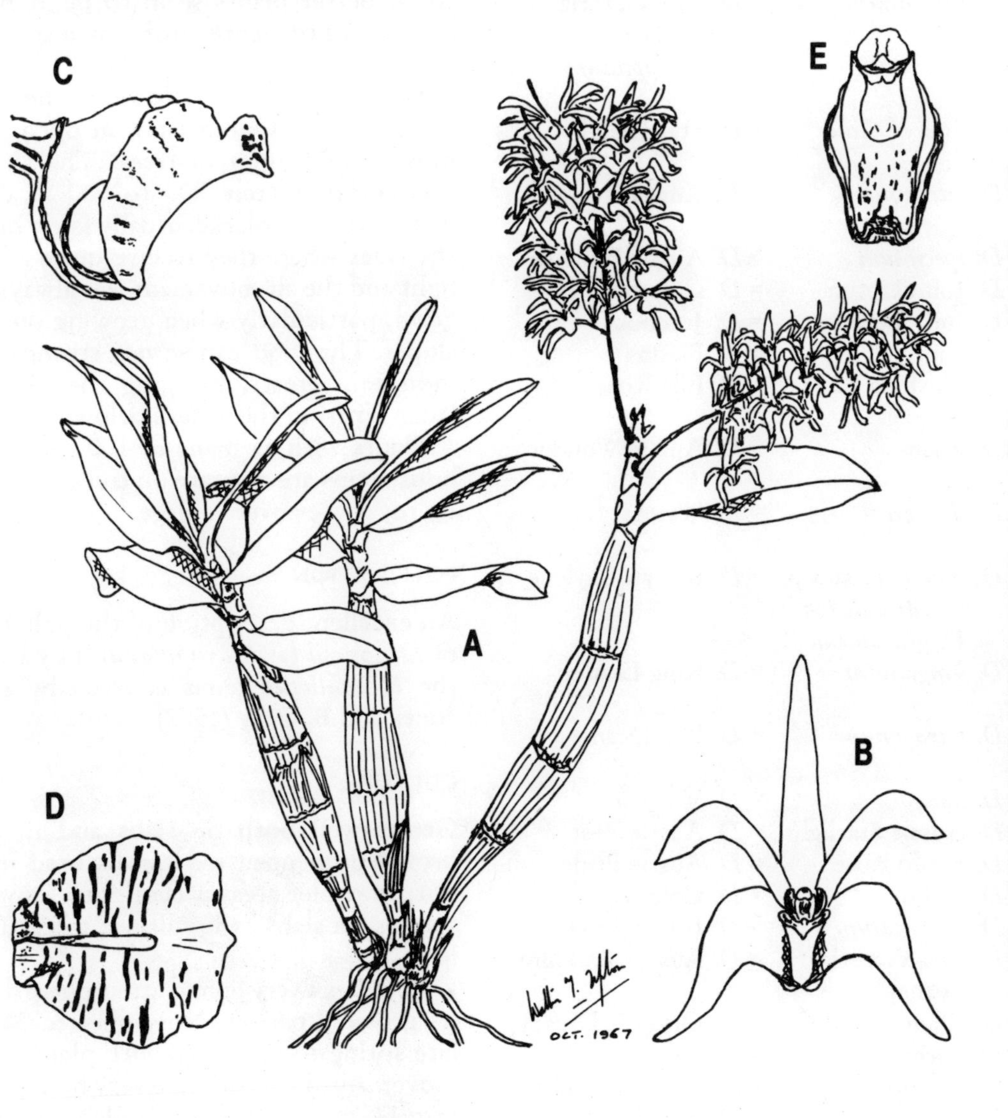

A Plant 1/3X
B Flower, front-view 2X
C Column and labellum, side-view 4X
D Labellum, pressed flat 4X
E Column, front-view 4X

HYBRIDS

D. jonesii (as *D. ruppianum*)	X *D.* Aussie Ira	= *D.* Aussie Nave
D. jonesii (as *D. ruppianum*)	X *D.* Ella Victoria Leaney	= *D.* Aussie Merit
D. jonesii (as *D. ruppianum*)	X *D.* Ellen	= *D.* Colin
D. jonesii (as *D. ruppianum*)	X *D.* Eureka	= *D.* Aussie Ecru
D. jonesii (as *D. ruppianum*)	X *D. fleckeri*	= *D.* Eureka
D. jonesii (as *D. ruppianum*)	X *D. macropus* subsp. *gracilicaule* (as *D. gracilicaule*)	= *D.* Golden Cascades
D. jonesii (as *D. ruppianum*)	X *D.* Hastings	= *D.* Aussie Gypsy
D. jonesii (as *D. ruppianum*)	X *D.* Hilda Poxon	= *D.* Ira Butler
D. jonesii (as *D. ruppianum*)	X *D. kingianum*	= *D.* Ella Victoria Leaney
D. jonesii (as *D. ruppianum*)	X *D. speciosum*	= *D. x ruppiosum*
D. jonesii (as *D. ruppianum*)	X *D.* Star of Gold	= *D.* Star Delight
D. jonesii (as *D. ruppianum*)	X *D. tetragonum*	= *D.* Kungara
D. Aussie Gypsy	X *D. speciosum*	= *D.* Aussie Frontier
D. Colin	X *D.* John Upton	= *D.* Coljohn
D. Colin	X *D.* Judy Leroy	= *D.* Jombock
D. Colin	X *D.* Kim	= *D.* Fiesta
D. Ella Victoria Leaney	X *D.* Blushing Rose	= *D.* Ella Rose
D. Ella Victoria Leaney	X *D.* Ellen	= *D.* Aussie Wonder
D. Ella Victoria Leaney	X *D. fleckeri*	= *D.* Wyuna
D. Ella Victoria Leaney	X *D. macropus* subsp. *gracilicaule* (as *D. gracilicaule*)	= *D.* Robyn Kaylene
D. Ella Victoria Leaney	X *D. kingianum*	= *D.* King Leaney
D. Ella Victoria Leaney	X *D. tetragonum*	= *D.* Phil Deane
D. Eureka	X *D. adae*	= *D.* Warooa
D. Eureka	X *D.* Aussie Starlight	= *D.* Aussie Zest
D. Eureka	X *D.* Bardo Rose	= *D.* Aussie Pride
D. Eureka	X *D.* Colin	= *D.* Colreka
D. Eureka	X *D. x delicatum*	= *D.* Aussie Love
D. Eureka	X *D.* Ella Victoria Leaney	= *D.* Aussie Treasure
D. Eureka	X *D.* Ellen	= *D.* Ocker 's Dream
D. Eureka	X *D. falcorostrum*	= *D.* Nooro
D. Eureka	X *D.* Gracious Falcon	= *D.* Aussie Delight
D. Eureka	X *D.* Hastings	= *D.* Arcadia

FLOWERING PERIOD

July to November (later winter to spring).

DISTRIBUTION

From the ranges to the north-west of Mackay (21° 05′S) in north-eastern Queensland, northward to the vicinity of the Iron Range in Cape York Peninsula (12°40′S) and a very doubtful recording in New Guinea.

HABITAT

On trees or rocks in sunny situations in the ranges and lowland gorges, from 150 to 1400 m. Seems to be happiest in open forests on the outskirts of rainforests, although often seen high up on rainforest trees. Better plants seem to be at higher altitudes. Frequently grows on a species of *Casuarina.*

I have seen it in full flower in the Coane Range, north Queensland, in open forest mostly on *Casuarina* trees. The altitude was approximately 700 to 900 m. Some plants were very large, usually fairly high in the trees where they receive quite a lot of light and the air movement was always very good, particularly when growing on steep slopes. The wind can be very strong on the mountain crests, bringing the moisture laden air from the valleys below.

Plants with as many as 100 pseudobulbous stems are not rare on trees, but those on rocks are never as large.

POLLINATION

An excellent description of the pollination of *D. jonesii* (as *D. ruppianum*) by bees of the *Homalictus* genus is given by D. L. Jones and B. Gray (1977).

CULTURE

Grows well both on slabs and in pots, provided an open medium is used in the pot. In cooler areas it is safer to grow this species on slabs as it will not tolerate wet feet.

Requires a very humid atmosphere during its growing period (November to March, late spring to summer), with plenty of air movement. Does not like very high temperatures, over 30°C. Needs plenty of water in the growing season, but should be kept on the dry side during the resting period (winter). Minimum winter temperature of 8°C.

In Sydney, NSW, I hang it up high in the greenhouse in the middle of winter so it gets lots of light and can be kept dry. The rest of the year it is in a shade-house under 50 per cent shade cloth.

IN HYBRIDIZATION

Not a very dominant species. Some of the good characteristics passed on to its progeny are the number of flowers on the raceme, its floriferousness, its habit of raceme, and the broad mid-lobe to the lip. It does not appear to be dominant in its vegetative characteristics.

HYBRIDS (con't)

D. Eureka	X *D.* Hilda Poxon	= *D.* Aussie Magic
D. Eureka	X *D.* Kingrose	= *D.* Aussie Hero
D. Eureka	X *D. kingianum*	= *D.* Yondi
D. Eureka	X *D. macropus* (as *D. gracilicaule*)	= *D.* Nugget
D. Eureka	X *D.* Peter	= *D.* Aussie Sunshine
D. Eureka	X *D. speciosum*	= *D.* Lynette Banks
D. Eureka	X *D. tetragonum*	= *D.* Nunkumbil
D. Eureka	X *D.* Wyuna	= *D.* Barnard
D. Kungara	X *D.* John Upton	= *D.* Wakehurst
D. Nunkumbil	X *D. adae*	= *D.* Whitehart
D. Tully	X *D.* Carawah	= *D.* Omega
D. Wyuna	X *D. speciosum*	= *D.* Nerang
D. Wyuna	X *D.* Sunglow	= *D.* Tully
D. Yondi	X *D.* Kingrose	= *D.* Jenni
D. Yondi	X *D. speciosum*	= *D.* Jesmond Gem

Dendrobium sp. aff. *jonesii*

Fig. 38, Plate 35

BASIONYM

Dendrobium fusiforme (F. M. Bail.) F. M. Bail., var. *fusiforme* forma *magnificum* Dockr., *North Q/ld. Nat.* 24(114):24 (1956)

SYNONYM

Dendrobium ruppianum A. D. Hawkes var. *ruppianum* forma *magnificum* (Dockr.) Dockr., *Aust. Pl.* 3:224 (1965)

DESCRIPTION

Clements is researching this taxon, and in this work it is unnamed at specific rank (Pers. Comm. M. Clements, 1986). Originally collected by J. H. Wilkie at Mount Bellenden Ker, north Queensland in October 1951.

Plants usually quite robust. Flowers of this taxon are larger than *D. jonesii* and are white when first open. Labellum has purple markings or sometimes without any markings at all. Flowers from 3 to 8 cm in diameter and, like *D. jonesii*, turn yellow with age (often only after 3 to 4 days). I have a clone of this form that has flowers 9 cm in diameter when flattened out; the flowers are very crowded on the raceme. It frequently carries 35 or more fragrant flowers. I have another clone with no purple markings on the lip whatsoever and the flowers are 8 cm in diameter; it frequently carries over 50 flowers on the crowded raceme.

FLOWERING PERIOD

October to November (spring).

DISTRIBUTION

From the Burdekin River (20° 00′S) to the Annan River (15° 30′S).

HABITAT

As for *D. jonesii*, on rocks or trees in sunny situations in the ranges and lowland gorges, although it is seldom found below 900 m. On Evelyn Tableland at approximately 1200 m both *D. jonesii* and this species are found growing together; here is a pure white flower form with no purple markings on the labellum. The plant cannot be distinguished from *D. jonesii* except when in flower. On the Evelyn Tableland they receive up to 305 cm of rain per year but the temperature seldom rises above 27°C.

CULTURE

Similar to *D. jonesii*, but with slightly more light through the winter to ensure it flowers well.

HYBRIDS

D. sp. aff. *jonesii* (as *D. ruppianum*)	X *D.* Ellen	= *D.* Colin
D. sp. aff. *jonesii* (as *D. ruppianum*)	X *D. fleckeri*	= *D.* Eureka
D. sp. aff. *jonesii* (as *D. ruppianum*)	X *D.* Hilda Poxon	= *D.* Ira Butler
D. sp. aff. *jonesii* (as *D. ruppianum*)	X *D. kingianum*	= *D.* Ella Victoria Leaney
D. sp. aff. *jonesii* (as *D. ruppianum*)	X *D. tetragonum*	= *D.* Kungara
D. Colin	X *D.* Judy Leroy	= *D.* Jombock
D. Colin	X *D.* Kim	= *D.* Fiesta
D. Ella Victoria Leaney	X *D. kingianum*	= *D.* King Leaney
D. Eureka	X *D. tetragonum*	= *D.* Nunkumbil

IN HYBRIDIZATION

Same remarks as for *D. jonesii* apply here; except maybe it is more dominant in its vegetative characteristics.

This taxon has been grown and used under the name *D. ruppianum* var. *ruppianum* forma *magnificum* for many years, so any hybrids using this taxon would have been registered under *D. ruppianum.*

Although I have made the following hybrids using this taxon they have been registered under the name *D. ruppianum* (now *D. jonesii*), as taxons below the ranks of species are not recognized in hybrid parentage. No doubt many other hybridists have used this taxon because of its large and superior flowers.

Fig. 38 *Dendrobium* sp. aff. *jonesii* From near Tinaroo Falls Dam, north Queensland

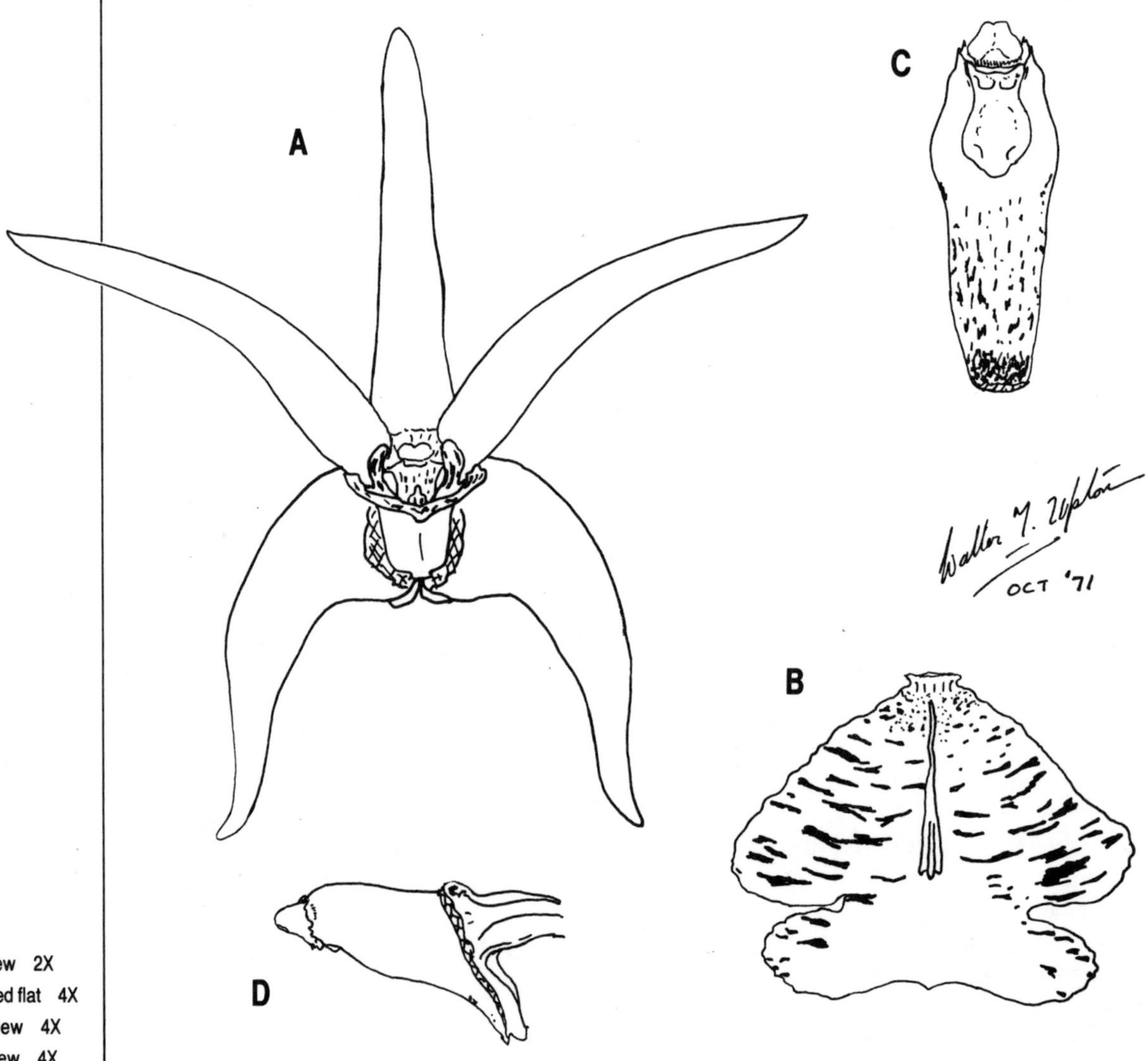

A Flower, front-view 2X
B Labellum, pressed flat 4X
C Column, front-view 4X
D Column, side-view 4X

Dendrobium kingianum Bidw.,

Edwards's Bot. Reg. 30: misc 11 (1844), *Edwards's Bot. reg.* 31: t.61 (1845)

Figs 39 and 40, Plates 3, 37, 38, 39 and 40

DESCRIPTION

It is generally understood that *D. kingianum* was found by John Carne Bidwill in 1839 on the Buckets, a mountain just to the west of Gloucester, NSW (M. K. Maxwell, 1967). Named after Captain (later Rear-Admiral) Phillip Parker King who was the Royal Navy commander of many expeditions which surveyed the coast of Australia. He was the son of Governor Philip Gidley King. Common name is pink rock lily, and in early literature known as Capt. King's Dendrobium. In *Orchidei v. Kultura* by T. M. Tcherevtchenko and G. P. Kushnir, chapter 6 mentions the use of orchids in the Russian space program and the interesting data produced. *D. kingianum* was one of the species used.

It is safe to say that this species is the most variable of all Australian dendrobes, mainly in flower form and colour, but also in vegetative form. The species varies from area to area. Generally plants in the southern area of its distribution are quite small; plants around the border country between NSW and Queensland are largest; and, furthest north of its distribution and in Carnarvon Gorge, Queensland, plants become smaller again. But, like all orchids, only a generalized statement can be made.

In the Tamborine area of Queensland (27° 55′S), Figure 40, I have seen plants 55 cm long, and in some cases with one or two aerials off the one stem making a total length of 125 cm. On Alum Mountain at Bulahdelah, NSW, Figure 39, the stems of mature plants are often only 3 cm long with few if any aerial growths. I have seen numerous plants of this species on the Glasshouse Mountains in Queensland, and the small flowers hardly opened up at all. I think some of the best *D. kingianum* flowers are found in the Gloucester to Hastings River area (32°S to 31°S). They are often large, widely opening and good colour forms.

This species has two varieties and the descriptions for each are given separately.

Fig. 39 *Dendrobium kingianum* Bidw.
B, C, D and F *D. kingianum* 'Albert West'
A and E Collected Alum Mountain, Bulahdelah

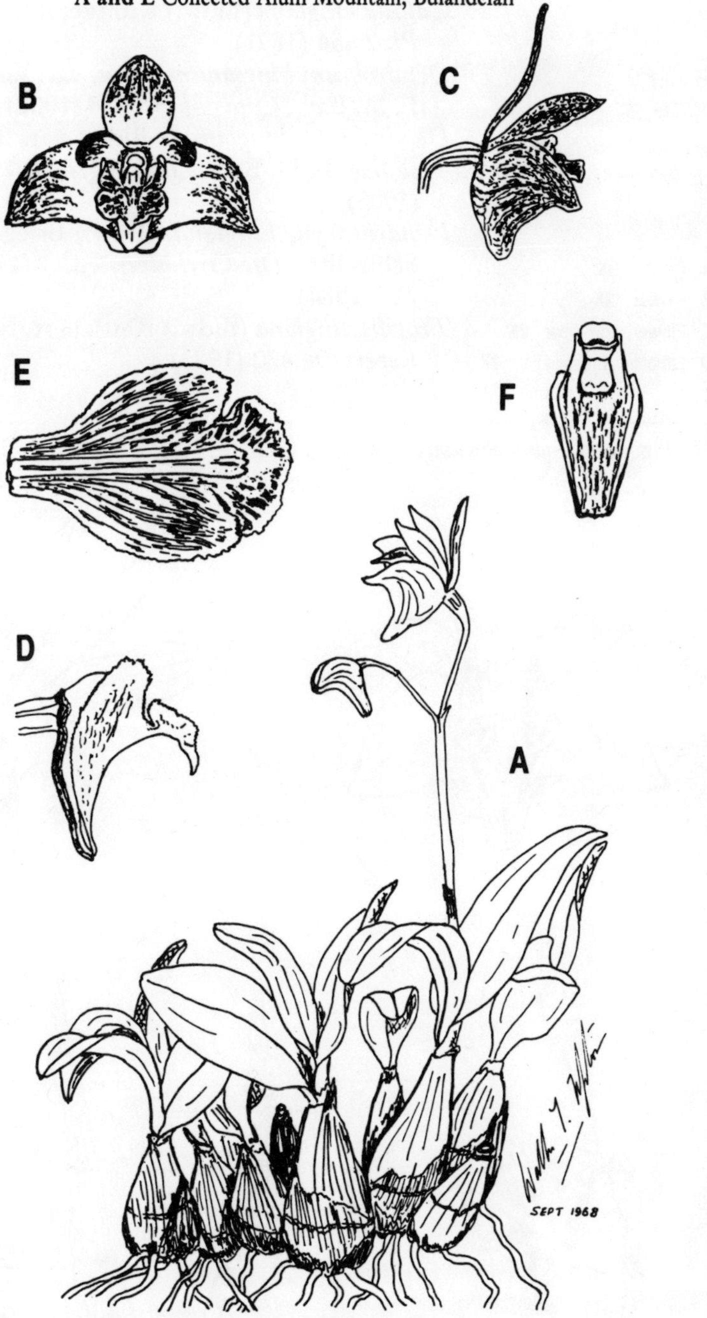

A Plant 1X
B Flower, front-view 1X
C Flower, side-view 1X
D Column and labellum, side-view 2X
E Labellum, pressed flat 2X
F Column, front-view 2X

D. kingianum Bidw. var. *kingianum*

SYNONYMS

Dendrobium kingianum Bidw. var. *pallidum*, F. M. Bail., *Proc. Roy. Soc. Q/ld.* 1:11 (1884)

A Plant $^1/_2$X
B Flower 2X
C Flower, side-view 2X
D Labellum, pressed flat 2X

Dendrobium kingianum Bidw. subvar. *pallidum*, Veitch, *Man. Orchid Pl.* 1(3):52 in obs. (1888)
Callista kingiana (Bidw.) Kuntze, *Rev. Gen. Pl.* 2:654 (1891)
Dendrobium kingianum Bidw. var. *silcockii* F. M. Bail., *Q/ld. Fl.* 5:1528 (1902)
Dendrobium kingianum Bidw. var. *aldersoniae*, F. M. Bail., *Q/ld. Agric. J.* 15:781 (1905)
Dendrocoryne kingiana (Bidw.) Brieger, in Schlechter *Die Orchideen* ed. 3 (11/12): 724 (1981)
Tropilis kingiana (Bidw.) Rauschert, *Feddes Repert.* 94:470 (1983)

DESCRIPTION

Stems up to 55 cm. Aerial growths frequently produced in some clones, mostly the elongated stem forms. Up to six or more leaves at the top of the stem.

Racemes up to approximately 18 cm long but most frequently about 12 cm, carrying from two to 15 flowers up to 3 cm in diameter. Racemes upright to pendulous, most frequently about six flowers produced on an arching raceme. Flowers vary from pure white through various shades of pink to deep mauve. Labellum is from pure white in the albino form to very heavily marked with red or purple. Many combinations of

Fig. 40 *Dendrobium kingianum* Bidw. Collected Mount Tamborine, Queensland

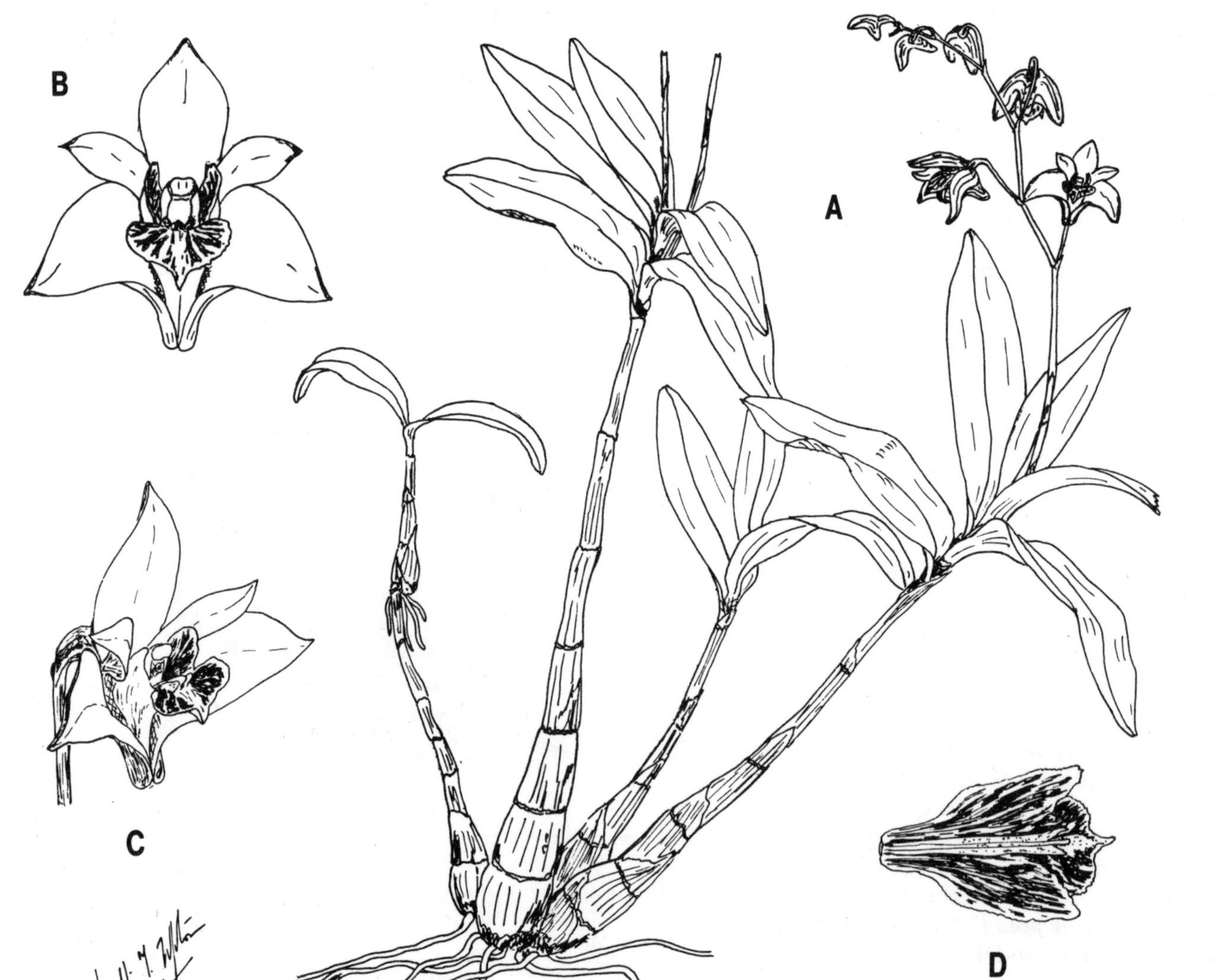

43 *D. linguiforme* var. *nugentii*

44 *D. lichenastrum*

45 *D. lobbii*

46 *D. luteocilium*

47 *D. macropus* subsp. *macropus* (*D. brachypus* form)

48 *D. macropus* subsp. *gracilicaule*

49 *D. macropus* subsp. *howeanum*

50 *D. mortii*

51 *D. nindii*

52 *D. pugioniforme* var. *alba*

53 *D. mirbelianum*

54 *D. pugioniforme*

55 *D. prenticei*

56 *D. racemosum*

colour exist — white with a dark-purple lip, mauve with a white mid-lobe to lip, pink with a pink lip, pink with a dark-purple lip, purple tips to pink segments, etc. (I have over 80 different colour forms in my collection.) Has a faint but pleasing scent. Individual flowers last for 2 to 4 weeks in good condition.

FLOWERING PERIOD

August to November (late winter to spring). Some plants flower at various other times.

DISTRIBUTION

From just north of the Hunter River, NSW (32° 30′S) to just below the Tropic of Capricorn (23° 30′S) and Carnarvon Gorge (25° 00′S by 148° 15′E) approximately 450 km inland. There have been reports of isolated plants being found near Ourimbah, Buc ketty and Burragorang Valley, but these cannot be accepted as a valid southern extension of its range.

HABITAT

On rocks, very rarely on trees, in open forests from sea level to 1200 m. I have seen it growing on an extremely large rock, about 600 m across, in the Gloucester area of NSW. The rock was surrounded by a belt of trees but the orchids were in absolutely full sun for most of the day. Many of the pseudobulbous stems were almost purple. There were thousands of plants, all flowering well.

Near Port Macquarie (31° 30′S) it grows on cliffs overlooking the ocean, receiving full morning sun only. I have seen it in the McPherson Range on the NSW/Queensland border at approximately 900 m, growing on rocks in quite heavily timbered areas, with plants nearby in full sun on a steep cliff, and some in the ground in large tufts of grass where they would hardly see the sun (all looked healthy), but there was a continuous movement of air coming up the cliffs from the valley below and through the heavily timbered area. Humidity was high. A personal communication from John Power refers to his sighting *D. kingianum* in Carnarvon Gorge in late August 1973 — 'Pale blush pink, broad petals and sepals, crystalline, not opening flat, good perfume, seen in gorges, only on ledges and sheer vertical sandstone cliffs, mostly no humus debris, very abundant in specialised areas only — about 100,000 plants'. Near the Bellinger River a few kilometres to the north of Bellingen, NSW, I have seen *D. kingianum* growing in a damp and heavily shaded gully, it had exceptionally fine growth and some of the pseudobulbous stems were over 60 cm high.

The average rainfall in its habitat would be in the vicinity of 100 cm per year. An excellent habitat description of *D. kingianum* growing in Carnarvon Gorge, Queensland is given by A. L. Willersdorf (1968).

D. kingianum var. *pulcherrimum*

Rupp, *Orchids New South Wales* 116 (1943)

Name comes from *pulchellus*, meaning beautiful. Originally collected from the Upper Hastings River area by a Mr Fieldsend in 1937.

The original description reads: 'Pseudobulbs crowded, all short, not exceeding 9 cm. Leaves as long as the pseudobulbs, tinged with red. Flowers comparatively large, the sepals and petals deep mauve. Labellum white on the under surface and the margins'.

Generally this variety carries only one to three flowers.

FLOWERING PERIOD

September (spring).

DISTRIBUTION

Between Hunter River (32° 30′S) and the Macleay River (31° 00′S).

Colour forms

Although I have followed Clement's (1982) treatment and included the following varieties as synonyms of *D. kingianum* var. *kingianum*, I am including them here purely for the horticulturist, as they are only colour forms.

var. *pallidum* F. M. Bail.

Weak slender stems and pale lilac stained flowers. In Veitch (1888) it states 'A subvar. called *pallidum* has white flowers with a few purple stripes on the labellum'.

HYBRIDS

D. kingianum	X *D. adae*	= *D.* Tooloom
D. kingianum	X *D. aemulum*	= *D.* Emmy
D. kingianum	X *D.* Andrew Persson	= *D.* Alan Printer
D. kingianum	X *D.* Aussie Bonanza	= *D.* Aussie Sequins
D. kingianum	X *D.* Aussie Freckles	= *D.* Fiona Banks
D. kingianum	X *D.* Bardo Rose	= *D.* Kingrose
D. kingianum	X *D. bigibbum*	= *D.* Esme Poulton
D. kingianum	X *D.* Blushing Rose	= *D.* Red Baron
D. kingianum	X *D.* Dainty Cascades	= *D.* Elfin
D. kingianum	X *D. x delicatum*	= *D.* Gillian Leaney
D. kingianum	X *D.* Double Two	= *D.* Nathan Wall
D. kingianum	X *D.* Ella Victoria Leaney	= *D.* King Leaney
D. kingianum	X *D.* Ellen	= *D.* Kathryn Banks
D. kingianum	X *D.* Eureka	= *D.* Yondi
D. kingianum	X *D. falcorostrum*	= *D.* Bardo Rose
D. kingianum	X *D. fleckeri*	= *D.* Hastings
D. kingianum	X *D.* Goblin	= *D.* King Goblin
D. kingianum	X *D. x gracillimum*	= *D.* Penny Ann
D. kingianum	X *D.* Graham Hewitt	= *D.* Julatten
D. kingianum	X *D.* Ku-Ring-Gai	= *D.* Zip
D. kingianum	X *D.* Hastings	= *D.* Rutherford Surprise
D. kingianum	X *D.* Hilda Poxon	= *D.* Telekon
D. kingianum	X *D. jonesii* (as *D. ruppianum*)	= *D.* Ella Victoria Leaney
D. kingianum	X *D.* Kingrose	= *D.* Joy Wray
D. kingianum	X *D. macropus* subsp. *gracilicaule* (as *D. gracilicaule*)	= *D. x suffusum*
D. kingianum	X *D.* Mini Pearl	= *D.* Berry
D. kingianum	X *D. monophyllum*	= *D.* Karani
D. kingianum	X *D.* Peter	= *D.* Kim
D. kingianum	X *D. speciosum*	= *D. x delicatum*
D. kingianum	X *D. speciosum*	= *D.* Specio-kingianum
D. kingianum	X *D.* Star of Gold	= *D.* Star of Dawn
D. kingianum	X *D. x suffusum*	= *D.* Cawongla
D. kingianum	X *D.* Sunglow	= *D.* Gulginni
D. kingianum	X *D.* Susan	= *D.* Kit Murdoch
D. kingianum	X *D. teretifolium*	= *D.* Pink Pencil
D. kingianum	X *D. tetragonum*	= *D.* Ellen
D. kingianum	X *D.* Wonga	= *D.* King-Wong
D. Aussie Bonanza	X *D.* Aussie Mist	= *D.* Aussie Jewel
D. Aussie Bonanza	X *D.* Aussie Freckles	= *D.* Aussie Flame
D. Aussie Bonanza	X *D.* Golden Fleck	= *D.* Margery Angeleri
D. Aussie Freckles	X *D. falcorostrum*	= *D.* Aussie Majestic
D. Aussie Glow	X *D.* Golden Fleck	= *D.* Tambar
D. Aussie Gypsy	X *D. speciosum*	= *D.* Aussie Frontier
D. Aussie Ira	X *D. bigibbum*	= *D.* Aussie Angel
D. Aussie Ira	X *D. falcorostrum*	= *D.* Aussie Bonanza

var. *silcockii* F. M. Bail.

Named after P. Silcock. Flowers pure white with a mauve lip. These flowers appear in all forms of this species with varying degrees of mauve to purple markings in the lip.

White forms

The pure white form is often incorrectly referred to as *D. kingianum* var. *album* Williams, but from the original description in Williams Orchids Growers Manual, it is not *D. kingianum*

Pure white flowers appear in all growth forms and the degree of whiteness varies, many so-called albino forms have a slight palish mauve in the lip.

CULTURE

D. kingianum is, in my opinion, the easiest of all the native species of *Dendrobium* to grow. It can be grown in pots, on slabs, on trees, in properly-prepared garden beds, in rockeries, in shade, in full sun and in just about any way you want, providing a few basic, common sense cultural practices are observed. It does not like to be overpotted or to have wet feet. Plenty of air movement and a medium to high degree of humidity is essential to produce a healthy plant; these factors are always found in its natural habitat.

Shallow terracotta saucers, with lots of drainage holes drilled in the bottom, suit best. It will grow well in an ordinary terracotta or plastic pot, provided it is the smallest pot possible that will allow for no more than 2 years growth, and it is in a good open medium with perfect drainage.

Repot immediately after flowering as the new lead will then be commencing its growth. Liberal feeding in its growing season is beneficial. Poultry manure and blood and bone sprinkled on top of the growing medium two or three times during this season, as well as a regular once a week half-strength inorganic feed, produces strong plants. Ample water while in growth with just enough to stop the pseudobulbous stems from shrivelling at other times. Minimum winter temperature of 2°C.

Medium to high light intensity will ensure it flowers well. Generally if the leaves have a slight purplish tinge to them, they are receiving the right amount of light.

IN HYBRIDIZATION

A very dominant species in size, shape and colour. In some of its third-generation progeny the colour is still coming through.

Because *D. kingianum* is such a variable species, the dominant characteristics of each clone have to be known to predict with any degree of certainty its particular dominant characteristics. Most of the albino forms do not, in my experience, generally produce whites, except in very minor numbers. It appears that the colour genes are generally recessive in these forms.

It should not be difficult to maintain *D. kingianum* shape and colour by back crossing to this species every third or fourth generation, although we could lose size and floriferousness.

Plant form generally favours *D. kingianum.*

HYBRIDS (con't)

D. Aussie Ira	X *D. fleckeri*	= *D.* Aussie Glow
D. Aussie Ira	X *D.* Hastings	= *D.* Aussie Gift
D. Aussie Ira	X *D. jonesii* (as *D. ruppianum*)	= *D.* Aussie Nave
D. Aussie Ira	X *D.* Kingrose	= *D.* Aussie Lust
D. Aussie Ira	X *D.* Peter	= *D.* Aussie Ember
D. Aussie Ira	X *D. speciosum*	= *D.* Aussie Starlight
D. Aussie Ira	X *D.* Star of Gold	= *D.* Aussie Parade
D. Aussie Mist	X *D. x delicatum*	= *D.* Aussie Sandy
D. Aussie Mist	X *D.* Star of Gold	= *D.* Aussie Luck
D. Aussie Starlight	X *D. x delicatum*	= *D.* Aussie Reward
D. Aussie Starlight	X *D.* Eureka	= *D.* Aussie Zest
D. Aussie Starlight	X *D.* Gillian Leaney	= *D.* Aussie Springtime
D. Aussie Starlight	X *D.* Kingrose	= *D.* Aussie Utmost
D. Bardo Rose	X *D.* Andrew Persson	= *D.* Ultravox
D. Bardo Rose	X *D.* Eureka	= *D.* Aussie Pride
D. Bardo Rose	X *D. falcorostrum*	= *D.* Aussie Bounty
D. Bardo Rose	X *D.* John Upton	= *D.* John Rose
D. Bardo Rose	X *D.* Kingrose	= *D.* Tawhai
D. Bardo Rose	X *D.* Peter	= *D.* Aussie Mist
D. Bardo Rose	X *D. speciosum*	= *D.* Dilgry
D. Bardo Rose	X *D.* Star of Gold	= *D.* Aquarius
D. Bardo Rose	X *D. tetragonum*	= *D.* Silvester
D. Blushing Rose	X *D. speciosum*	= *D.* Regal Rose
D. Blushing Star	X *D.* Hilda Poxon	= *D.* Kim Heinze
D. Blushing Star	X *D.* Sunglow	= *D.* Blushing Sun
D. Colin	X *D.* Eureka	= *D.* Colreka
D. Colin	X *D.* Gillian Leaney	= *D.* Narraweena
D. Colin	X *D.* John Upton	= *D.* Coljohn
D. Colin	X *D.* Judy Leroy	= *D.* Jombock
D. Colin	X *D.* Kim	= *D.* Fiesta
D. x delicatum	X *D. adae*	= *D.* Kevin Banks
D. x delicatum	X *D.* Andrew Persson	= *D.* Tom Jupp
D. x delicatum	X *D.* Eureka	= *D.* Aussie Love
D. x delicatum	X *D. falcorostrum*	= *D.* Delicate Falcon
D. x delicatum	X *D. x gracillimum*	= *D.* Sun Sprite
D. x delicatum	X *D.* Kingrose	= *D.* Debra Wray
D. x delicatum	X *D.* Peter	= *D.* Memoria Eric Gordon
D. x delicatum	X *D. speciosum*	= *D.* Jane Leaney
D. x delicatum	X *D.* Star of Gold	= *D.* Star One
D. x delicatum	X *D. x superbiens*	= *D.* Elsie Cox
D. x delicatum	X *D. tetragonum*	= *D.* Ku-Ring-Gai
D. Ella Victoria Leaney	X *D.* Blushing Rose	= *D.* Ella Rose
D. Ella Victoria Leaney	X *D.* Eureka	= *D.* Aussie Treasure
D. Ella Victoria Leaney	X *D. fleckeri*	= *D.* Wyuna
D. Ella Victoria Leaney	X *D. jonesii* (as *D. ruppianum*)	= *D.* Aussie Merit

HYBRIDS (con't)

D. Ella Victoria Leaney	X *D. macropus* subsp. *gracilicaule* (as *D. gracilicaule*)	= *D.* Robyn Kaylene
D. Ella Victoria Leaney	X *D. tetragonum*	= *D.* Phil Deane
D. Ellen	X *D.* Aussie Bonanza	= *D.* Dot Sheen
D. Ellen	X *D. x delicatum*	= *D.* Delicate Ellen
D. Ellen	X *D.* Ella Victoria Leaney	= *D.* Aussie Wonder
D. Ellen	X *D.* Emmy	= *D.* Goblin
D. Ellen	X *D.* Eureka	= *D.* Ocker's Dream
D. Ellen	X *D. falcorostrum*	= *D.* Warrambool
D. Ellen	X *D. fleckeri*	= *D.* Aussie Dream
D. Ellen	X *D.* Gillian Leaney	= *D.* Aussie Charm
D. Ellen	X *D.* Graham Hewitt	= *D.* Pinterry
D. Ellen	X *D. jonesii* (as *D. ruppianum*)	= *D.* Colin
D. Ellen	X *D.* Kim	= *D.* John Upton
D. Ellen	X *D. macropus* (as *D. gracilicaule*)	= *D.* Ellen Grace
D. Ellen	X *D.* Penny Ann	= *D.* Ferg Martin
D. Ellen	X *D.* Peter	= *D.* Aussie Treat
D. Ellen	X *D.* Peewee	= *D.* Lorikeet
D. Ellen	X *D. speciosum*	= *D.* Graham Hewitt
D. Ellen	X *D.* Star of Gold	= *D.* Kookaburra
D. Ellen	X *D.* Susan	= *D.* Sweet Ellen
D. Ellen	X *D. tetragonum*	= *D.* Aussie Ira
D. Emma	X *D. x delicatum*	= *D.* Delemma
D. Emmy	X *D. speciosum*	= *D.* Craig Hewitt
D. Gillian Leaney	X *D.* Graham Hewitt	= *D.* Aussie Quest
D. Gillian Leaney	X *D.* Peter	= *D.* Gilpet
D. Gillian Leaney	X *D. tetragonum*	= *D.* Emma
D. Graham Hewitt	X *D. tetragonum*	= *D.* Elinjaa
D. Gulginni	X *D. tetragonum*	= *D.* Coral Sun
D. Hastings	X *D.* Eureka	= *D.* Arcadia
D. Hastings	X *D. falcorostrum*	= *D.* Aussie Spice
D. Hastings	X *D.* Golden Fleck	= *D.* Peach Glow
D. Hastings	X *D.* Hilda Poxon	= *D.* Peter Hewitt
D. Hastings	X *D. jonesii* (as *D. ruppianum*)	= *D.* Aussie Gypsy
D. Hastings	X *D. speciosum*	= *D.* Wonga
D. Hastings	X *D. tetragonum*	= *D.* Peach Star
D. Hastings	X *D.* Wonga	= *D.* Rutherford Supreme
D. John Upton	X *D.* Jamie Upton	= *D.* Colsboys
D. John Upton	X *D.* Kungara	= *D.* Wakehurst
D. John Upton	X *D.* Ku-Ring-Gai	= *D.* Ring of Fire
D. John Upton	X *D. speciosum*	= *D.* Ingleside
D. John Upton	X *D. tetragonum*	= *D.* Bright Spark
D. Kevin Banks	X *D. tetragonum*	= *D.* Winston Hills
D. Kingrose	X *D.* Eureka	= *D.* Aussie Hero
D. Kingrose	X *D. falcorostrum*	= *D.* King Falcon
D. Kingrose	X *D.* Yondi	= *D.* Jenni

HYBRIDS (con't)

D. Ku-Ring-Gai	X *D.* Hilda Poxon	= *D.* Warrior
D. Ku-Ring-Gai	X *D. speciosum*	= *D.* Northmead
D. Ku-Ring-Gai	X *D.* Star of Gold	= *D.* Star Child
D. Penny Ann	X *D.* Bardo Rose	= *D.* Penny Rose
D. Penny Ann	X *D. falcorostrum*	= *D.* Twinkle
D. Penny Ann	X *D.* Goblin	= *D.* Minuet
D. Penny Ann	X *D. tetragonum*	= *D.* Fairy Floss
D. Pink Pencil	X *D. speciosum*	= *D.* Wesley Rose
D. x suffusum	X *D.* Bardo Rose	= *D.* Blushing Rose
D. x suffusum	X *D.* Ellen	= *D.* Pixie
D. x suffusum	X *D. falcorostrum*	= *D.* Aussie Freckles
D. x suffusum	X *D. speciosum*	= *D.* Harold Hirsch
D. x suffusum	X *D. tetragonum*	= *D.* Blushing Star
D. Tully	X *D.* Carawah	= *D.* Omega
D. Wonga	X *D.* Hilda Poxon	= *D.* Vida Banks
D. Wonga	X *D.* Jamie Upton	= *D.* Karen Nair
D. Wonga	X *D. speciosum*	= *D.* Woko
D. Wyuna	X *D.* Eureka	= *D.* Barnard
D. Wyuna	X *D. speciosum*	= *D.* Nerang
D. Wyuna	X *D.* Sunglow	= *D.* Tully
D. Yondi	X *D. speciosum*	= *D.* Jesmond Gem

Dendrobium lichenastrum (F. Muell.) Krzl. emend. Dockr.

Nth. Q/ld. Nat. 24(115):20 (1956)

Fig. 41, Plate 44

BASIONYM

Bulbophyllum lichenastrum F. Muell., *Fragm. Phyt. Aust.* 7:60 (1869) as 'bolbophyllum'

SYNONYMS

Dendrobium lichenastrum (F. Muell.) Krzl., *Engl. Pflanzenreich* 45:289 (1910)

Dockrillia lichenastrum (F. Muell.) Brieger, in Schlechter *Die Orchideen* ed. 3 (11/12): 745 (1981)

DESCRIPTION

A plant of this species was received by the Botanic Gardens in Edinburgh in 1836 from Earl Fitzwilliam of Wentworth. The type species was collected by J. Dallachy near Townsville in 1869 and named by Mueller as '*Bolbophyllum*' *lichenastrum*. The name comes from *lichen*, meaning a cryptogamic plant, and *astrum*, meaning a kind of—referring to the creeping habit resembling a lichen.

This species varies considerably in leaf form but not in flower form. It has a creeping, branching rhizome, tightly packed with leaves of varying shapes, often forming dense patches. Oval to round, almost button-shaped leaves varying in colour from green to reddish depending on the amount of light received.

Single flowers produced on a short stem about 1 cm long. Flowers approximately 4 to 7 mm in diameter and vary in colour from greenish to cream to pinkish with reddish striations. The well-proportioned segments are short and broad, and the rounded, prominent labellum varies from yellow to orange. Flowers last from 2 to 3 weeks and have a slightly musty odour when in the sun.

The opening of the capsule is considerably different from other members of the genus in Australia, except *D. prenticei*, see Plates 206, 207. At the commencement of the opening, the carpels break away from each other and open all the way down the suture join. After 2 to 4 days the carpels are fully open and splayed right back. By this time most, if not all, of the seed has been discharged. (Five seed capsules, two on *D. lichenastrum* and three on *D. prenticei* were observed to open this way.)

Fig. 41 *Dendrobium lichenastrum* (F. Muell.) Krzl. emend. Dockr.
From Mt. Lewis, north Queensland

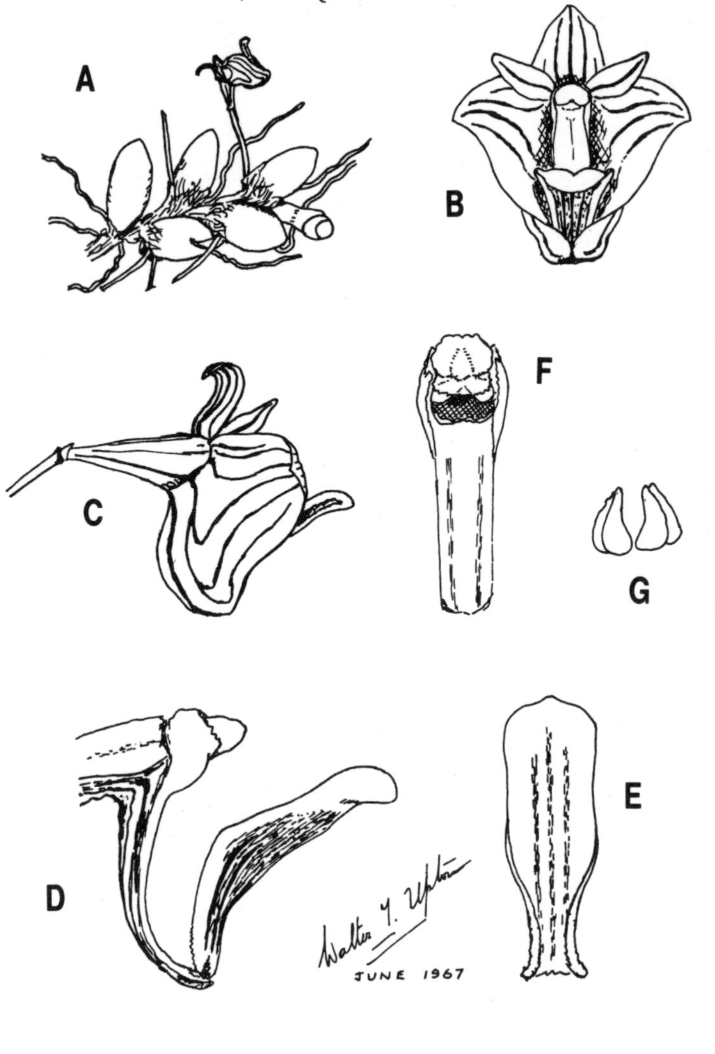

A Plant 1X
B Flower, front-view 4X
C Flower, side-view 4X
D Column and labellum, side-view 6X
E Labellum, pressed flat 6X
F Column, front-view 6X
G Pollinia greatly enlarged

FLOWERING PERIOD

All the year round.

DISTRIBUTION

Coastal ranges of north Queensland from approximately (21° 00′S) northward to (17° 00′S).

HABITAT

On rainforest trees or rocks, in fairly open forests. I have seen it growing on moss-covered branches of trees overhanging a steep, rocky creek in fairly bright light in the Coane Range at about 600 m. Humidity was very high.

POLLINATION

In nature this species is pollinated by a native bee, *Trigonia carbonaria* (R. S. Smythe, 1970c).

CULTURE

Slab culture suits best. Requires lots of water, and must not dry out when grown on slabs. Needs 50 per cent shade. Minimum winter temperature of 6°C.

IN HYBRIDIZATION

No hybrids registered to date.

Dendrobium linguiforme Sw. *Kongl. Vetensk. Acad. Nya Handl. Stockh.* 21:247 (1800)

Fig. 42

DESCRIPTION

Name comes from *linguae*, meaning tongue, and *formae*, meaning shape. This was the first Australian species of the genus *Dendrobium* to be named. I quote the original description by Olaf Swartz: '*D. linguaeforme*: caule radicante, foliis sessilibus ovatis carnosis; scapise basi foliorum multifloris. Ex insulis maris pacifici. Communicavit cel. SMITH'. Following Swartz's description in 1800, a detailed description, in English was given by Sir James Edward Smith in his *Exotic Botany* in 1804, together with a detailed drawing by James Sowerby. His description begins 'Found upon rocks along the seashore at Port Jackson'. As Surgeon-General John White sent many botanical specimens, including orchids, to both Smith and Banks, it is presumed that the plant described by Swartz was actually collected by White on the shores of Port Jackson.

Many writings in Australia have quoted *D. linguiforme* as the type species for the genus; this is not so as *D. linguiforme* was not one of the original 19 species named by Swartz in 1799 when he created the genus. The type species is *D. moniliforme* (Lindley) Swartz., a Japanese species which was included in the original list.

The common names are tongue orchid

and thumbnail orchid. Admiral William Bligh is understood to have also introduced this species into England in 1810.

D. linguiforme has three varieties which are each described separately.

D. linguiforme Sw. var. *linguiforme*

Plate 41

SYNONYMS

Callista linguiformis (Sw.) Kuntze, *Rev. Gen. Pl.* 2:655 (1891)

Dockrillia linguiforme (Sw.) Brieger, *Die Orchideen* ed. 3 (11/12):745 (1981)

DESCRIPTION

Tough, creeping, branched rhizome with fleshy leaves conspicuously ribbed longitudinally, frequently growing into large mats. Leaves vary considerably in shape and size but mostly oval. Colour of leaves can vary from a light green through to dark green with a purple tinge, depending on the amount of light they receive. Has dark purple blotches in the leaves when grown in full sun.

Racemes are approximately 6 to 15 cm long and bear up to 20 fragrant white to cream (very rarely pale yellow) flowers up

Fig. 42 *Dendrobium linguiforme* Sw. Collected Gloucester area, NSW

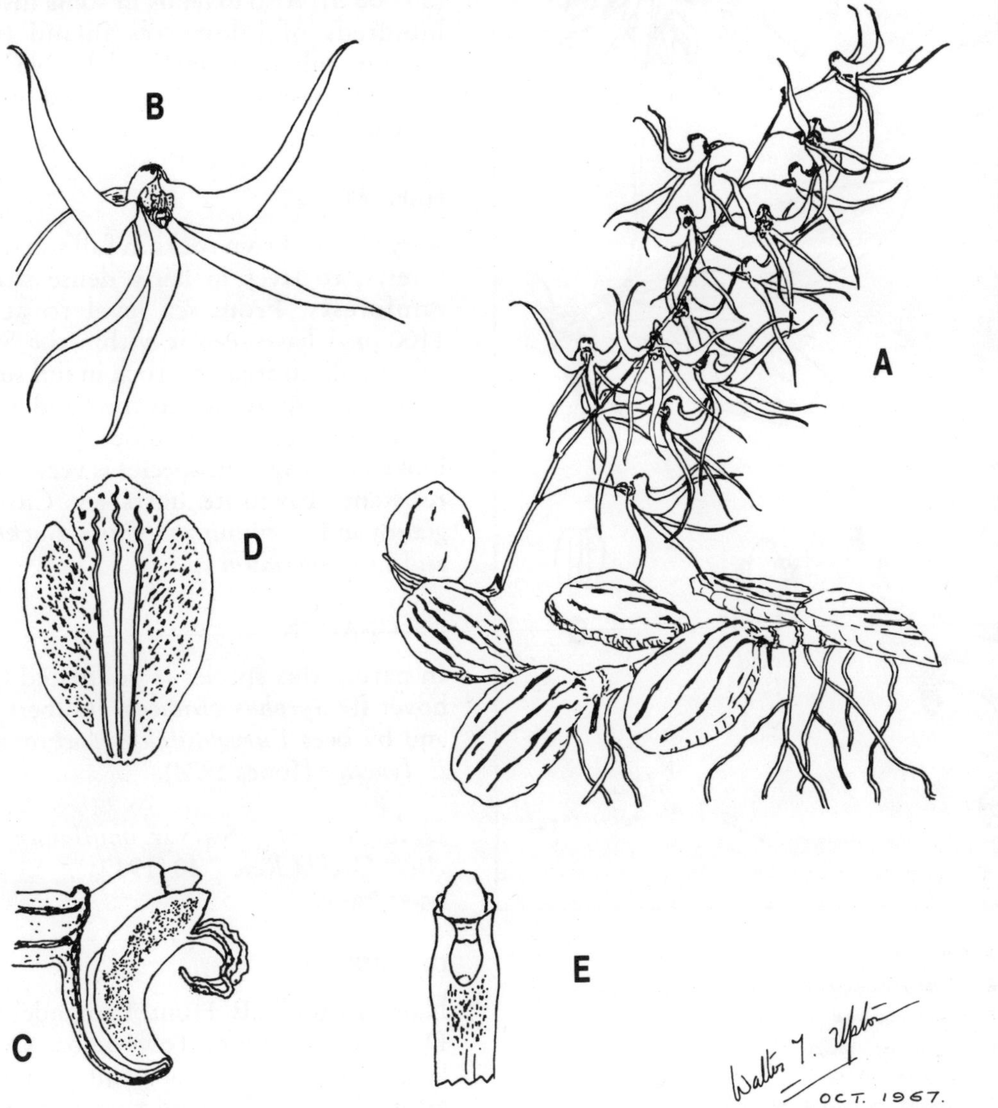

A Plant 1X
B Flower, front-view 2X
C Column and labellum, side-view 8X
D Labellum, pressed flat 8X
E Column, front-view 8X

Fig. 43 *Dendrobium linguiforme* Sw. var. *huntianum* Rupp
Collected from Ipswich District of south- eastern Queensland

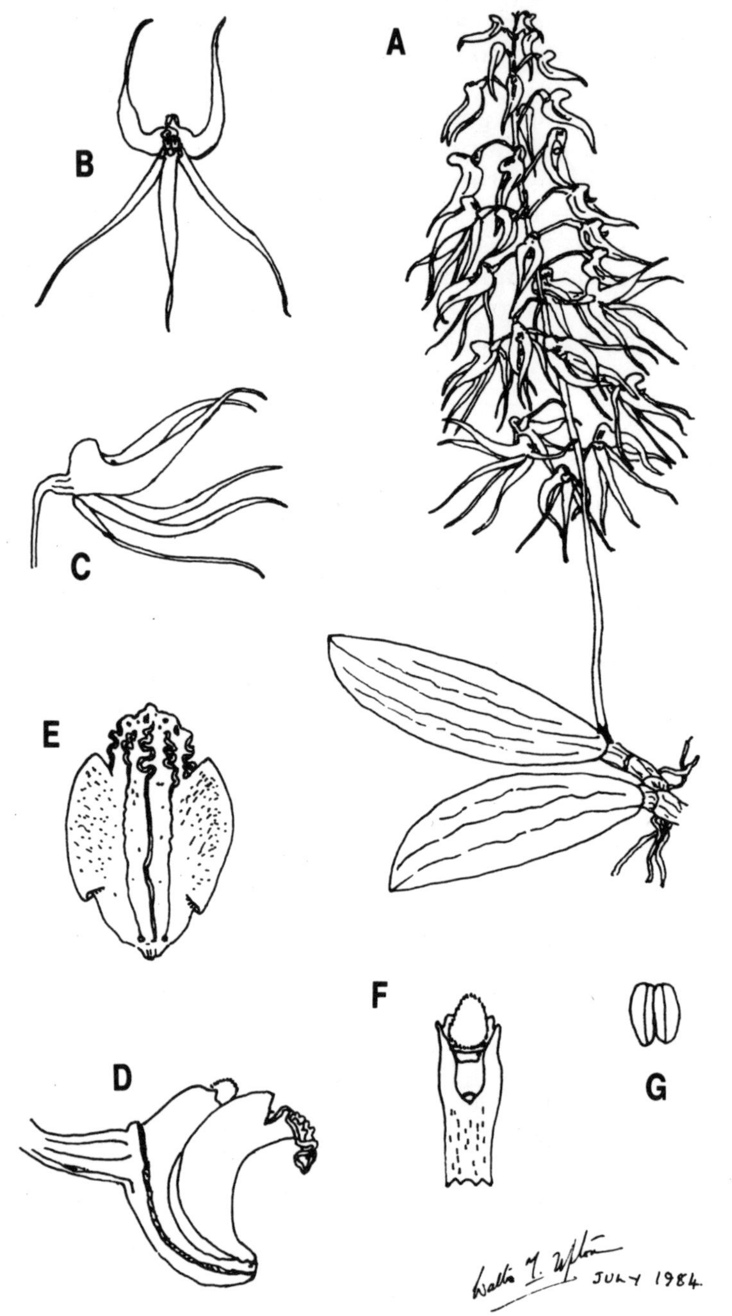

A Plant 1X
B Flower, front-view 2X
C Flower, side-view 2X
D Column and labellum, side-view 8X
E Labellum, pressed flat 8X
F Column, front-view 8X
G Pollinia greatly enlarged

to 5 cm in diameter. Generally plants produce about 10 flowers that are approximately 3 cm in diameter. Labellum is short in relation to other segments and marked in varying degrees with purple. Flowers last about 3 weeks. On warm days has a faint but pleasant perfume.

FLOWERING PERIOD

From August to November (winter to spring).

DISTRIBUTION

From at least as far south as Narooma (36° 15′S) in south-eastern NSW to the Burdekin River in north-eastern Queensland (20° 00′S). Also extends in some instances hundreds of kilometres inland (about 250 km inland near Coolah, NSW, although mainly coastal. Found in New Caledonia.

HABITAT

Very varied. From rocks in full sun in open forests, to trees in fairly dense shade in rainforests. From sea level to at least 1100 m. I have seen it within the Sydney metropolitan area on a rock in full sun, the rock being quite hot to the hand and the leaves almost a purple colour, yet the plant looked healthy. The species is very drought resistant. Favourite hosts are *Casuarina glauca* and *C. cunninghamiana, Backhousia* and *Leptospermum* sp.

POLLINATION

In nature this species is pollinated by the hover fly *Syrphus viridiceps* (Gilbert 1958) and by bees *Euryglossina hypochroma* and *E. lynettae* (Jones 1974).

D. linguiforme Sw. var. *huntianum* Rupp

Aust. Orchid Rev. 7:40 (1942)
Fig. 43, Plate 42

DESCRIPTION

Named after T. E. Hunt. I am indebted to Dr Noel Grundon, Towoomba, Queensland, as he obtained the plant for me after asking at a local orchid society if anyone knew where this variety was originally collected. A member's father knew the location and collected two plants which were then forwarded to me by Dr Grundon. Both plants continue to flower in June. I find it difficult to accept this as a distinct variety although it does have a distinct look about it.

The plant form is similar to the type variety although the flowers have a distinctive look, and in many regards are near to var. *nugentii.* The racemes are up to 18 cm long and the four racemes on my two plants carried 31, 27, 43 and 34 flowers and averaged 16 cm long. The fragrant flowers are off-white to cream.

FLOWERING PERIOD

June to July (mid-winter).

DISTRIBUTION

Ipswich district of south-eastern Queensland (27° 35′S).

HABITAT

Occurs sparingly on the paperbarks (*Melaleuca* sp.) of the lowlands.

D. linguiforme Sw. var. *nugentii* F. M. Bail.

Queensland Fl. 6:1533 (1902)

Fig. 44, Plate 43

DESCRIPTION

Named after L. J. Nugent. Habit of plant similar to var. *linguiforme.* Leaves generally larger, thicker, broader and more robust. Longitudinal furrows usually more pronounced and surface rougher.

Flowers usually have much shorter segments and are more evenly spaced on the raceme. Petals mostly longer than sepals. Side lobes of the labellum frequently quite heavily marked with red-purple.

FLOWERING PERIOD

June to August (winter).

DISTRIBUTION

North-eastern Queensland in both the highlands (where it is more common) and some areas of the lowlands. From south of Townsville (say 20° 00′S) to about as far north as Cooktown (15° 30′S).

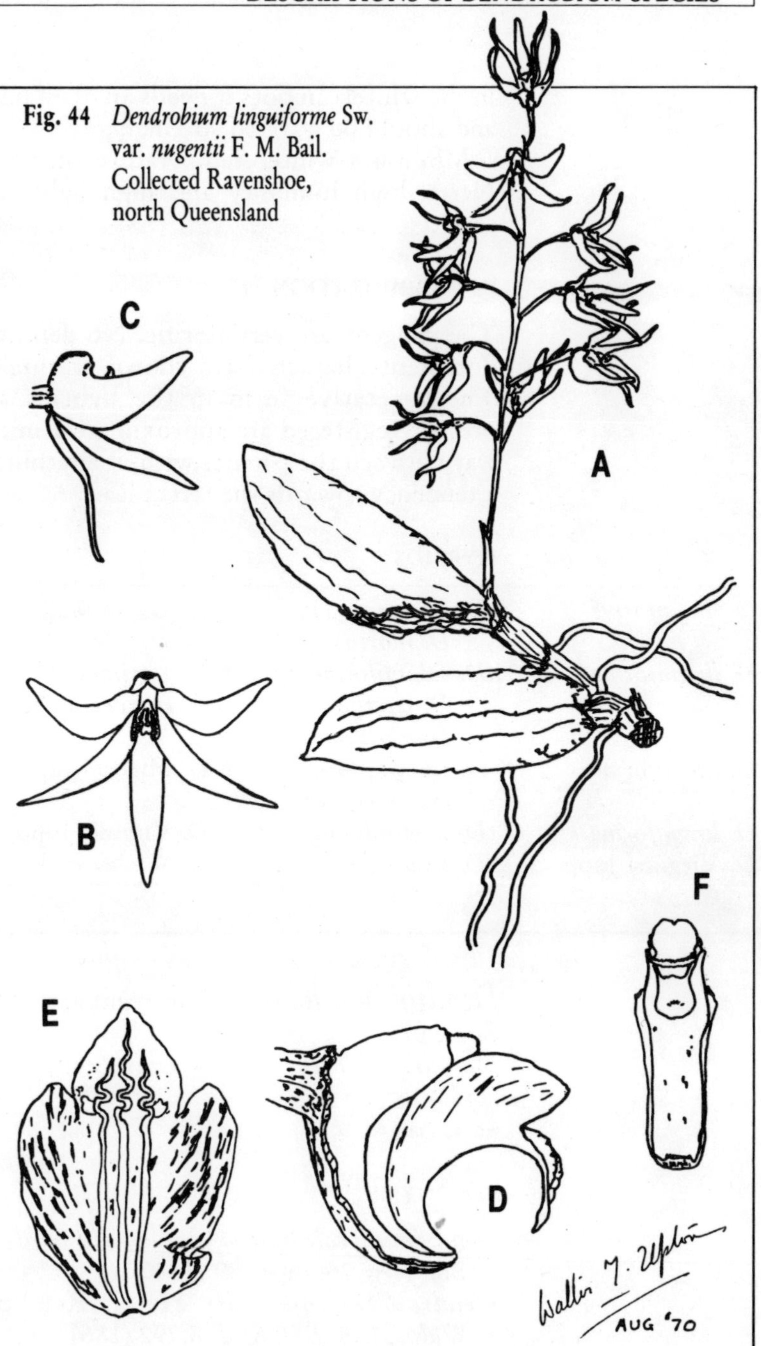

Fig. 44 *Dendrobium linguiforme* Sw. var. *nugentii* F. M. Bail. Collected Ravenshoe, north Queensland

A Plant 1X
B Flower, front-view 2X
C Flower, side-view 2X
D Column and labellum, side-view 8X
E Labellum, pressed flat 8X
F Column, front-view 8X

HABITAT

Similar to type variety although shows a preference for trees, particularly trees in dry creeks.

CULTURE

D. linguiforme grows well on hardwood or cork slabs, or tied to trees in the garden. *Jacaranda* or frangipani trees are quite suitable. On slabs it needs to be watered every evening in the summer and in the mornings in the winter. In pots it needs an open mix and should be watered sparingly.

Minimum winter temperature of 1°C. Needs high humidity and high light intensity.

IN HYBRIDIZATION

The progeny are very florific. No definite dominant characteristic shown through. The vegetative form in the hybrids at present registered are approximately midway between the parents with, if anything, a tendency towards the terete leaf.

HYBRIDS

D. linguiforme	X *D. bowmanii* (as *D. mortii*)	= *D.* Glenn Wall
D. linguiforme	X *D. calamiforme* (as *D. teretifolium* var. *fasciculatum*)	= *D. x grimesii* (natural hybrid)
D. linguiforme	X *D. schoenimum* (as *D. beckleri*)	= *D.* Michael Jupp
D. linguiforme	X *D. teretifolium*	= *D.* Virginia Jupp
D. Virginia Jupp	X *D. teretifolium*	= *D.* Wesley Pink

Dendrobium lobbii Teijsmann et Binnend.

Natuurk. Tijdschr. Ned.—Indie 5:491 (1853)

Fig. 45, Plate 45

SYNONYMS

Dendrobium calcaratum Lindley, *Edwards's Bot. Reg.* 26: misc. 89 (1840)
Dendrobium conostalix H. G. Reichb. *Walp. Ann. Bot. Syst.* 6:292 (1861)
Callista conostalix (H. G. Reichb.) Kuntze, *Rev. Gen. Pl.* 2:654 (1891)
Callista lobbii (Teijsmann et Binnend.) Kuntze, *Rev. Gen. Pl.* 2:655 (1891)
Conostalix calcarata (Lindley) Brieger, in Schlechter, *Die Orchideen*, ed. 3 (11/12): 659 (1981)
Conostalix lobbii (Teijsmann et Binnend.) Rauschert, *Feddes Repert.* 94:444 (1983)

DESCRIPTION

First recorded collection in Australia was in 1948 by Dr L. J. Brass in the Iron Range area of North Queensland. This species is terrestrial.

Erect stems up to 1 m tall, very slender, wiry, slightly pseudobulbous at the base. Many leaves with long sheathing bases, throughout most of the stem.

Flowers mostly face downwards, borne singly or in pairs. Sepals and petals greenish-yellow to yellow with orange veins (not always pronounced), labellum white. Flowers are quite long lasting.

FLOWERING PERIOD

July to September (winter to spring), but can occur at other times.

DISTRIBUTION

Originally collected by Dr L. J. Brass in the Iron Range area of north Queensland in 1948 and then in 1968 Ian Walters found a plant near Cardwell, North Queensland in the foothills of the Cardwell Range. It was again located in 1976 by Dr P. Lavarack and party at Tozer's Gap and Brown Creek; and in 1983 Dr Lavarack and party found this species in the hills above Carron Valley approximately 12° 24′S. No doubt as future expeditions are made to Cape York this orchid will be found in localities between Cardwell (18° 00′S) and the Carron Valley (12° 24′S) and further north. Also in New Guinea, Sumatra, Borneo, Malaya and Thailand.

HABITAT

In wet open swampy areas, among grasses and reeds where it receives a fair amount of light. Plants hard to see as they merge so well into the grasses and reeds. Water is seeping through the area continuously Also found by running creeks.

CULTURE

I find this species extremely difficult to grow. Pot in a coarse sand mixture with plenty of leaf-mould and peatmoss. Stand pot on pebbles in a tray of water, just allowing the pot to touch the water. Do not allow the water to become stagnant, change it frequently. The mixture must be kept just moist.

Requires high humidity and high light intensity. Minimum winter temperature of 12°C (54°F).

IN HYBRIDIZATION

No hybrids have been recorded to date.

Fig. 45 *Dendrobium lobbii* Teijsmann et Binnend. Collected near Cardwell, north Queensland

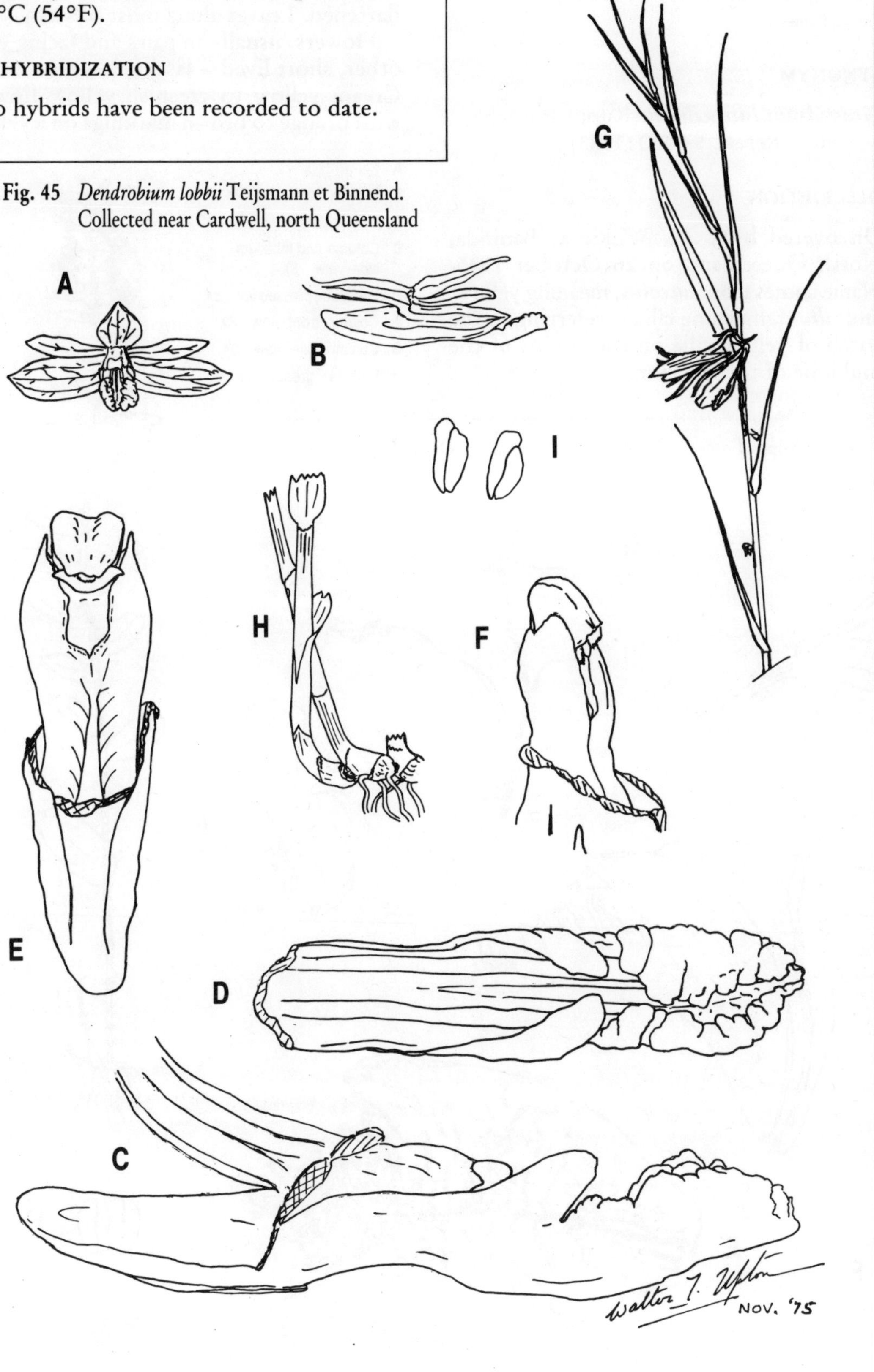

A Flower, front-view 3X
B Flower, side-view 3X
C Column and labellum, side-view 10X
D Labellum from above 10X
E Column, front-view 12X
F Column, side-view 12X
G Top of stem 1X
H Base of stems and new growth 2X
I Pollinia 14X

Dendrobium luteocilium Rupp

Nth. Q/ld. Nat. 13 (77):1 (1945)

Fig. 46, Plate 46

SYNONYM

Grastidium luteocilium (Rupp) Rauschert, *Feddes Repert.* 94:450 (1983)

DESCRIPTION

Discovered by J. H. Wilkie at Babinda, North Queensland on 26 October 1945. Name comes from *luteolus*, meaning yellow, and *cilium*, meaning cilia—referring to the patch of yellow cilia on the lamina of the mid-lobe of the labellum.

A tall plant with stems up to 200 cm, commonly about 1 m long. Stems slightly flattened. Leaves along most of stem.

Flowers, usually in pairs and facing each other, short lived—lasting less than a day. Cream yellow to greenish yellow flowers with orange to brown markings on a yellow

A Plant 1/9X
B Flower, front-view 2X
C Flower, side-view 2X
D Column and labellum, side-view 8X
E Labellum, pressed flat 8X
F Column, front-view 8X
G Column, side-view 8X
H Pollinia greatly enlarged

Fig. 46 *Dendrobium luteocilium* Rupp From the McIlwraith Ranges, north Queensland

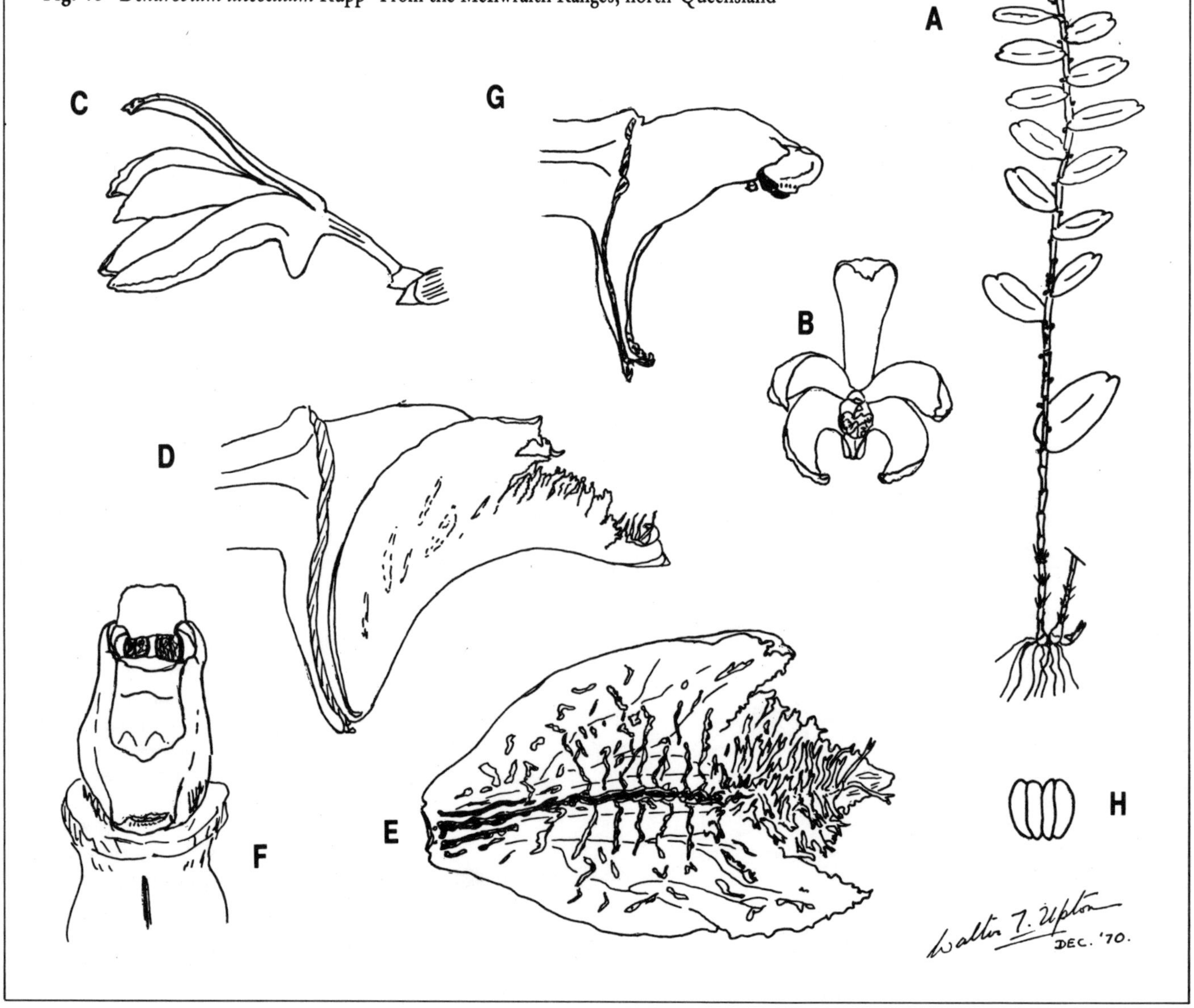

labellum; the ridge down the centre of the labellum bright orange, the cilia yellow. Frequently all the flowers in the one area open on the same day emitting a strong honey perfume. Plant in Fig. 46 had 14 flowers, they all opened at approximately 7 am and the majority of them closed by 4 pm. It is interesting to note that the cilia possibly assist in self-pollination as the labellum seems to furl up as the plant dies and the anther and pollinia are dislodged by cilia. Most of the dying flowers, when forced open, had the pollinia lodged in the cilia; the flowers left on the plant produced seed capsules.

FLOWERING PERIOD

October to February (spring to summer).

DISTRIBUTION

From the Tully River (18° 00'S) northward to the tip of Cape York Peninsula and also in Papua New Guinea.

HABITAT

On rocks and trees mostly in the lowlands although it is found at higher altitudes. In the lowland rainforests, in mangrove swamps and in open areas on rocks and occasionally trees at higher elevations. Areas of high humidity, good light and lots of air movement. Frequently growing into large clumps and on trees overhanging the creeks.

CULTURE

Pot culture is quite satisfactory with an open medium. About 50 per cent shade in the summer and no shade in mid-winter. It prefers to be kept moist all year round but does not like 'wet feet'.

Weak fertilizer all year except the coldest months of winter when it is not growing. Minimum winter temperature of 12°C.

IN HYBRIDIZATION

No hybrids have been registered to date.

Dendrobium macropus (Endl.) H. G. Reichb. ex Lindley

J. Proc. Linn. Soc. Bot. 3:9 (1859)

BASIONYM

Thelychiton macropus Endl., *Prod. Florae Norfolkicae* 33: (1833)

DESCRIPTION

A species named from material originally collected in Norfolk Island and subsequently found in Australia, Lord Howe Island, Fiji and New Caledonia.

In a personal communication from the late (Mrs) Pat Ralston of Norfolk Island, dated 7/9/65, she stated 'Incidentally, Dr. Garay said that both *D. brachypus* and *D. macropus* are only forms of *D. gracilicaule* F. Muell'. At that time, although it was realized these species were one and the same, no action was taken to change any names because both *D. brachypus* and *D. macropus*, the earlier named species, were abnormal forms (being peloric and in most instances cleistogamous) and the International Code of Botanical Nomenclature outlawed names based on abnormalities. But in 1975 the article in the International Code of Botanical Nomenclature restricting the use of abnormal species in name priority was deleted, leaving the way clear to bring all varieties of this one species under the one name. The name changes have been carried out by P. S. Green, Honorary Research Associate, Royal Botanic Gardens, Kew in the *Journal of the Arnold Arboretum*, USA.

The name comes from *macropus*, meaning long footed — referring to the length of the pseudobulbous stems. It is a complex species of three subspecies, each described separately.

D. macropus (Endl.) H. G. Recihb. ex Lindley subsp. *gracilicaule* (F. Muell.) P. S. Green

J. of Arnold Arbor. 67:115 (1986)

Fig. 47, Plate 48

BASIONYM

Dendrobium gracilicaule F. Muell., *Fragm. Phyt. Aust.* 1:179 (1859)

Fig. 47 *Dendrobium macropus* (Endl.) H. G. Reichb. subsp. *gracilicaule* (F. Muell.) P. S. Green
From Gloucester area, NSW

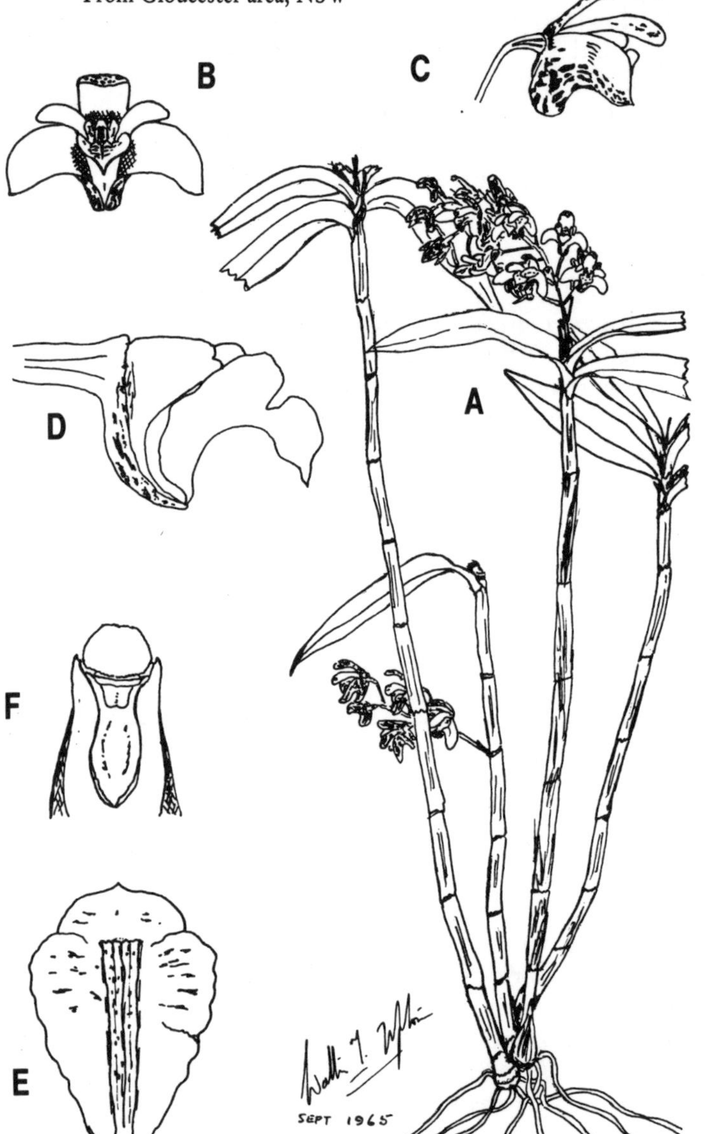

A Plant 1/3X
B Flower, front-view 2X
C Flower, side-view 2X
D Column and labellum, side-view 4X
E Labellum, pressed flat 4X
F Top of column, front-view 8X

SYNONYMS

Dendrobium elongatum A. Cunn., *Edwards's Bot. Reg.* 25: misc. 33 (1839)
Dendrobium brisbanense H. G. Reichb., *Walp. Ann. Bot. Syst.* 6:299 (1861)
Callista gracilicaulis (F. Muell.) Kuntze, *Rev. Gen. Pl.* 2:654 (1891)
Dendrobium comptonii Rendle, *J. Linn. Soc.* 45:247 (1920)
Tropilis gracilicaulis (F. Muell.) Rauschert, *Feddes Repert.* 94:470 (1983)

DESCRIPTION

From the synonyms you will notice that Cunningham's name *D. elongatum* was applied in 1839, but this was invalid because of an earlier *D. elongatum* named by Lindley in 1830 and applying to an entirely different species. Mueller's name of *D. gracilicaule*, although not applied until 1859, had priority until *D. macropus* was eligible due to the deletion of article 71 from the International Code of Botanical Nomenclature.

Mueller described this species from a plant collected by W. Hill at Moreton Bay. The first record of it being grown in England is of a plant sent to Kew by J. F. Roberts of Melbourne in 1883. The name comes from *gracilis*, meaning slender, and *caulis*, meaning stalk—referring to the slender pseudobulbous stems.

Very slender pseudobulbous stems up to 90 cm long, commonly about 30 to 40 cm long. Leaves lanceolate, up to six near-terminal.

Racemes commonly near-terminal in young stems while older stems flower further down. Racemes bear from a few to 40 very fragrant flowers, approximately 1 to 2 cm across. Flowers from green-yellow to bright yellow, most frequently dull yellow, with varying amounts of red-brown blotching on the outside. Occasionally flowers yellow to almost gold with no markings whatsoever. Labellum mostly dull yellow with few red-brown markings. Individual flowers generally last from 3 to 4 weeks. Whole raceme lasts over a month.

FLOWERING PERIOD

August to October (late winter to spring).

DISTRIBUTION

From Kiama (34° 40′S) in the south of NSW to the Bloomfield River (16° 00′S) in the south-eastern Cape York Peninsula in Queensland. Quite common from southern Queensland (25° 00′S) to the Hawkesbury River in NSW (33°30′S).

HABITAT

Mostly on trees but occasionally on rocks, palms, etc. Habitat quite varied, from coastal scrub and open forest, frequently in very sunny situations, to rainforest (perhaps its most common occurrence). In the most northerly part of its distribution it is only found in the ranges and tablelands.

D. macropus (Endl.) H. G. Reichb. ex Lindley subsp. *howeanum* (Maiden) P. S. Green
J. Arnold Arbor. 67:115 (1986)
Fig. 48, Plate 49

BASIONYM

Dendrobium gracilicaule F. Muell. var. *howeanum* Maiden, *Proc. Linn. Soc. N.S.W.* 24:382 (1889)

DESCRIPTION

Named after Lord Howe Island.

Pseudobulbous stems usually thicker than the type variety and seldom above 40 cm long. Leaves generally slightly larger and longer.

Raceme generally shorter. Flowers golden yellow to pale yellow and sometimes cream; usually slightly larger than subsp. *gracilicaule.* Labellum cream and marked with red-pink. Flowers fragrant but different from the type variety.

FLOWERING PERIOD

September to October (spring).

DISTRIBUTION

Lord Howe Island. There have been unconfirmed reports of this variety occurring on the mainland at Mangrove Mountain, near Gosford, NSW.

HABITAT

Common on trees and widespread throughout the island. Large plants frequently seen. When plants fall from the trees they continue to grow quite happily on the ground or rocks beneath.

In New Caledonia and Fiji there are forms of *D. macropus* which are possibly nearer to subsp. *howeanum* than subsp. *gracilicaule.*

Fig. 48 *Dendrobium macropus* (Endl.) H. G. Reichb. ex Lindley subsp. *howeanum* (Maiden) P. S. Green From Lord Howe Island

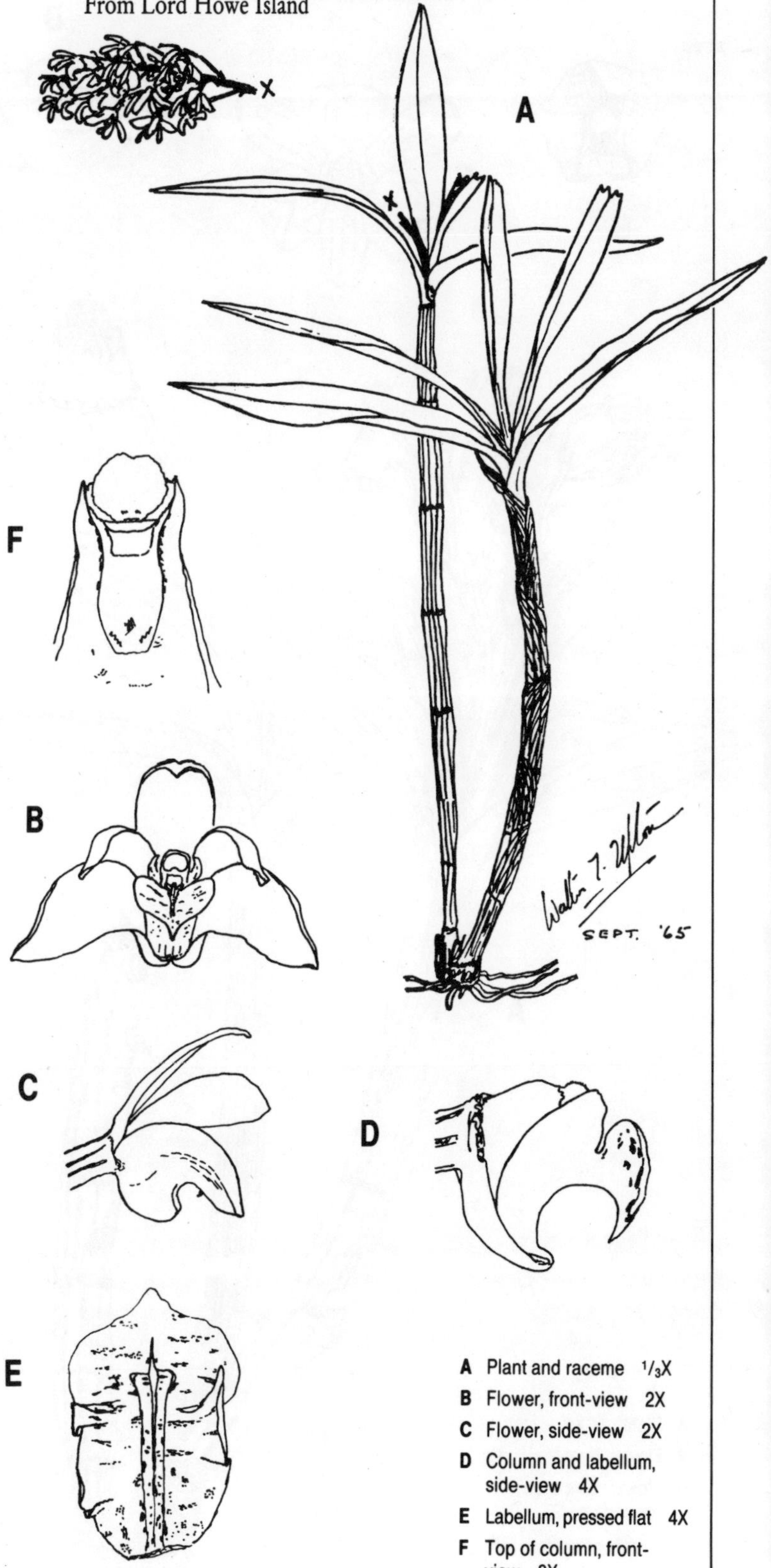

A Plant and raceme $^1/_3$X
B Flower, front-view 2X
C Flower, side-view 2X
D Column and labellum, side-view 4X
E Labellum, pressed flat 4X
F Top of column, front-view 8X

Fig. 49 *Dendrobium macropus* (Endl.) H. G. Reichb. ex Lindley subsp. *macropus* (Endl.) P. S. Green
Collected by Pat Ralston near Mt Pitt

A Plant ²/₃X
B Flower, front-view 2X
C Flower, front-view, opened up 2X
D Flower side-view 2X
E Column, front-view 4X

D. macropus (Endl.) H. G.Reichb. ex Lindley subsp. *macropus* (Endl.) P. S. Green

J. Arnold Abor. 67:116 (1986)
Fig. 49, Plate 47

SYNONYMS

Thelychiton brachypus Endl., *Prod. Florae Norfolkicae* 33 (1833)
Dendrobium brachypus (Endl.) H. G. Reichb., Linnaea xli., 42 (1877)

DESCRIPTION

Pseudobulbous stems, mainly yellowish-green, from 12 to 25 cm long and mostly grooved longitudinally. Three to six leaves borne on the upper part of the pseudobulbous stems.

Raceme (in all the plants I have seen) terminal, but no doubt also flowers laterally from the upper nodes. Raceme bears eight to ten yellow-green to creamy flowers. Sepals and petals about 12 mm long but the sepals slightly broader. Flowers do not appear to open fully. Flowers are peloric; the labellum being indistinguishable from the other petals. Column is short and column foot absent. As in the form *D. brachypus*, flowers appear to be frequently self-pollinating, therefore the ovary always appears large, no doubt due to this.

FLOWERING PERIOD

Mainly August to September (late winter to spring), but can flower much earlier.

DISTRIBUTION

Endemic to Norfolk Island where it is common on the pines (*Araucaria*) and other trees of the Mount Pitt Reserve; but less common elsewhere on the island.

HABITAT

On trees and very occasionally on rocks in heavily-wooded areas up to 300 m.

D. brachypus Plate 47

I cannot agree with the conclusions arrived at in an article 'The orchids of Norfolk Island' in *Orchadian*, September 1985, that from studying Ferdinand Bauer's original drawings *D. macropus* had been based on a

57 *D. monophyllum*

58 *D. moorei*

59 *D. rigidum*

60 *D. schoeninum*

61 *D. schneiderae*

62 *D. schneiderae* var. *major*

63 *D. striolatum*

64 *D. stuartii*

65 *D. speciosum* var. *speciosum* 'Buckety'

66 *D. speciosum* var. *hillii*

67 *D. speciosum* var. *grandiflorum*

68 *D. speciosum* var. *capricornicum* type plant, photo S. Clemesha

69 *D. speciosum* var. *pedunculatum*

70 *D. speciosum* var. *curvicaule*

71 *D. speciosum* var. *pedunculatum*

fully grown plant and *D. brachypus* on younger material in which the pseudobulbous stems had not matured. I have studied copies of the original drawings and have my own drawings.

In September 1965, Mrs Pat Ralston sent me plants of both *D. brachypus* and *D. macropus*, which I have grown and flowered ever since. She stated in the letter that 'some plants are long and thin and some are short and fat'. There is quite a distinction between the two in plant form, although they grew together in many instances.

D. brachypus has never increased the size of its pseudobulbous stems and *D. macropus* has, if anything, got longer. I selfed the *D. macropus* in September 1981 and brought numerous seedlings to maturity. At no time did they look like *D. brachypus* having quite quickly (1985) grown longish stems. Clements (Pers. Comm., 1986) suggested that the difference may be in the chromosome number.

I am giving another description here for the taxon previously called *D. brachypus*, as no doubt it should be either a separate subspecies or at least a variety of *D. macropus* var. *macropus.*

DESCRIPTION

Name comes from *brachypus*, meaning short footed—referring to the short pseudobulbous stems.

Pseudobulbous stems crowded together, 13 to 55 mm long, swollen in the middle and taper gradually to the top. Three to four leaves borne from the top of the stems; leaves getting smaller with the top leaf approximately half the size of the first leaf.

Off-white flowers borne in a different manner from any member of the Australian native *Dendrobium* species. Generally from three to four approximately 12 mm flowers produced on a 40 to 50 mm raceme. Each flower emerges from the axil of a roundish bract. Pedicel of the first flower much longer than that of the last flower, so that the inflorescence has an almost capitulate appearance. Bracts also become smaller towards the top of the raceme. Petals slightly narrower than sepals. Flowers are peloric, labellum being indistinguishable from petals. Column short and column foot absent. Flowers appear to be frequently self-pollinating. I have opened many immature flowers and the pollinia is already in position in the stigmatic cavity. Ovary always appears large, but this is probably the reason.

FLOWERING PERIOD

Mainly August to September (late-winter to spring), but can flower much earlier.

DISTRIBUTION

Endemic to Norfolk Island. Common in the forests of Mount Pitt Reserve and less common elsewhere on the island.

HABITAT

On trees and very occasionally on rocks in heavily-wooded areas.

CULTURE

Either pot or slab culture seem satisfactory. Drainage in a pot must be excellent as *D. macropus* has fine roots which will deteriorate quickly if overwatered, particularly during the winter months.

Requires good light, good air movement, good humidity, regular feeding during its growing period and must be allowed to dry out between waterings. Minimum winter temperature of approximately 2°C with a little more warmth for some of the Pacific Island forms.

IN HYBRIDIZATION

A very dominant species in shape, size and colour; both in plant and flower form. When crossed with the dominant *D. kingianum*, that species is dominant for colour.

HYBRIDS

D. macropus subsp. *gracilicaule* (as *D. gracilicaule*)	X *D. adae*	= *D.* Gracia
D. macropus subsp. *gracilicaule* (as *D. gracilicaule*)	X *D. aemulum*	= *D.* Liela Gilbert
D. macropus subsp. *gracilicaule* (as *D. gracilicaule*)	X *D.* Ella Victoria Leaney	= *D.* Robyn Kaylene

HYBRIDS (con't)

D. macropus subsp. *gracilicaule* (as *D. gracilicaule*)	X *D.* Ellen	= *D.* Ellen Grace
D. macropus subsp. *gracilicaule* (as *D. gracilicaule*)	X *D.* Eureka	= *D.* Nugget
D. macropus subsp. *gracilicaule* (as *D. gracilicaule*)	X *D. falcorostrum*	= *D.* Susan
D. macropus subsp. *gracilicaule* (as *D. gracilicaule*)	X *D. fleckeri*	= *D.* Golden Fleck
D. macropus subsp. *gracilicaule* (as *D. gracilicaule*)	X *D. x gracillimum*	= *D.* Shan Leaney
D. macropus subsp. *gracilicaule* (as *D. gracilicaule*)	X *D. kingianum*	= *D. x suffusum*
D. macropus subsp. *gracilicaule* (as *D. gracilicaule*)	X *D. jonesii* (as *D. ruppianum*)	= *D.* Golden Cascades
D. macropus subsp. *gracilicaule* (as *D. gracilicaule*)	X *D. speciosum*	= *D. x gracillimum*
D. Allyn Star	X *D. tetragonum*	= *D.* Tarantula
D. Aussie Freckles	X *D.* Aussie Bonanza	= *D.* Aussie Flame
D. Aussie Freckles	X *D. falcorostrum*	= *D.* Aussie Majestic
D. Aussie Freckles	X *D. kingianum*	= *D.* Fiona Banks
D. Blushing Rose	X *D.* Ella Victoria Leaney	= *D.* Ella Rose
D. Blushing Rose	X *D. kingianum*	= *D.* Red Baron
D. Blushing Rose	X *D. speciosum*	= *D.* Regal Rose
D. Blushing Star	X *D.* Hilda Poxon	= *D.* Kim Heinze
D. Blushing Star	X *D.* Sunglow	= *D.* Blushing Sun
D. Double Two	X *D. kingianum*	= *D.* Nathan Wall
D. Double Two	X *D. speciosum*	= *D.* Ronald Banks
D. x gracillimum	X *D. adae*	= *D.* Double Two
D. x gracillimum	X *D. x delicatum*	= *D.* Sun Sprite
D. x gracillimum	X *D. falcorostrum*	= *D.* Gracious Falcon
D. x gracillimum	X *D. kingianum*	= *D.* Penny Ann
D. x gracillimum	X *D. speciosum*	= *D.* Gracious Cascades
D. x gracillimum	X *D. tetragonum*	= *D.* Allyn Star
D. Gracious Cascades	X *D. tetragonum*	= *D.* Rutherford Sunstar
D. Gracious Falcon	X *D.* Eureka	= *D.* Aussie Delight
D. Gracious Falcon	X *D. speciosum*	= *D.* Karloo
D. Gracious Falcon	X *D.* Sunglow	= *D.* Kalang
D. Gracious Falcon	X *D. tetragonum*	= *D.* Gracious Star
D. Golden Fleck	X *D.* Aussie Bonanza	= *D.* Margery Angeleri
D. Golden Fleck	X *D.* Aussie Glow	= *D.* Tambar

HYBRIDS (con't)

D. Golden Fleck	X *D.* Hastings	= *D.* Peach Glow
D. Golden Fleck	X *D.* Hilda Poxon	= *D.* Hot Shot
D. Golden Fleck	X *D.* Peter	= *D.* Whispering Jack
D. Golden Fleck	X *D. tetragonum*	= *D.* Jamie Upton
D. Jamie Upton	X *D.* Golden Fleck	= *D.* Green Star
D. Jamie Upton	X *D.* John Upton	= *D.* Colsboys
D. Jamie Upton	X *D. speciosum*	= *D.* Lemon Glow
D. Jamie Upton	X *D.* Wonga	= *D.* Karen Nair
D. Penny Ann	X *D.* Bardo Rose	= *D.* Penny Rose
D. Penny Ann	X *D.* Ellen	= *D.* Ferg Martin
D. Penny Ann	X *D. falcorostrum*	= *D.* Twinkle
D. Penny Ann	X *D.* Goblin	= *D.* Minuet
D. Penny Ann	X *D. tetragonum*	= *D.* Fairy Floss
D. x suffusum	X *D.* Bardo Rose	= *D.* Blushing Rose
D. x suffusum	X *D.* Ellen	= *D.* Pixie
D. x suffusum	X *D. falcorostrum*	= *D.* Aussie Freckles
D. x suffusum	X *D. kingianum*	= *D.* Cawongla
D. x suffusum	X *D. speciosum*	= *D.* Harold Hirsch
D. x suffusum	X *D. tetragonum*	= *D.* Blushing Star
D. Susan	X *D. adae*	= *D.* Green Goddess
D. Susan	X *D. kingianum*	= *D.* Kit Murdoch
D. Susan	X *D.* Ellen	= *D.* Sweet Ellen

Dendrobium malbrownii Dockr.

Aust. Pl. 4:134 (1967)

Fig. 50

SYNONYMS

Monanthos malbrownii (Dockr.) Rauschert, *Feddes Repert.* 94:455 (1983)

Fig. 50 *Dendrobium malbrownii* Dockr. From the McIlwraith Range in Cape York Peninsula

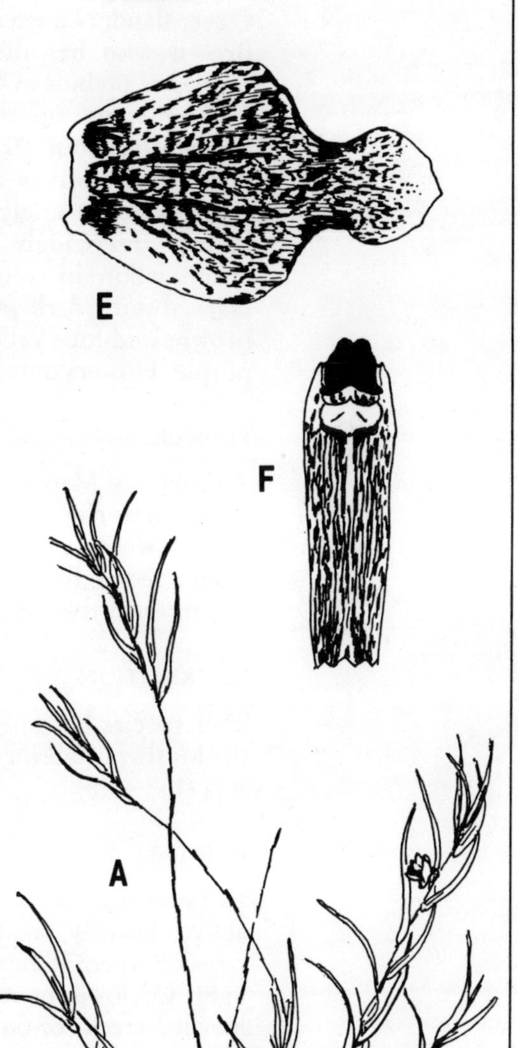

A Plant 1/3X
B Flower, front-view 4X
C Flower, side-view 4X
D Column and labellum, side-view 8X
E Labellum, pressed flat 8X
F Column, front-view 8X
G Pollinia greatly enlarged

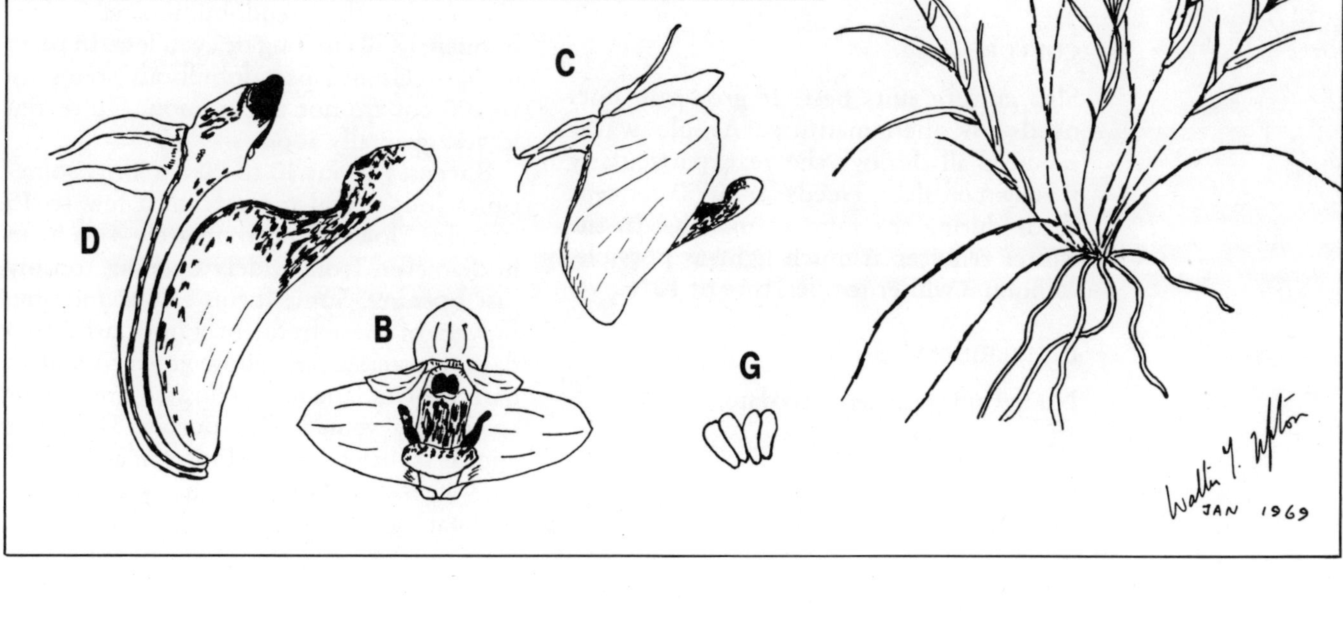

DESCRIPTION

Discovered on 4 March 1967 by Malcolm Brown in the McIlwraith Range, north Queensland. Named in honour of Malcolm Brown who has discovered many new species of orchids in Cape York Peninsula.

Wiry stems up to 35 cm long, most frequently about 20 to 25 cm long in a dense mass; leaves along most of stem. Flowers borne singly, about 8 mm in diameter, and widely opening. Cream to cream-yellow in colour. Labellum heavily marked with dark purple to almost dark brown; mid-lobe yellow. Anther very dark purple. Flowers only last about 2 to 3 days.

FLOWERING PERIOD

January to March (summer to autumn) seems to be the main flowering period, but also flowers at other times. One of my plants seems to have one or two flowers out most of the time.

DISTRIBUTION

This species has only been recorded from the McIlwraith Range in Cape York Peninsula (13° 45′S).

HABITAT

In rainforests at altitudes of 400 m and above. Usually on the upper branches of trees or where it experiences fairly intense light. On lower branches it is usually overhanging creeks or on the edges of forests.

CULTURE

Slab culture suits best. If grown in pots needs an open medium. Ample water required all through the year, particularly if grown on slabs. Needs about 50 per cent shade during the summer months. In the winter requires as much light as possible. Minimum winter temperature of 10°C.

IN HYBRIDIZATION

No hybrids registered to date.

Dendrobium mirbelianum Gaud.

Freycinet Voy. Uranie 423, t.38 (1826)

Fig. 51, Plate 53

SYNONYMS

Dendrobium rosenbergii Teijsmann et Binnend., *Natuurk. Tijdschr. Ned.—Indie* 24:317 (1862)

Dendrobium polycarpum H. G. Reichb., *Gard. Chron.* (2):492 (1883)

Dendrobium giulianettii F. M. Bail., *Q/ld. Agric. J.* 3:158 (1898)

Dendrobium aruanum Krzl., *Engl. Pflanzenreich Heft* 45:242 (1910)

Dendrobium wilkianum Rupp, *Nth. Q/ld. Nat.* 10(67):4 (1941)

DESCRIPTION

This species was originally found in Australia in 1934 by Gillie Bates, of the Gordonvale Sugar Experiment Station, and Bill Kearns, a cane farmer. In October 1941, Jack H. Wilkie, who in his early days discovered many of our native orchids, found a plant at Babinda, north Queensland, and sent it to the Rev. H. M. R. Rupp, who described it as *D. wilkianum.*

The type plant was collected at Rawak on the Island of Waigeo off the north-west coast of New Guinea. It is named after Professor Mirbel.

In the Australian form the pseudobulbous stems are generally up to 90 cm long, although in the southern part of its distribution the pseudobulbous stems are frequently 30 cm long or even less. In parts of New Guinea pseudobulbous stems up to 200 cm are not uncommon. Up to ten leaves, generally about six.

Racemes about 10 to 20 cm long, sometimes longer and carry from a few to 15 flowers. Flowers approximately 4 to 5 cm in diameter, from widely opening to only just opening. Some forms self-pollinating, flowers of these frequently only last 2 to 3 days. Flowers from yellow-green to quite a dark brown; labellum yellowish-green conspicuously veined with purple. Does not appear to have a scent. Flowers last from 2 to 3 weeks in the forms that are not self-pollinating.

FLOWERING PERIOD

Mainly August to October (late winter to spring), but often occurs much earlier and later than this.

DISTRIBUTION

Eastern Cape York Peninsula from the Johnstone River (17° 30′S) northwards and into New Guinea, westward to the Moluccas, and eastward to the Bismarck Archipelago and the Solomon Islands.

HABITAT

On trees in lowland forests and coastal swamps where it gets strong light. Also seen in a fair amount of shade from the canopy of tall, straight-trunk trees that are unbranched for about 12 m, growing happily at about 10 m. Area always very humid, being but a short way from a river or mangrove swamp.

CULTURE

Very similar culture to *D. discolor* except seems to prefer a minimum winter temperature of 12°C.

IN HYBRIDIZATION

Appears to be a typical *Spatulata* section parent, although in its primary hybrids the petals are not quite as twisted.

All the early hybrids using *D. mirbelianum* were obviously with the larger and more spectacular New Guinea forms. With the exception of *D.* Sauterne, all hybrids in the list below were made prior to 1966 and only *D.* Gloucester Dawn and *D.* Princess Stephanie were made in Australia.

A Plant and raceme $1/_3$X
B Flower, front-view 1X
C Flower, side-view $1^1/_3$X
D Column and labellum 2X
E Labellum, pressed flat 3X
F Column, front-view 4X
G Column, side-view 2X

Fig. 51 *Dendrobium mirbelianum* Gaud. Purchased from Limberlost Nursery, Cairns as *D. wilkianum*

HYBRIDS

D. mirbelianum	X *D. antennatum* (as *D. d'albertisii*)	= *D.* Moses Mark
D. mirbelianum	X *D. bigibbum*	= *D.* Princess Stephanie
D. mirbelianum	X *D. canaliculatum*	= *D.* Gloucester Dawn
D. mirbelianum	X *D. discolor* (as *D. undulatum*)	= *D.* Champagne
D. mirbelianum	X *D. x superbiens* (as *D. goldiei*)	= *D.* Moira Stewart
D. mirbelianum	X *D. johannis*	= *D.* Kaipu
D. Champagne	X *D.* Dang Toi	= *D.* Rajah
D. Champagne	X *D. discolor* ((as *D. undulatum*)	= *D.* Yellow Curls
D. Champagne	X *D. johannis*	= *D.* Sauterne
D. Yellow Curls	X *D.* Verninha	= *D.* Wee Cheow Lay

Dendrobium monophyllum F. Muell.

Fragm. Phyt. Aust. 1:189 (1859)
Fig. 52, Plate 57

SYNONYMS

Dendrobium tortile A. Cunn., *Edwards's Bot. Reg.* 25:misc. 33 (1839): nomen nudum non Lindley (1847)
Callista monophylla (F. Muell.) Kuntze, *Rev. Gen. Pl.* 2:655 (1891)
Australorchis monophylla (F. Muell.) Brieger in Schlechter, *Die Orchideen* ed. 3 (11/12):741 (1981)

DESCRIPTION

Named from a plant sent to Mueller by W. Hill, Queensland Colonial Botanist, and collected from the woods at Moreton Bay, Queensland. Name comes from the Greek *mono*, meaning single, and *phyllon*, meaning leaf. The original description only refers to one leaf, no doubt the specimen from which it was described had only one leaf. It is not unusual for *D. monophyllum* to have two leaves. Common name is lily-of-the-valley orchid.

Creeping rhizome with erect, almost conical pseudobulbous stems set close together. Often growing into large masses, rhizomes not infrequently growing back over the older ones and forming a number of layers. Leaves solitary or occasionally two.

Raceme is erect and up to 18 cm high, carrying from six to 20 or more (generally around ten), fragrant, dull to bright-yellow flowers about 8 mm in diameter. Labellum yellow. Flowers partly nodding and do not generally open widely. Individual flowers last over 2 months in good condition in some clones.

FLOWERING PERIOD

Main season August to October (late winter to spring). I have a plant that is seldom out of flower.

DISTRIBUTION

From the Clarence River (28° 30′S) in northern NSW (there have been reports of its existence near Gloucester (32°S), but I have searched for it in vain in this area) to north Queensland, at least to the Annan River (15° 30′S).

HABITAT

On rocks and trees, frequently exposed to the full sunlight, on cliff faces, in open forests and on the high branches of rain-forest trees. Always where it gets plenty of light and air movement. I have seen it growing on trees on a headland in south Queensland where it receives almost continuous salt spray from the sea. Generally a lowland plant in the southern portion of its range; it is not uncommon in both the coastal swamps and on the ranges in southern Queensland. In north Queensland it appears to be confined to the highlands and seldom seen below 600 m.

Trees are often seen with their branches completely matted with this species and many other species of orchids, growing happily together.

CULTURE

Slab culture suits best. It is essential to tie it tightly to its host as it is slow to establish itself in cultivation. Generally a slow grower.

Requires excellent ventilation, plenty of light and frequent watering, although it is necessary to dry plants out between waterings. Minimum winter temperature of 2°C, but some plants seem to need a little warmth in the coldest winter months; no doubt these are plants collected in Queensland at lower altitudes or in the coastal swamps.

IN HYBRIDIZATION

Only five hybrids have been registered to date. The most important characteristic that is being passed on by *D. monophyllum* is the length of time the flowers last. So far the shape appears to be reasonably dominant. In *D.* Orana and *D.* Golligosh the yellow colour has dominated.

Flowers of the progeny are delightful and have a *D. monophyllum* look about them. *D. monophyllum* as a parent has a great future. I have at least five more hybrids using this species which are nearing flowering size and the plant form in most of them is so far favouring *D. monophyllum*.

HYBRIDS

D. monophyllum	X *D. adae*	= *D.* Orana
D. monophyllum	X *D. canaliculatum*	= *D.* Collette Fleur Simpson
D. monophyllum	X *D. kingianum*	= *D.* Karani
D. monophyllum	X *D. schneiderae*	= *D.* Golligosh
D. monophyllum	X *D. speciosum*	= *D.* Aussie Choice

Fig. 52 *Dendrobium monophyllum* F. Muell. Collected in north Queensland

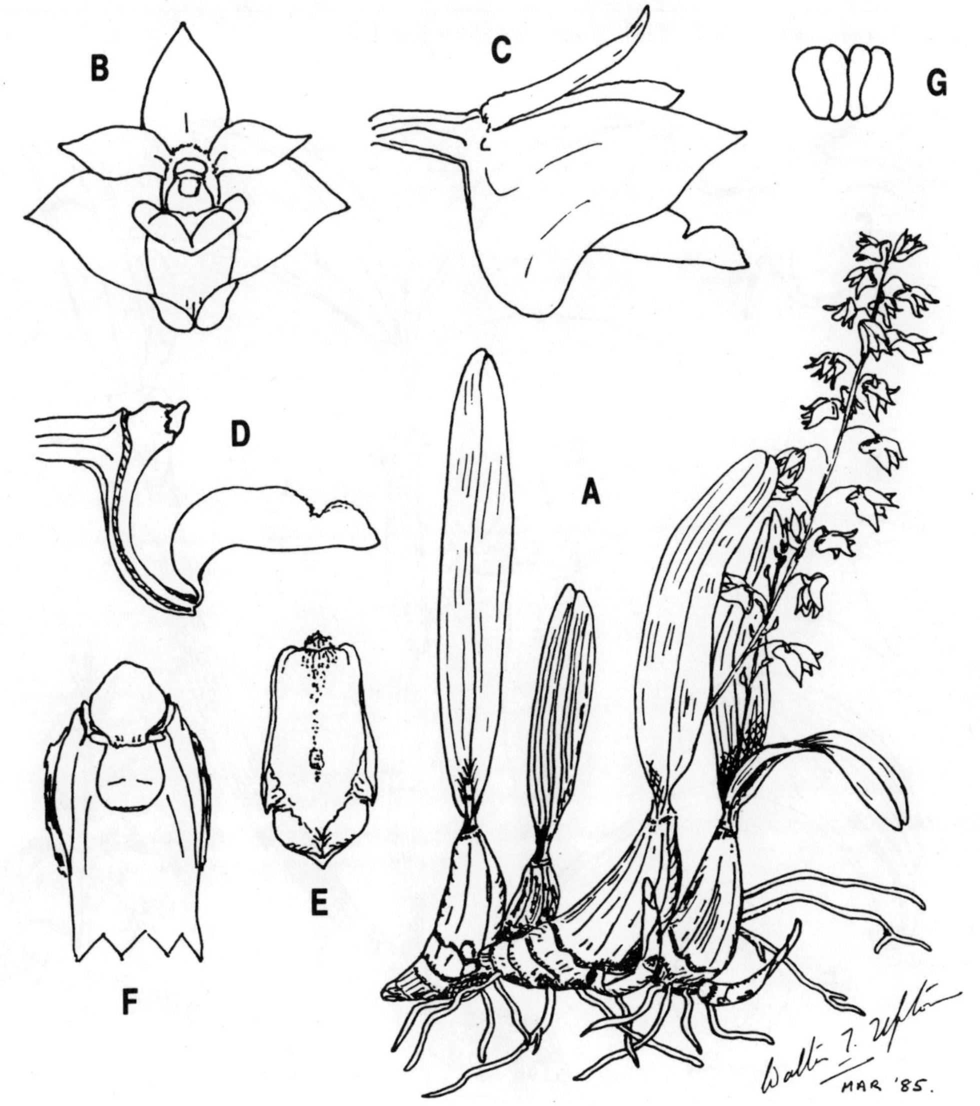

A Plant ²/₃X
B Flower, front-view 4X
C Flower, side-view 4X
D Column and labellum, side-view 4X
E Labellum viewed from above 4X
F Top of column, front-view 8X
G Pollinia greatly enlarged

Dendrobium moorei F. Muell.

Fragm. Phyt. Aust. 7:29 (1869)
Fig. 53, Plate 58

SYNONYM

Tropilis moorei (F. Muell.) Rauschert, *Feddes Repert.* 94:470 (1983)

DESCRIPTION

Discovered by R. D. Fitzgerald in 1869 on Lord Howe Island, growing 'on precipices among the mountains, and on the mossy branches of trees which hang over the cliffs'. Named in honour of Charles Moore, who at the time was Director of the Sydney Botanic Gardens.

Pseudobulbous stems up to 25 cm long, varying from quite slender to quite stout; two to six leaves. Racemes up to 15 cm long (usually about 10 cm), bearing up to 20 flowers. Flowers a pure glistening white but do not generally open very wide. Labellum also pure white. Flowers spotted with light purple are occasionally seen. The stigmatic cavity in all the flowers I have inspected contains a large globule of clear liquid, which makes the cavity appear convex instead of concave. When pollinating the flowers, the pollinia sinks deeply into the clear liquid. Flowers which are long lived do not appear to be perfumed.

Of this plant, Fitzgerald (1888) said 'When seen adhering to the black basalt of

Fig. 53 *Dendrobium moorei* F. Muell. From Lord Howe Island

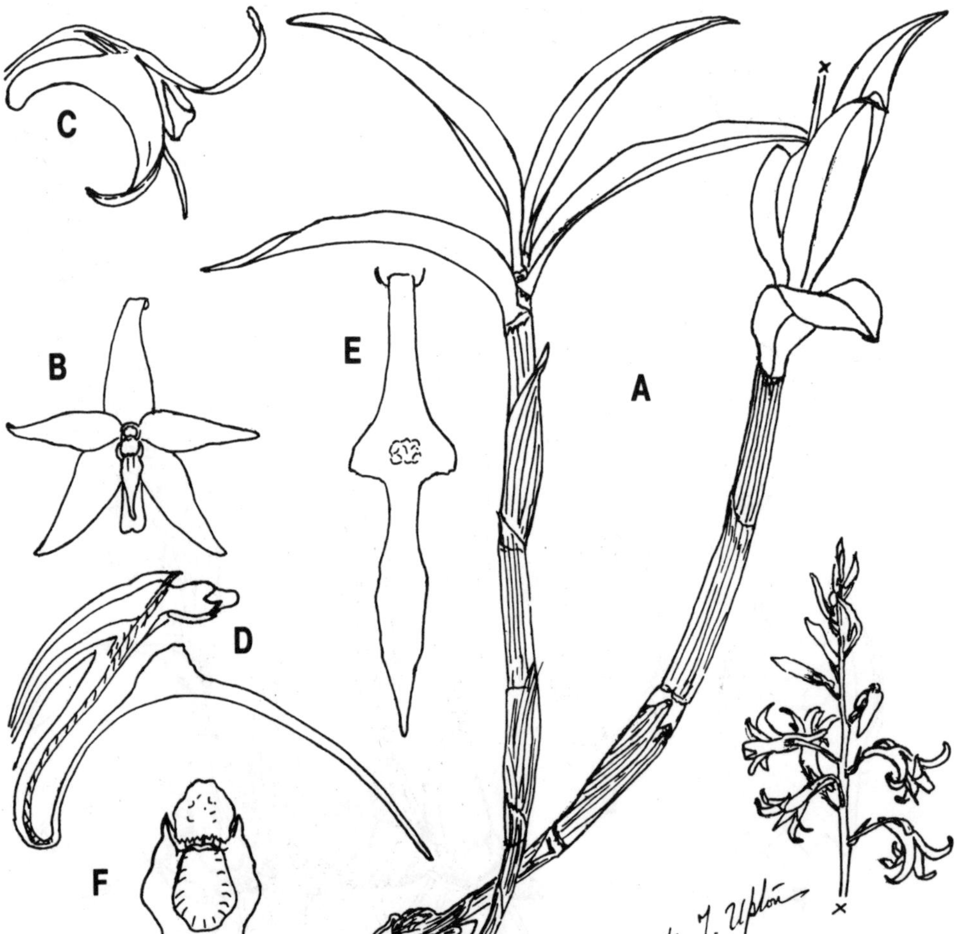

A Plant and raceme 2/3X
B Flower, front-view 2X
C Flower, side-view 2X
D Column and labellum, side-view 4X
E Labellum, pressed flat 4X
F Column, front-view 8X

the chasms, the white waxy flowers, more like hyacinths than orchids, make a beautiful contrast to the wildness of the scene. The flower is like a snowflake in the very act of melting'.

FLOWERING PERIOD

Mainly March to July (autumn to winter). Racemes take approximately 4 months from first showing to maturing. A well-grown, happy plant will seldom be out of flower.

DISTRIBUTION

Endemic on Lord Howe Island (31° 33'S and 150° 4'E). Lord Howe Island is about 702 km north-east of Sydney and is approximately 11 km long and 1 km wide. The climate is mild with moderate humidity. The normal temperature range is between 12° and 30°C.

HABITAT

C. A. Jaede (1936) states 'I originally found this species growing on Mt. Gower from 1000 ft above sea level . . . They grow on rocks, tree ferns, scrub and appear to do particularly well in the soil. The best plants were those upon tree ferns . . . On my last trip to the island I located *D. moorei* at the other end of the island growing in a narrow vertical strip and coming down to within 100 ft above sea level. Here it was that the orchid was growing particularly well in the soil.'

D. moorei also grows on figs and palms. Grows fairly low down on the trees in a fair amount of light. Is quite abundant.

CULTURE

Although I have seen many plants of this species growing happily on slabs, I find it is best grown in pots in an open mix. When in full growth, spring to early autumn, needs plenty of water and regular fertilizing. Produces aerial growths (keikis) very easily, particularly if overwatered. Should dry out between watering to help prevent aerial growths.

In the Sydney area of NSW it grows easily in an open bush-house in about 50 per cent shade. Some clones are shy at flowering unless given a little warmth in winter, say a minimum of 12°C.

IN HYBRIDIZATION

In the two hybrids I have seen, *D. moorei X D. kingianum* and (*D. moorei X D. kingianum*) *X D. tetragonum*, it appears to pass on the very delicate and glistening characteristics of the flowers to its progeny. Also it is fairly dominant in its labellum shape. Neither of the hybrids have been registered.

Dendrobium mortii F. Muell.

Fragm. Phyt. Aust. 1:214 (1859)

Fig. 54, Plate 50

SYNONYMS

Callista mortii (F. Muell.) Kuntze, *Rev. Gen. Pl.* 2:655 (1891)

Dendrobium tenuissimum Rupp, *Proc. Linn. Soc. N.S.W.* 52:570 (1927)

Dendrobium robertsii F. Muell. ex Rupp, *Q/ld. Nat.* 9:52 in obs. (1934) nom. invalid

Dockrillia mortii (F. Muell.) Rauschert, *Feddes Repert.* 94:446 (1983)

Dockrillia tenuissima (Rupp) Rauschert, *Feddes Repert.* 94:447 (1983)

DESCRIPTION

The plant was collected from woods near the Hastings River by Dr Beckler and sent to Mueller. It is named after T. S. Mort, outstanding promoter of horticulture in NSW.

M. Clements has recently sighted the type sheet for *D. mortii* in Melbourne and determined that the name *D. tenuissimum* has been misapplied to this taxon (Pers. Comm. M. Clements, 1986). *D. tenuissimum*, it would appear, was originally collected by John Hopson and the Rev. H. M. R. Rupp from the Upper Allyn River in October 1925. *D. robertsii* is from the sources of the Condamine River.

In plant form, *D. mortii* is the daintiest of the fleshy-leaved members of the genus *Dendrobium*. It resembles a miniature *D. teretifolium* and has very slender, thread-like stems. It is pendulous and plants are

Fig. 54 *Dendrobium mortii* F. Muell. Collected in the Gloucester area, NSW

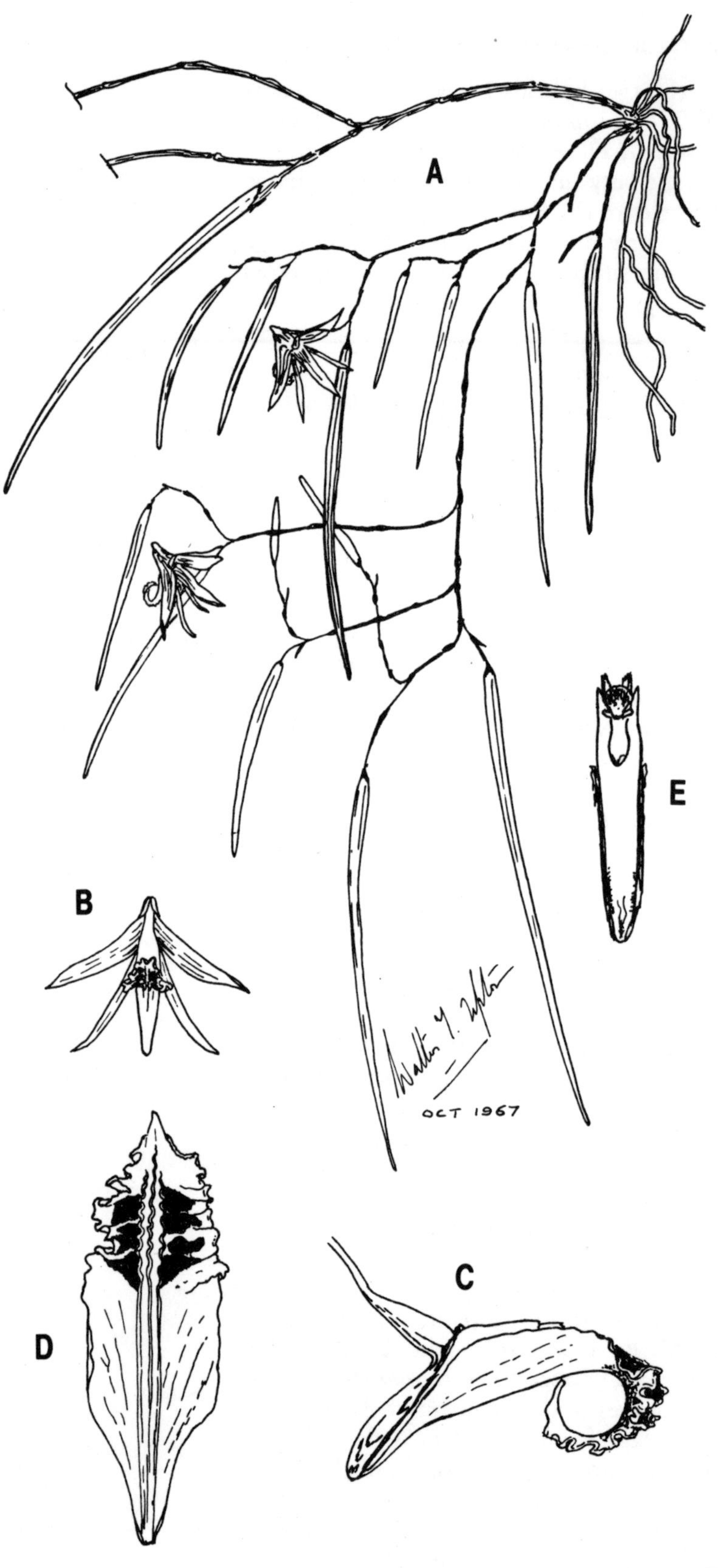

A Plant ²/₃X
B Flower, front-view 1X
C Column and labellum, side-view 3X
D Labellum, pressed flat 3X
E Column, front-view 3X

generally about 30 cm long, although I have seen plants well over 60 cm long. The roots are very conspicuous as they have red growing tips. Leaves pendulous, cylindrical and about 4 to 12 cm long.

Normally single flowered, approximately 2 to 3 cm in diameter and widely opening. Vary in colour from a tawny-brown to greenish-brown through to a strong dark-green frequently with rich-purple striae on the outside of the segments and bright-purple markings on the white labellum. Pleasantly fragrant.

CHROMOSOME NUMBER

2n = 38 (K. Y. Lim, 1985).

FLOWERING PERIOD

September to October (spring).

DISTRIBUTION

From the southern foothills of Barrington Tops (say 32° 30′S) in NSW to the ranges in the extreme south-east of Queensland (say 28° 00′S).

HABITAT

In rainforests on moss-covered trees where dew and mists are commonplace, at altitudes of 300 to 900 m. Occasionally in more open forests. I can recall seeing hundreds of plants of this species on the trees that had been left after logging had been carried out in their rainforest home. They hung from the uppermost branches and trunks like dainty flimsy lace garments belonging to the fairies, dancing in the slightest breeze. But with the large tree cover gone, how long would they exist in this changed environment with its unwelcome intense light and burning sun.

CULTURE

Slab culture suits well because of the plant's pendulous nature. Similar culture to *D. striolatum*, except that it requires slightly more shade, say 60 to 70 per cent. Minimum winter temperature of 4°C.

This plant can be easily propagated as it makes aerial roots quite freely.

IN HYBRIDIZATION

Only six hybrids have been registered to date. Four within its own section, *Rhizobium*, and two with a species of the section *Dendrocoryne*. The dainty appearance of this species is passed on to its progeny. All hybrids to date have been registered under the species name *D. tenuissimum.*

HYBRIDS

D. mortii (as *D. tenuissimum*)	X *D. falcorostrum*	= *D.* Amaroo
D. mortii (as *D. tenuissimum*)	X *D. pugioniforme*	= *D.* Dainty Cascades
D. mortii (as *D. tenuissimum*)	X *D.* Rosemary Jupp	= *D.* Super Ruffles
D. mortii (as *D. tenuissimum*)	X *D. schoeninum* (as *D. beckleri*)	= *D.* Ida Mary
D. mortii (as *D. tenuissimum*)	X *D. striolatum*	= *D.* Julie Skillicorn
D. Dainty Cascades	X *D. kingianum*	= *D.* Elfin

Dendrobium nindii W. Hill

Votes Proc. Legis. Ass. Q/ld. Session 2:871 (1874)

Fig. 55, Plate 51

SYNONYMS

Dendrobium tofftii F. M. Bail., *Syn. Q/ld. Fl.* suppl. 3:71 (1890)

Dendrobium ionoglossum Schltr., *Repert. Spec. Nov. Regn. Veg. Beih.* 1:551 (1912)

DESCRIPTION

Originally collected by P. H. Nind in September 1873 at Nind Creek off the Moresby River (17° 40′S). Mr Nind had joined the North Eastern Coast Exploring Expedition, of which Walter Hill, Colonial Botanist, Director and Founder of the Brisbane Botanic Gardens, was a member. Hill named the species *D. nindii*, after Mr Nind in his report of the Brisbane Gardens presented to the Legislative Assembly of Queensland in 1874 (Collins, 1979). Surveyor Tofft also collected this species in the 1880s and it was described by F. M. Bailey as *D. tofftii.* Common name is blue orchid. In 1986 this species appeared on a 55 cent postage stamp of Australia.

Pseudobulbous stems up to 300 cm tall, but generally about 100 to 120 cm in cultivation. Usually about 10 to 15 leaves in the distal half of the stem.

Racemes from the upper nodes of the stem, erect or arching and usually about 40 cm long, carrying from eight to 25 widely-opening flowers, approximately 6 cm in diameter. Racemes with over 50 flowers have been recorded. The fragrant flowers, which have slightly twisted perianth segments, are extremely beautiful. Sepals generally white to cream to pale lilac, petals erect or almost erect and white with a few pale-violet veins sometimes quite strongly marked with violet. Labellum pale violet-mauve with darker veins. An alba form has been recorded (Tierney, 1946). Flowers long-lasting.

FLOWERING PERIOD

Mainly July to September (winter to spring).

DISTRIBUTION

Northern Queensland. From Johnstone River (17° 30′S), northward into New Guinea. Uncommon.

HABITAT

Prefers to grow on trees in the coastal swamps, but occasionally found in low altitude gorges with high light intensity. Grows on mangroves (*Avicennia*), *Barringtonia* sp., Leichhardt (*Nauclea orientalis*) and the river cherry (*Wormia alata*). Some plants grow out in the full sun. It is probably safe from most collectors as it is frequently found in swamps that are crocodile infested with large pythons in the trees. The swamps are slimy, stinking and steamy.

HYBRIDS

D. nindii (as *D. tofftii*)	X *D. affine* (as *D. dicuphum*)	= *D.* Gresford
D. nindii (as *D. tofftii*)	X *D. bigibbum*	= *D.* Australia
D. nindii (as *D. tofftii*)	X *D.* Dang Toi (as *D.* Albanense)	= *D.* Frances Henton
D. nindii (as *D. tofftii*)	X *D. discolor* (as *D. undulatum*)	= *D.* Peter Petersen
D. nindii (as *D. tofftii*)	X *D. johannis*	= *D.* Tusco
D. nindii (as *D. tofftii*)	X *D. x superbiens*	= *D.* Blue Gloucester
D. Peter Petersen	X *D. canaliculatum*	= *D.* Son of Gloucester
D. Peter Petersen	X *D. carronii*	= *D.* Cherub
D. Peter Petersen	X *D. discolor* (as *D. undulatum*)	= *D.* Camp Island
D. Peter Petersen	X *D. x superbiens*	= *D.* Bluey

Fig. 55 *Dendrobium nindii* W. Hill. From Daintree River area

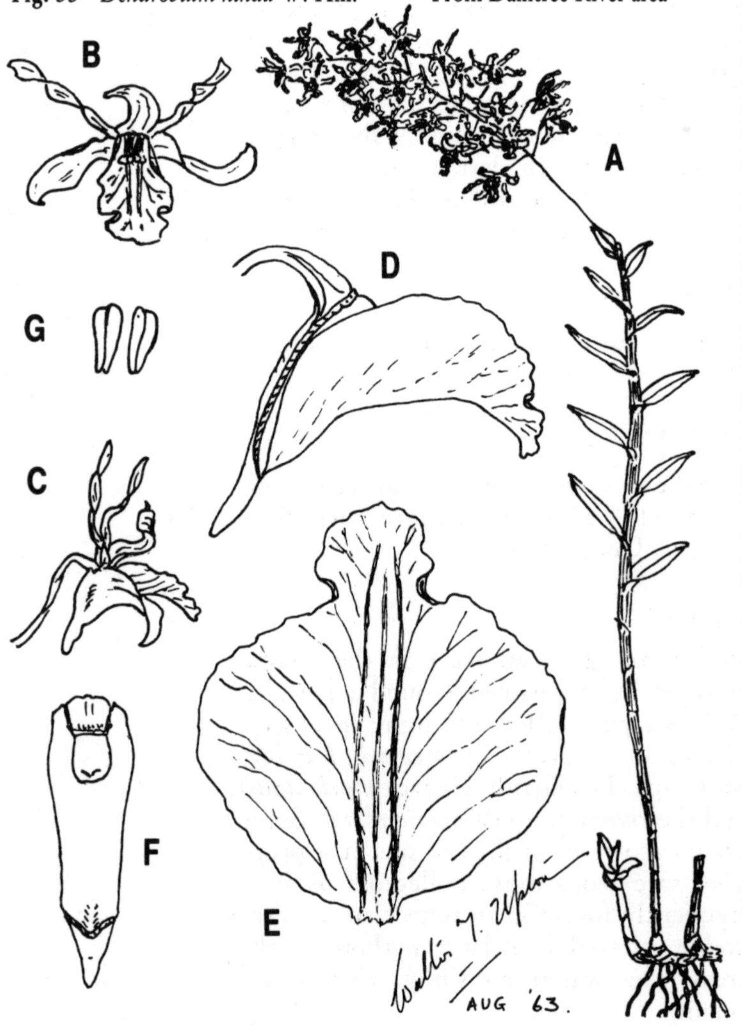

A Plant $^1/_{10}$X
B Flower, front-view $^1/_2$X
C Flower, side-view $^1/_2$X
D Column and labellum, side-view 2X
E Labellum, pressed flat 2X
F Column, front-view 2X
G Pollinia greatly enlarged

CULTURE

Grows well when affixed to a sheet of cork (Portuguese cork preferred) about 30 cm long by 25 cm wide. Tie it on tightly and place staghorn peat or spaghnum moss over the roots to get it established quickly. Requires heavy watering during the heat of the summer, but when the inflorescence appears reduce watering. Do not allow the plant to become completely dry. Its natural habitat is in a swamp, therefore its roots must always be just damp—reminiscent of the steamy atmosphere in nature.

Give it full sunlight but keep its roots shaded from the direct rays of the sun. In colder areas take care when watering, maybe twice a week should be ample most of the year to keep roots moist. Minimum winter temperature of 15°C.

IN HYBRIDIZATION

Appears to be fairly dominant in plant size and mostly in flower size. The mauve lip comes through in many of its progeny.

Dendrobium prenticei (F. Muell.) Nicholls

Nth. Q/ld. Nat. 7 (56):2 (1938)

Fig. 56, Plate 55

BASIONYM

Bulbophyllum prenticei F. Muell., *Sth. Sci. Rec.* 1:173 (1881)

SYNONYMS

Bulbophyllum lichenastrum Fitzg., *Aust. Orch.* 2(5):(t.18) 1893, non F. Muell., (1869)

Dendrobium variabile Nicholls, *Nth. Q/ld. Nat.* 7(56):2 (1938)

Dendrobium aurantiaco-purpureum Nicholls, *Nth. Q/ld. Nat.* 10(68):2 (1942)
Dendrobium lichenastrum (F. Muell.) Krzl. var. *prenticei* (F. Muell.) Dockr., *Nth. Q/ld. Nat.* 24(115):20 (1956)
Dendrobium lichenastrum (F. Muell.) Krzl. var. *prenticei* (F. Muell.) Dockr. forma *aurantiaco-purpureum* (Nicholls) Dockr., *Nth. Q/ld. Nat.* 24(115):20 (1956)

DESCRIPTION

In recent research Clements determined the change in status for this taxon and the earlier specific name is again recognized. The taxon *D. lichenastrum* var. *prenticei* forma *aurantiaco-purpureum* is to be reduced to a synonym of *D. prenticei* (Pers. Comm., M. Clements, 1986).

Originally found on Bellenden Ker Range in 1881. Named after Dr Charles Prentice.

An extremely variable plant, leaves from oval shaped to terete or, in some instances, almost cylindrical. Vary in length from approximately 1 to 6 cm. Leaf from prostrate to completely erect.

Flowers variable in size, shape and colour, being approximately 4 to 9 mm in diameter. Generally flowers open well and frequently segments furl back, though some hardly open at all. Sepals much broader and longer than petals. Colour varies considerably from a translucent dirty cream through cream to a pale translucent pink; some flowers quite heavily striate with mauve to purple. The degree of striation varies greatly. In some cases these colours suffused throughout. Labellum from pale yellow to orange in colour. Flowers last about 15 to 20 days, and have a musty odour.

Plates 206, 207 show the opening of the seed capsule of *D. prenticei*, which is considerably different from other members of the genus in Australia, except *D. lichenastrum* (refer to this species). This species is more common than *D. lichenastrum.*

FLOWERING PERIOD

All the year round.

DISTRIBUTION

From the Burdekin River (20° 00′S) to the Annan River (15° 30′S).

Fig. 56 *Dendrobium prenticei* (F. Muell.) Nicholls
From the Atherton Tableland

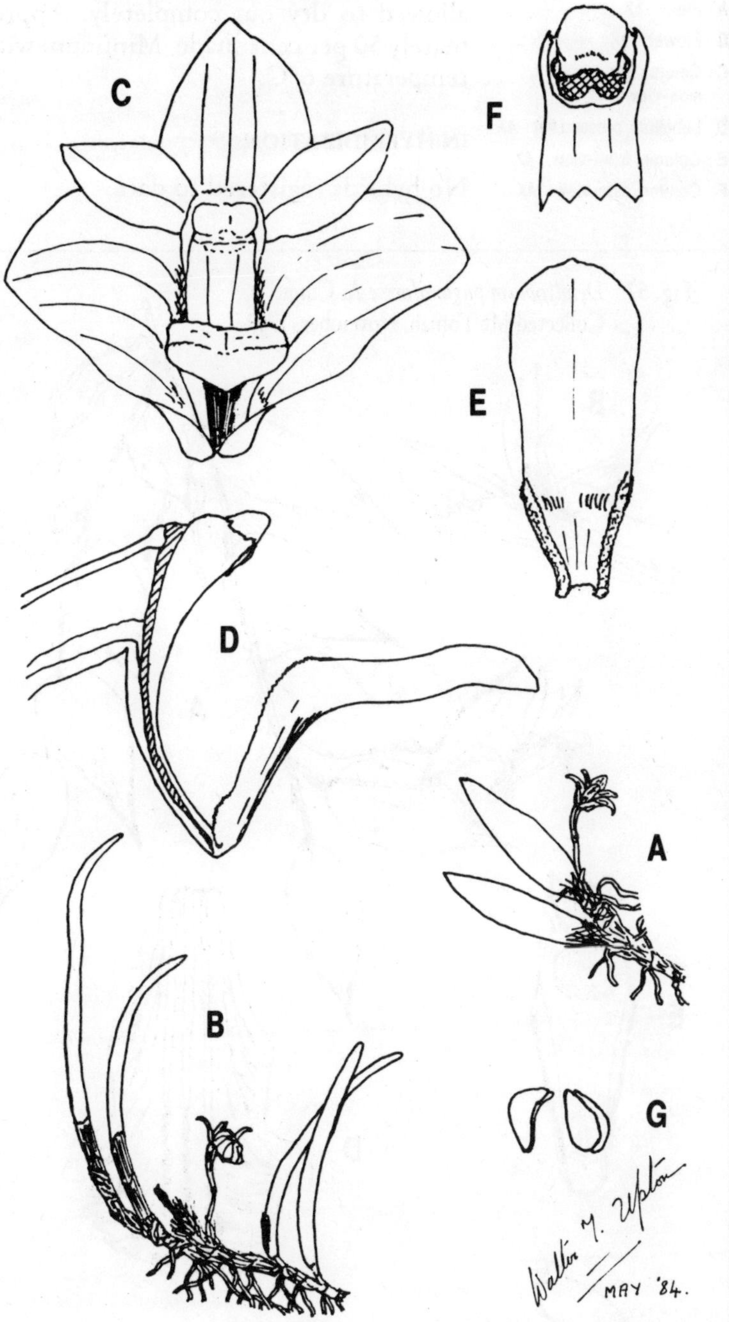

A & B Different plant and leaf forms 1X
C Flower, front-view 8X
D Column and labellum, side-view 10X
E Labellum, pressed flat 8X
F Column, front-view 10X
G Pollinia greatly enlarged

HABITAT

In fairly open forests on trees or rocks. Frequently growing with other orchid species, and on moss-covered branches overhanging creeks. From the ranges almost down to sea level.

A Plant 1X
B Flower 2X
C Column and labellum, side-view 4X
D Labellum, pressed flat 4X
E Column, front-view 4X
F Column, side-view 4X

CULTURE

Slab culture suits best, but must not be allowed to dry out completely. Approximately 50 per cent shade. Minimum winter temperature 6°C.

IN HYBRIDIZATION

No hybrids registered to date.

Fig. 57 *Dendrobium pugioniforme* A. Cunn.
Collected Mt Tomah, November, 1959

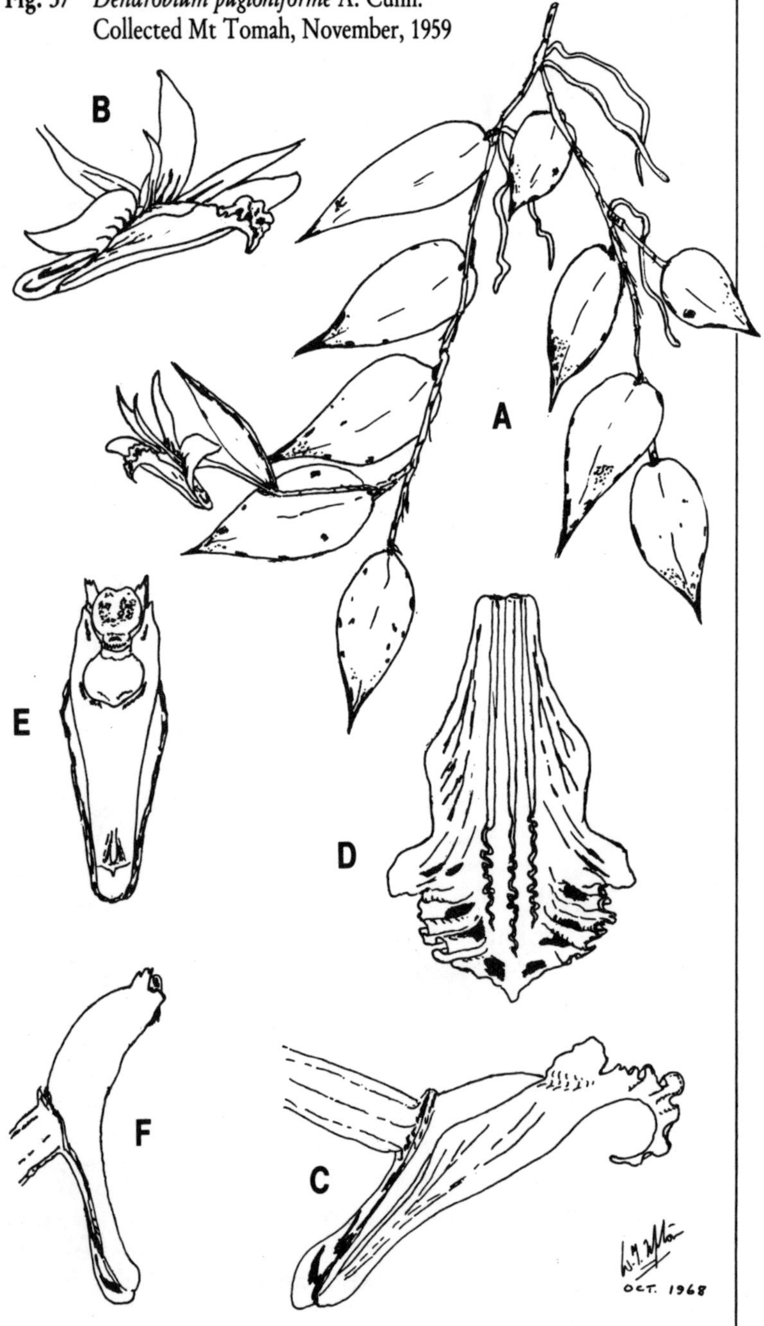

Dendrobium pugioniforme A. Cunn.

Edwards's Bot. Reg. 25:misc. 33 (1839)
Fig. 57, Plates 52 and 54

SYNONYMS

Dendrobium pungentifolium F. Muell., *Fragm. Phyt. Aust.* 1:189 (1859)
Callista pugioniformis (A. Cunn.) Kuntze, *Rev. Gen. Pl.* 2:655 (1891)
Dockrillia pugioniformis (A. Cunn.) Rauschert, *Feddes Repert.* 94:446 (1983)

DESCRIPTION

Originally found in the Illawarra district of NSW by Allan Cunningham. Name comes from *pugionis*, meaning dagger, and *formae*, meaning shape—referring to the shape of the leaves. Common name is dagger orchid.

Pendulous habit. Not uncommon to see large plants up to 4 m in length hanging precariously from the outer branches of trees in highland rainforests. Roots mostly thin and wiry, frequently developing at the nodes of the branches. Makes aerial roots very freely. Stems slender but woody. Leaves numerous, leathery, thick, flattened or occasionally three-sided from round to dagger shaped (up to 8 cm long), often with a needle-sharp point.

Flowers fragrant, 2 to 2·5 cm, generally borne singly on the inflorescence but occasionally in twos and threes. Light green to yellowish with purple to red markings on the whitish labellum. A form without any purple markings on the labellum is occasionally encountered. Flowers last about 14 days.

FLOWERING PERIOD

September to November (spring).

DISTRIBUTION

From the vicinity of the Bega district of south-eastern NSW (36° 40′S) to south-eastern Queensland (26° 20′S).

HABITAT

On trees and rocks from the coastal lowlands (in southern NSW) to the ranges up to about 1300 m. It prefers the rainforests of the mountainous ranges, where damp mossy and heavily-shaded conditions exist. I have seen it growing near Mt Tomah, NSW (150° 25′E and 33° 30′S) at approximately 680 m in a rainforest, many large plants hanging from the trunks and branches of many different trees, large masses up to 4 m long and 1 m wide, with moss hanging from the trees all around it. It was a beautiful sight. In the Kangaroo Valley District (34° 45′S) of NSW, I have seen mats of this species covering large rocks and hanging over the edge on most sides. Large rainforest trees were all around so the light was not very good and it appeared to be continuously moist if not wet, but air movement was good.

CULTURE

Grows well on tree limbs in a shady, sheltered position but with continuous gentle air movement and high humidity. Will also grow on cork or similar slabs but must be kept moist at all times. Minimum winter temperature of 2°C.

IN HYBRIDIZATION

Fairly dominant in its leaf form, although as only four hybrids have been registered it is too early to be definite. The flowers, so far, have been midway between the two parents.

HYBRIDS

D. pugioniforme	X *D. mortii* (as *D. tenuissimum*)	= *D.* Dainty Cascades
D. pugioniforme	X *D. schoeninum* (as *D. beckleri*)	= *D.* Aussie Cascade
D. pugioniforme	X *D. striolatum*	= *D.* Duffy
D. Dainty Cascades	X *D. kingianum*	= *D.* Elfin

Dendrobium racemosum (Nicholls) Clemesha et Dockr.

Orchadian 1:52 (1964)

Fig. 58, Plate 56

BASIONYMS

Dendrobium beckleri F. Muell. var. *racemosum* Nicholls, *Nth. Q/ld. Nat.* 4(45): 35 (1936)

Dockrillia racemosa (Nicholls) Rauschert, *Feddes Repert.* 94:446 (1983)

DESCRIPTION

Sent to W. H. Nicholls by Mr H. G. Henson of Cairns. Name comes from *racemosum*, meaning in the form of a raceme—referring to the inflorescence.

Plant form similar to *D. schoeninum*, but has more robust leaves, which do not get smaller as the stem grows longer to anything like the degree in *D. schoeninum*. Stems also more robust. Plant often grows to 90 cm, although it seems to have a remarkably small root system. Leaves are up to 20 cm long, but generally about 10 cm.

Raceme is 8 cm long and carries eight to 15 fragrant flowers approximately 2 to 3 cm in diameter. Flowers cream to yellowish to almost brownish in colour, with the column and anther purplish. Flowers last about 10 days.

FLOWERING PERIOD

September and October (spring), although may occur at any time of the year.

DISTRIBUTION

North Queensland. It has only been found in a few areas—the Atherton and Evelyn Tablelands, and the lower Russell and Johnstone Rivers (all roughly around 17° 30′S).

HABITAT

Main habitat seems to be above 760 m, where it receives a high rainfall and is subject to frequent mists. Also found in one area on the coastal lowlands. Grows on rainforest trees in very humid and sunny places.

CULTURE

Slab culture suits best, cork or hardwood blocks. About 50 per cent shade. Needs plenty of moisture at all times. In the very hot months frequent mistings are beneficial. High humidity and good air circulation necessary. Minimum winter temperature of 5°C.

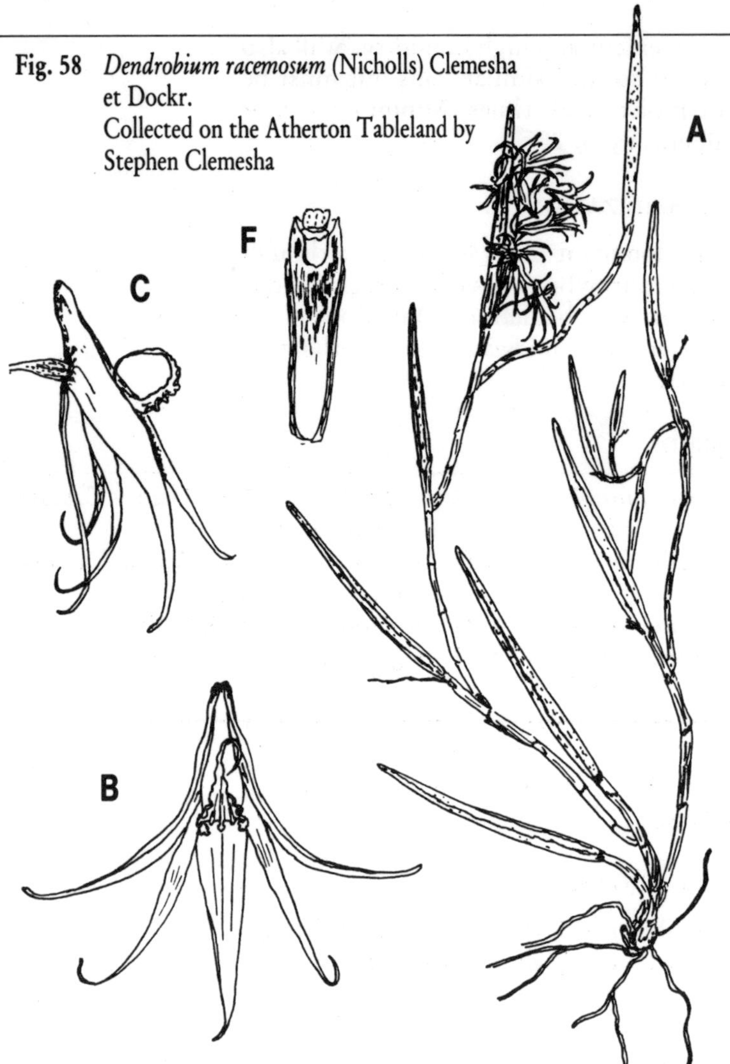

Fig. 58 *Dendrobium racemosum* (Nicholls) Clemesha et Dockr.
Collected on the Atherton Tableland by Stephen Clemesha

A Plant 1/3X
B Flower, front-view 2X
C Flower, side-view 2X
D Column and labellum, side-view 4X
E Labellum, pressed flat 4X
F Column, front-view 4X
G Column, side-view 4X

IN HYBRIDIZATION

No hybrids registered to date.

Dendrobium rigidum R.Br.

Prodr. Fl. Novae Holl. 333 (1810)

Fig. 59, Plate 59

SYNONYMS

Callista rigida (R.Br.) Kuntze *Rev. Gen. Pl.* 2:655 (1891)
Dockrillia rigida (R.Br.) Rauschert, *Feddes Repert.* 94:447 (1983)

DESCRIPTION

This species was among the earliest Australian members of the genus *Dendrobium* to be described. It was discovered by Solander near the Endeavour River, North Queensland, in 1770. Solander's original sketch is in the British Museum. Name comes from *rigidus*, meaning stiff or hard—referring to the habit of the plant.

Branching, pendulous stems are wiry and numerous, occasionally growing into large plants up to 40 cm long. Leaves thick, almost cylindrical. Frequently curved and pointed and mostly of a grey-green colour, frequently spotted with purple. I have seen some plants with the leaves very heavily spotted all over with a dark purple.

Racemes 2 to 5 cm long with up to seven flowers, sometimes quite crowded on the raceme. Flowers from 10 to 20 mm across the lateral sepals, quite widely opening in some instances but not generally. Fragrant flowers very variable in size and markings, particularly in colour and markings on the labellum. Colour ranges from a cream to almost red, some well-spotted with a red-purple, others with only a little red spotting at the base of the segments. Petals usually less marked than sepals. Labellum colour ranges from all cream to nearly all red, but generally it is yellowish outlined in red to red-purple, the side-lobes all red to purple. Flowers last from 10 to 14 days.

CHROMOSOME NUMBER

2n = 38 (K. Y. Lim, 1985).

72 *D. speciosum* var. *speciosum* 'Jill'

73 *D. teretifolium*

74 *D. smillieae*

75 *D. teretifolium*

76 *D. toressae*

77 *D. tetragonum* var. *giganteum*

78 *D. tetragonum* var. *tetragonum*, Wyong, NSW

79 *D. tetragonum* var. *tetragonum* 'No. One'

80 *D. tetragonum* var. *giganteum*

81 *D. tetragonum* var. *giganteum*, pale green form

FLOWERING PERIOD

Anytime of the year and frequently several times a year. Main crop is in the spring.

DISTRIBUTION

Eastern Cape York Peninsula from about the Russell River (17° 25′S) northwards to New Guinea.

HABITAT

Very varied; from mangrove swamps (where it is frequently most common) to rather dry areas and on the low ranges sometimes a long way from the coast. It seems to require fairly high light intensity and is quite tolerant of very dry conditions. I found it to be very abundant on trees along the many creeks between the Wenlock and the Pascoe Rivers, Cape York Peninsula. Numerous plants (mostly smallish) on the trunks and branches of most trees lining the creeks or in some instances in the creeks and quite low down on trees that were almost horizontal (no doubt pushed over with the weight of the water in the rainy season). Some plants it would appear could even be in the water when the creeks were in flood.

CULTURE

This species will grow in the bush-house in Sydney, NSW, but is happier in a semi-shaded position in a warm greenhouse in the winter. Grows well on slabs of cork or hardwood if frequently misted in the hot summer months and kept just moist the rest of the year. If grown in pots, beware as it does not like 'wet feet' and should therefore dry out between waterings. Minimum winter temperature of 8°C.

IN HYBRIDIZATION

No hybrids registered to date, I have used this species but the progeny have not flowered yet. The plant form appears to favour *D. rigidum*. The natural hybrid *D. rigidum* X *D. teretifolium* = *D.* x *foederatum* is approximately midway between the parents.

Fig. 59 *Dendrobium rigidum* R.Br.
Purchased from Limberlost Nursery, Cairns

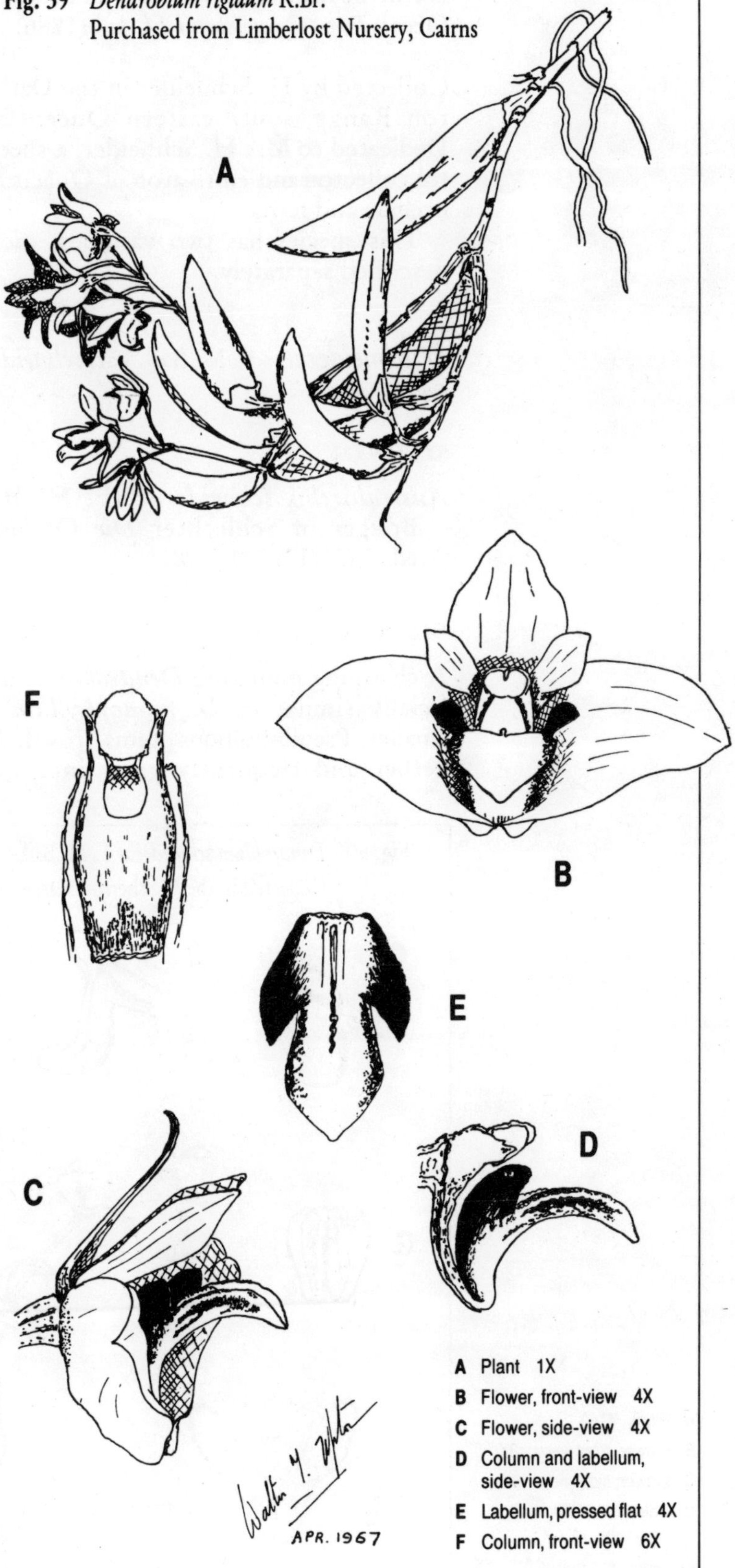

A Plant 1X
B Flower, front-view 4X
C Flower, side-view 4X
D Column and labellum, side-view 4X
E Labellum, pressed flat 4X
F Column, front-view 6X

Dendrobium schneiderae F. M. Bail.

Occas. Pap. *Queensland Fl.* 1:7 (1886)

Collected by H. Schneider in the Darlington Range, south-eastern Queensland. Dedicated to Mrs H. Schneider, a successful collector and cultivator of Queensland orchids and ferns.

This species has two varieties, each is described separately.

D. schneiderae F. M. Bail. var. *schneiderae*

Fig. 60, Plate 61

SYNONYM

Australorchis schneiderae (F. M. Bail.) Brieger in Schlechter *Die Orchideen* ed. 3 (11/12):741 (1981)

DESCRIPTION

A charming miniature *Dendrobium* superficially similar to *D. monophyllum* but smaller. Pseudobulbous stems crowded together and frequently grow into many pseudobulbous plants. One to three leaves, two being most common.

Raceme arched to drooping and approximately 8 cm long, bearing up to 20 flowers about 8 mm in diameter. Fragrant flowers, quite waxy in appearance and from cream to yellow-green, neatly outlined to a varying degree with dark maroon (with the exception of the lower half of the lateral sepals). Labellum very broad at the apex and bears a prominent callus plate. Stigmatic cavity also outlined (or edged) in maroon.

CHROMOSOME NUMBER

2n = 38 (K. Y. Lim, 1985).

FLOWERING PERIOD

February to April (summer to autumn).

DISTRIBUTION

North-eastern NSW and south-eastern Queensland, mostly restricted to the MacPherson Range on the border of NSW and Queensland.

Fig. 60 *Dendrobium schneiderae* F. M. Bail. var. *schneiderae*
Collected in the McPherson Range, south-eastern Queensland, 1960

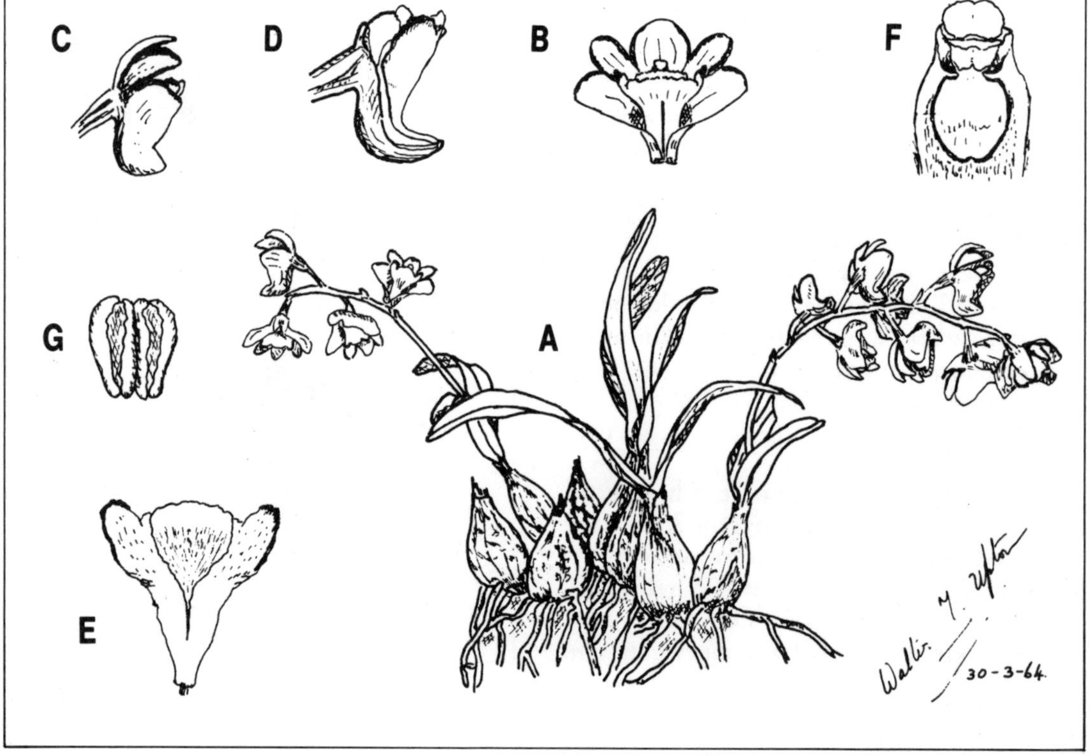

A Plant $1^1/_2$X
B Flower, front-view 3X
C Flower, side-view 3X
D Column and labellum, side-view $4^1/_2$X
E Labellum, pressed flat 6X
F Column, front-view 9X
G Pollinia greatly enlarged

HABITAT

Mostly growing high up on the branches of hoop pines (*Araucaria cunninghamii*), red bean trees (*Dysoxylum muelleri*) and other large rainforest trees, where it gets plenty of light. Has also been found growing on rocks. My wife and I found it in the MacPherson Range on 29 August 1960 at approximately 450 m on a very large red bean tree that we watched being felled. It was very steep country and a very exhausting climb to get to where the trees were being felled, but so rewarding to see so many different species of orchid high up in the branches of this one tree. There were, in addition to *D. schneiderae*, *D. monophyllum*, *D. schoeninum*, *D. macropus* subsp. *gracilicaule*, *D. teretifolium*, *D. tetragonum*, *D. speciosum* var. *hilli*, *Bulbophyllum elisae*, *B. bracteaum*, *B. aurantiacum*, *Cymbidium sauve* and *Sarcochilus falcatus*. They were all growing together, hundreds of plants on the one tree. The *D. schneiderae* was often on the underneath of the branches growing in the moss.

D. schneiderae F. M. Bail. var. *major* Rupp *Qld. Nat.* 11:3 (1939)

Fig. 61, Plate 62

This variety was discovered in 1937 by Dr C. P. Ledward in the Eungella Range near Mackay, North Queensland, some 1000 km north of the type variety.

Pseudobulbous stems generally more robust and raceme much longer and bears more flowers than var. *schneiderae*. The differences between the flowers of the two varieties are shown in the table (taken from S. Clemesha, 1969). Fig. 61 is from a raceme sent to me by Clemesha.

var. *schneiderae*	var. *major*
Flowers greenish-yellow; dorsal sepal, petals, basal part of upper margin of lateral sepals and stigma outlined in red	Flowers more yellow and all red markings on the petals and sepals absent, though stigma markings and that on upper margin of the lateral labellum lobes present as in var. *schneiderae*
Flowers with shorter segments than var. *major* and more widely expanding	Dorsal sepal more pointed
Labellum lateral lobes much better developed than those of var. *major* and about three times as large	
Labellum disc with faint red markings and clearly separated from the lamina all around and contracting into a keel extending towards the labellum base, to within a quarter of the labellum length	Labellum disc more yellow, devoid of the red markings present in var. *schneiderae* and separated from the lamina, except at the base, where it merges with it, keel absent
	Column longer and more slender

FLOWERING PERIOD

March to May (autumn).

DISTRIBUTION

Only known from the Eungella Range (21° 10′S) near Mackay, Queensland.

CULTURE

Slab culture suits *D. schneiderae* best, although it is slow to establish, so must be tied tightly on to its host. Requires excellent ventilation, plenty of light and frequent watering, though it is essential that plants dry out between waterings. Air around it should always be saturated with moisture. Minimum winter temperature of 6°C. Should the summer temperature reach over 30°C, continuous misting is recommended until the temperature goes down again.

A Flower, front-view 4½X
B Labellum, pressed flat 9X

Fig. 61 *Dendrobium schneiderae* F. M. Bail. var. *major* Rupp
Collected in the Eungella Range near Mackay

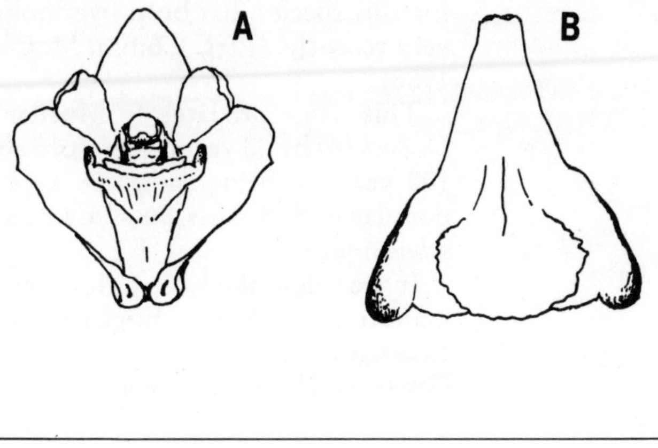

IN HYBRIDIZATION

Only one hybrid has been registered and it is approximately midway between its parents, although the flowers are possibly nearer to *D. schneiderae*. It is a real gem and the flowers are long lasting. Its name is *D.* Golligosh; its parents *D. schneiderae* X *D. monophyllum*. (How did it get its name? I showed it to a friend of ours, Mrs Joan Cook and said 'What name would you suggest for this delightful plant?' She exclaimed 'Oh, Golligosh', so I said 'That will do' and proceeded to fill in the registration form.)

HYBRIDS

D. schneiderae	X *D. monophyllum*	= *D.* Golligosh

Dendrobium schoeninum Lindley

Gard. Chron. 1:7 (1846)

Fig. 62, Plate 60

SYNONYMS

Dendrobium mortii Benth., *Fl. Aust.* 6:286 (1873) non F. Muell. (1859)
Dendrobium beckleri F. Muell., *Fragm. Phyt. Aust.* 7:59 (1869)
Dendrobium striolatum F. M. Bail., *Syn. Q/ld Fl.* 512 (1883) non H. G. Reichb. (1857)
Dendrobium striolatum H. G. Reichb. var. *beckleri* (F. Muell.) F. M. Bail. *Syn Q/ld. Fl. Suppl.* 1:55 (1886)
Callista beckleri (F. Muell.) Kuntze, *Rev. Gen. Pl.* 2:654 (1891)

The name *D. schoeninum*, the original name for this species, has been overlooked until very recently. (Pers. Comm. M. Clements, 1988).

This name predates F. Mueller's name *D. beckleri* by 23 years, therefore after over 100 years of being called *D. beckleri* this popular orchid must now be known as *D. schoeninum*.

It was described by J. Lindley from a plant imported into England by Messrs Loddiges. The name is possibly from the Greek word *schoenos* meaning reed or rush, no doubt referring to the shape of the leaves.

D. beckleri was described from a plant collected by R. D. Fitzgerald on the Macleay River, NSW and named after Dr Beckler a naturalist and collector in Australia.

Common name is pencil orchid.

DESCRIPTION

Stems slender, up to 90 cm long, with short, thick, conspicuously ribbed, terete leaves. Plant erect when short but as the stem elongates becomes more and more pendulous, although the terete leaves still remain more or less erect. Becomes an untidy, straggling plant due to its leaves and the end of each stem becoming progressively smaller.

Flowers generally produced singly or in pairs, but very occasionally as many as four seen with the pedicels very close together. Flowers open widely, sepals and petals greenish to cream and occasionally white or pale mauve. Outside of the perianth marked with a few purple striations. Labellum white, prominently frilled and bearing irregular, bright-purple markings. Very occasionally a form with a pure-white labellum, devoid of any markings, is found. Flowers quite fragrant and last about 2 weeks.

FLOWERING PERIOD

August to October (late winter to spring).

DISTRIBUTION

From the Hunter River, near Newcastle, NSW (33° 00'S) to the Burdekin River in north-eastern Queensland (20° 00'S). Exceedingly common in numerous areas.

HABITAT

On trees and rocks from the coastal lowlands to the foothills of the ranges, up to about 600 m. In rainforests and fairly open forests, areas of high humidity. Among its hosts are *Casuarina glauca*, *Callicoma serratifolia* and *Ehretia acuminata*.

CULTURE

Slab culture suits best, cork or hardwood blocks. Prefers to grow under 50 per cent shade cloth. Needs plenty of moisture, particularly in hot summer months. High humidity and good air circulation. Minimum winter temperature about 2°C.

IN HYBRIDIZATION

Generally the resultant hybrids appear to be midway between the parents. No obvious dominant characteristics coming through at this stage.

HYBRIDS

D. schoeninum (as *D. beckleri*)	X *D. cucumerinum*	= *D.* Burie
D. schoeninum (as *D. beckleri*)	X *D. linguiforme*	= *D.* Michael Jupp
D. schoeninum (as *D. beckleri*)	X *D. mortii* (as *D. tenuissimum*)	= *D.* Ida Mary
D. schoeninum (as *D. beckleri*)	X *D. pugioniforme*	= *D.* Aussie Cascade
D. schoeninum (as *D. beckleri*)	X *D. striolatum*	= *D.* Stephen
D. schoeninum (as *D. beckleri*)	X *D. teretifolium*	= *D.* Jiggi

Fig. 62 *Dendrobium schoeninum* Lindley Collected September, 1963, west of Gloucester, NSW

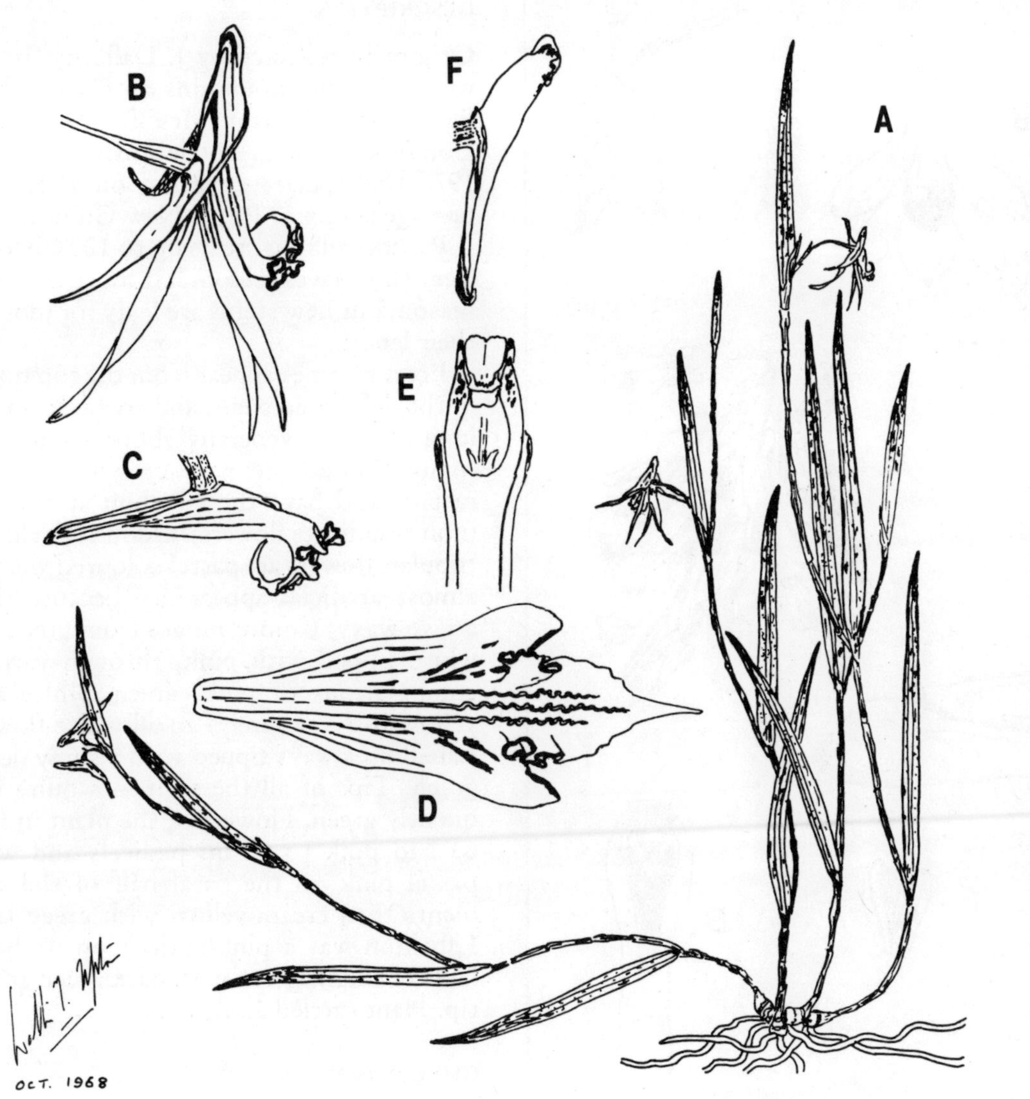

A Plant 1/3X
B Flower, side-view 2X
C Column and labellum, side-view 2X
D Labellum, pressed flat 3X
E Column, front-view 4X
F Column, side-view 3X

Fig. 63 *Dendrobium smillieae* F. Muell. From Daintree River area

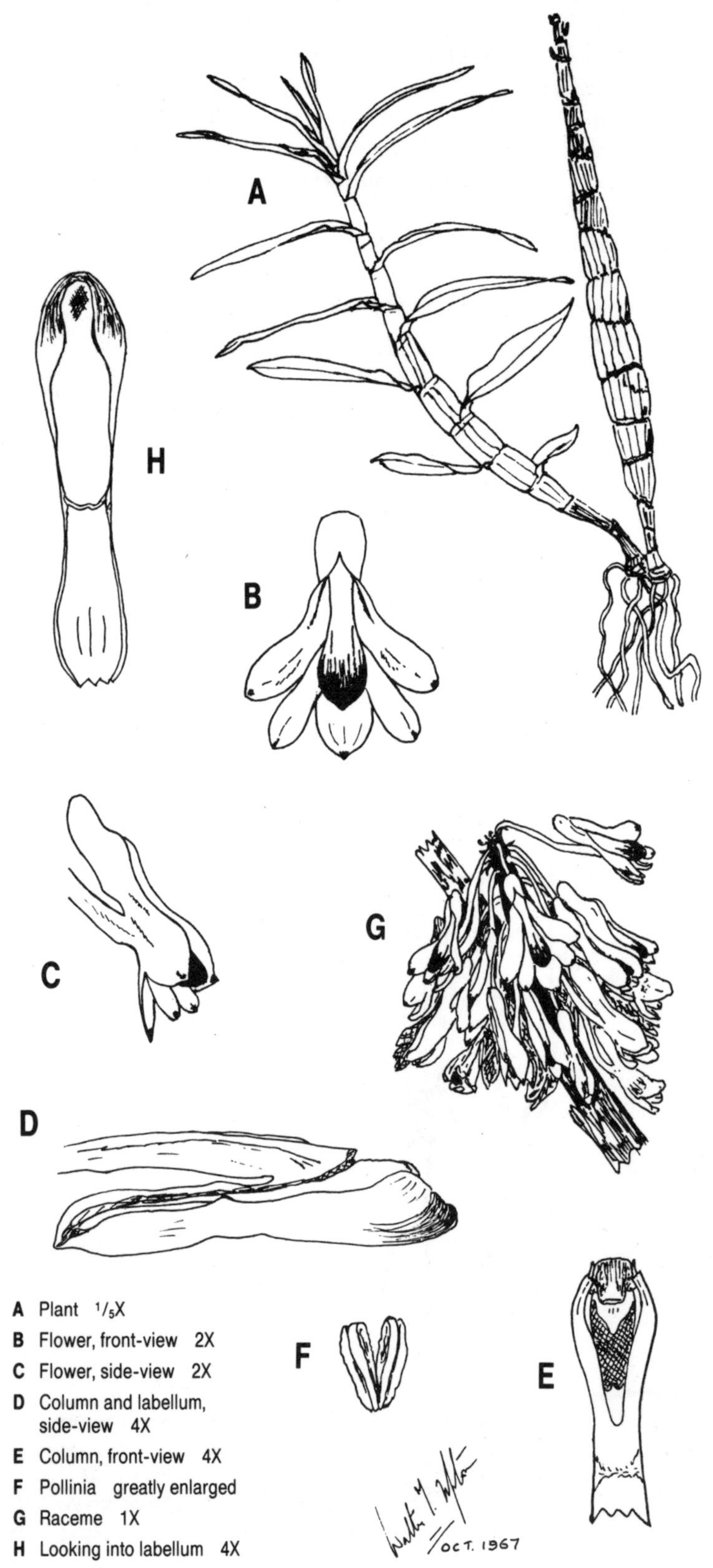

A Plant $^{1}/_{5}$X
B Flower, front-view 2X
C Flower, side-view 2X
D Column and labellum, side-view 4X
E Column, front-view 4X
F Pollinia greatly enlarged
G Raceme 1X
H Looking into labellum 4X

Dendrobium smillieae F. Muell.

Fragm. Phyt. Aust. 6:94 (1867)
Fig. 63, Plate 74

SYNONYMS

Dendrobium ophioglossum H. G. Reichb., *J. Proc. Linn. Soc. Bot.* 15:113 (1876)
Coelandria smillieae (F. Muell.) Fitzg., *Aust. Orch.* 1(7): (t.7) (1882)
Callista smillieae (F. Muell.) Kuntze, *Rev. Gen. Pl.* 2:655 (1891)
Dendrobium hollrungii Krzl. var. *australiense* Rendle, *J. Bot. Lond.* 37:339 (1899)
Pedilonum smillieae (F. Muell.) Rauschert, *Feddes Repert.* 94:464 (1983)

DESCRIPTION

Originally collected by J. Dallachy 'In the woods of the mountains at Rockingham Bay'. Named after Mrs E. J. Smillie. Common name is bottlebrush orchid. In 1975 this species appeared on a 20 cent postage stamp of Papua New Guinea.

Pseudobulbous stems up to 100 cm long. Very thin leaves seldom last more than one season, but new stems are leafy for most of their length.

Erect racemes appear from the top nodes of the defoliated stems and are up to 13 cm long although generally about 7 cm long. Many flowers are very crowded on the raceme and have the labellum at the top (non-resupinate flowers). Odourless, almost tubular, flowers are pastel coloured with an almost artificial appearance because they are so waxy. Colour ranges from greenish-white tipped with pink, through various combinations of pink, green, white and yellow to (very rarely) an all-white flower. Labellum always tipped with a shiny deep-green. Tips of all the segments quite frequently green. Flowers of the plant in Fig. 63 had pink to cream pedicels and were pastel pink on the basal half of the segments then cream-yellow with green tips. Labellum was a pink-apricot on its basal half then light green with dark green at the tip. Plant carried 22 flowers.

CHROMOSOME NUMBER

2n = 38 (K. Y. Lim, 1985).

FLOWERING PERIOD

August to November (winter to spring).

DISTRIBUTION

Quite common from the Burdekin River (20° 00′S) to the tip of Cape York Peninsula and the islands adjacent to the coast, and into Papua New Guinea.

HABITAT

From the coast up to about 600 m in the ranges. Seems to prefer the lowland areas although still quite plentiful in the lower ranges. Open forests bordering rainforest is its most common habitat. Some of its hosts are bloodwood (*Eucalyptus corymbosa*), swamp mahogany (*Tristania suaveolens*) and Leichhardt trees (*Sarcocephalus coadunatus*). In the rainforest, it will grow on tree ferns as well as trees. In some areas large plants will be seen growing happily on rocks in a fair amount of sun.

In the Daintree River area I have seen it growing on a large tree in the middle of a paddock, a very large plant in a fork of the tree at about 5 m above the ground. It received very little shade but no doubt the air was quite humid from the wide Daintree River about 200 m away. I have also seen it growing in the hollow trunks of trees with the roots well down into the tree. Those found growing on the low open country, close to the beach, are usually stunted. It seems to prefer hollow limbs or forks where the bark and leaves have collected and rotted to form a coarse soil.

POLLINATION

This species is adapted for pollination by birds. The pollinia are a slate grey to browny colour, no doubt so that they won't be noticed by the bird. Yellow pollinia would be noticed, and therefore less likely to be wiped off. Smythe (1970a) reports on the pollination of *D. smillieae* by the yellow honey-eater (*Meliphaga flava*).

CULTURE

Slab culture suits well. Requirements are a warm, damp, humid position that is fairly well shaded. The thin leaves burn easily in too much direct sunlight.

Plenty of water and regular feeding while in full growth, to bring up the new growth quickly. Needs a winter rest. Prefers to be hung up high in the greenhouse during this period with just enough water to ensure it doesn't dry out completely. Minimum winter temperature of 15°C.

IN HYBRIDIZATION

No hybrids registered to date.

Dendrobium speciosum Smith

Exotic Bot. 1:17, t.10 (1804)

DESCRIPTION

Name comes from *speciosus*, meaning showy, beautiful. Common names are NSW rock lily and Queensland king orchid.

Of all the species of *Dendrobium* in Australia, this is possibly the best known, and certainly the one species most commonly grown. A very large percentage of home gardens in Sydney have one in a pot, in a rockery or even planted in the garden soil. It is unrivalled for its ability to survive under most adverse conditions, for its sheer resistance to dryness, and for its ability to grow in full sunlight and (in some gardens) in very heavy shade. Many articles have been written about this species. The most comprehensive are by S. Clemesha in *Orchadian* Vol. 6 No. 11 and No. 12 and Vol. 7 No. 5, 1981–1982.

I quote from some interesting paragraphs from R. D. Fitzgerald's *Australian Orchids* Vol. 2, Part 4.

Dendrobium speciosum, about thirty years ago, was common on all the sandstone cliffs about Sydney, and on such outlying rocks as were large enough to afford safety from bush fires, none are now to be found except in gardens and every year the circle is enlarging within which it is disappearing never to return. There are exceptional years in which there is a general flowering of all the plants in a district, but *Dendrobium speciosum* never flowers freely in succeeding years.

The year 1889 was one in which all the Dendrobiums flowered well, and notes taken in that year will give an idea of the flowering and fertility of the genus in the open: *Dendrobium speciosum*, number of spikes on best plant, 35; number of flowers on best spike, 158; total of flowers on plant, on average of 120 flowers to each spike, 4200. Number of capsules set on plant, 7; number of spikes producing capsules, 5; largest number of capsule on spike, 3. Seven capsules to four thousand two hundred flowers.

Dendrobium hillii (form of *D. speciosum*), number of spikes on best plant, 95; number of flowers on best spike, 291; total of flowers on plant at average of 180 flowers to each spike, 9000. Number of capsules on plants, 9; number of spikes producing capsules, 6; largest number of capsules on a spike, 3. Nine capsules to nine thousand flowers.

North Queensland varieties grown in the cooler states of Australia, are much more difficult to flower than the southern varieties. Some north Queensland forms produce aerial growths quite regularly, whereas the southern forms seldom, if ever, do so. Pollination of the north Queensland form is by the native bee *Trigona caponaria* (R. Smythe, 1970b).

D. speciosum is an extremely variable species of six varieties, extending from north-eastern Victoria to the tip of Cape York Peninsula and possibly in Papua New Guinea, with a doubtful recording in New Caledonia. Each variety is described separately. (In geographical order, the most southern variety first).

D. speciosum Smith var. *speciosum*

Fig. 64, Plates 65 and 72

SYNONYMS

Callista speciosa (Smith) Kuntze, *Rev. Gen. Pl.* 2:655 (1891)
Dendrocoryne speciosum (Smith) Brieger, *Die Orchideen* ed. 3 (11/12):725 (1981)
Tropilis speciosa (Smith) Rauschert, *Feddes Repert.* 94:471 (1983)

DESCRIPTION

The type specimen is in Sir J. E. Smith's herbarium, held in the herbarium of the Linnean Society in London. On the type sheet is written 'Port Jackson, New South Wales'.

Pseudobulbous stems up to 60 cm tall, sometimes longer, thick at the base, up to 25 cm in circumference, and then tapering upwards. Average about 25 to 30 cm tall. Leaves flat or concave on the upper surface, rather thick, two to five at the top of the stem.

About 30 to 80 very fragrant flowers produced on a raceme about 30 to 45 cm long. Raceme erect to arching. (I have counted as many as 103 flowers on a 70 cm raceme.) Well-spaced flowers, from 2·5 to 5 cm in diameter on average, but flowers to 8 cm are not too uncommon. Colour varies from white through cream to yellow with some purple to red markings on the labellum. In dull weather and at night there is no perfume, but on a hot day they give off a strong sweet odour. Individual flowers last from 15 to 20 days.

FLOWERING PERIOD

August to October (late winter to spring). Flower buds start to swell noticeably about May.

DISTRIBUTION

From just south of Genoa (37° 40′S) in Victoria to Bulahdelah (32° 25′S) in central NSW and about 240 km inland to the Mudgee area (149° 40′E).

HABITAT

Mostly on rocks in open forest but in some rainforest areas on trees. I have seen it on large sandstone rocks in Victoria, a very short-stemmed form sitting up on top of the rocks like little sentinels; the flowers were a cream-yellow. The area was lightly timbered. On the Cambewarra Range just north of Nowra, NSW, it grows high up on rainforest trees; a very moist area but with plenty of air movement. This is a very long-stemmed form with smallish flowers.

In the ranges near Rylstone, NSW (Mudgee area), it is exposed to frost, snow and dry westerly winds in winter and fierce heat in summer.

It grows into large clumps. Some of the largest flower forms are found on rocks in the foothills of the Blue Mountains, NSW; on the eastern side in open forest. They have medium-sized pseudobulbous stems.

Fig. 64 *Dendrobium speciosum* Smith var. *speciosum* From near Bucketty, north of Sydney, NSW

To see a large rock face absolutely covered with *D. speciosum* is a sight not quickly forgotten, particularly when some of the individual clones consist of many hundreds of pseudobulbous stems.

A Plant $^{1}/_{3}$X
B Flower 1X
C Flower, side-view 1X
D Column and labellum, side-view 2X
E Labellum, pressed flat 3X
F Column, front-view, cut at centre 4X
G Column, side-view 4X

D. speciosum Smith var. *hillii* Masters
Gard. Chron. 7:112 (1877)
Fig. 65, Plate 66

SYNONYM

Dendrobium hillii Hook., *Bot. Mag.* 87 t.5261 (1861) nom. illeg. non F. Muell. (1859)

DESCRIPTION

Named after Walter Hill, Colonial Botanist, Director and Founder of the Brisbane Botanic Gardens. The type specimen, as *D. hillii*, is in the Kew herbarium; it was taken from a cultivated plant sent to Kew by Mr Walter Hill.

Pseudobulbous stems usually much longer than those of var. *speciosum*—up to 1 m tall. Usually straight and taper little throughout. Cultivated plants seldom more than 45 cm long. Flowers are crowded on the raceme and vary from 1·5 to 3 cm in diameter. It is quite common to have 100 or more flowers on the approximately 40 cm raceme. Flowers mostly white or pale cream. Sepals and petals generally narrower than the type variety. Labellum much smaller than either var. *speciosum* or var. *grandiflorum*.

FLOWERING PERIOD

Same as for var. *speciosum*.

Fig. 65 *Dendrobium speciosum* Smith var. *hillii* Masters Collected in the Wauchope area of NSW

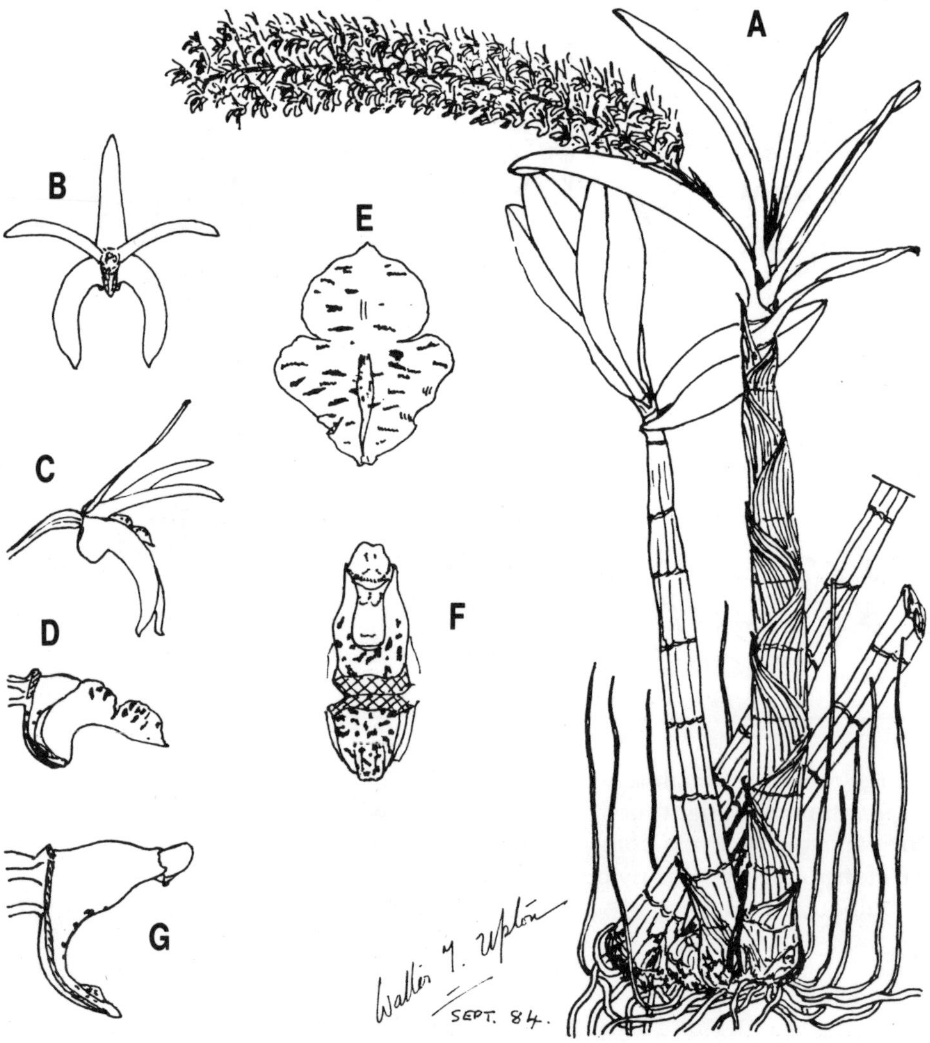

A Plant 1/5X
B Flower, front-view 1X
C Flower, side-view 1X
D Column and labellum, side-view 2X
E Labellum, pressed flat 3X
F Column, front-view, cut at centre 4X
G Column, side-view 4X

DISTRIBUTION

From Bulahdelah (32° 25′S) in central NSW to between Brisbane and Nambour (27° 00′S) in south-eastern Queensland. A few recordings in an area to the west and north of Wyong, NSW (Mangrove Mountain, Peats Ridge and Wattagan Mountains). (33° 00′S to 33° 20′S).

HABITAT

Similar to type variety although mostly on trees.

D. speciosum Smith var. *grandiflorum* F. M. Bail.

Bot. Bull. Dept. Agric. Qld. 14:12 (1896)

Fig. 66, Plate 67

SYNONYMS

Dendrobium speciosum var. *hillii* F. M. Bail., forma *grandiflorum* (F. M. Bail.) F. M. Bail., *Qld. Fl.* 5:1526 (1902)

Dendrobium speciosum forma *grandiflorum* (F. M. Bail.) H. G. Reichb., *Beitr. Syst. Pfl.* 271 (1871)

DESCRIPTION

Name comes from *grandiflorum*, meaning large flowered. Type specimen of this variety is in the Queensland Herbarium and the type sheet states 'Eumundi—September 1894'. Eumundi is approximately 120 km north of Brisbane.

Pseudobulbous stems to about 60 cm, quite slender, similar to var. *hillii.* Leaves generally longer and wider than var. *hillii.*

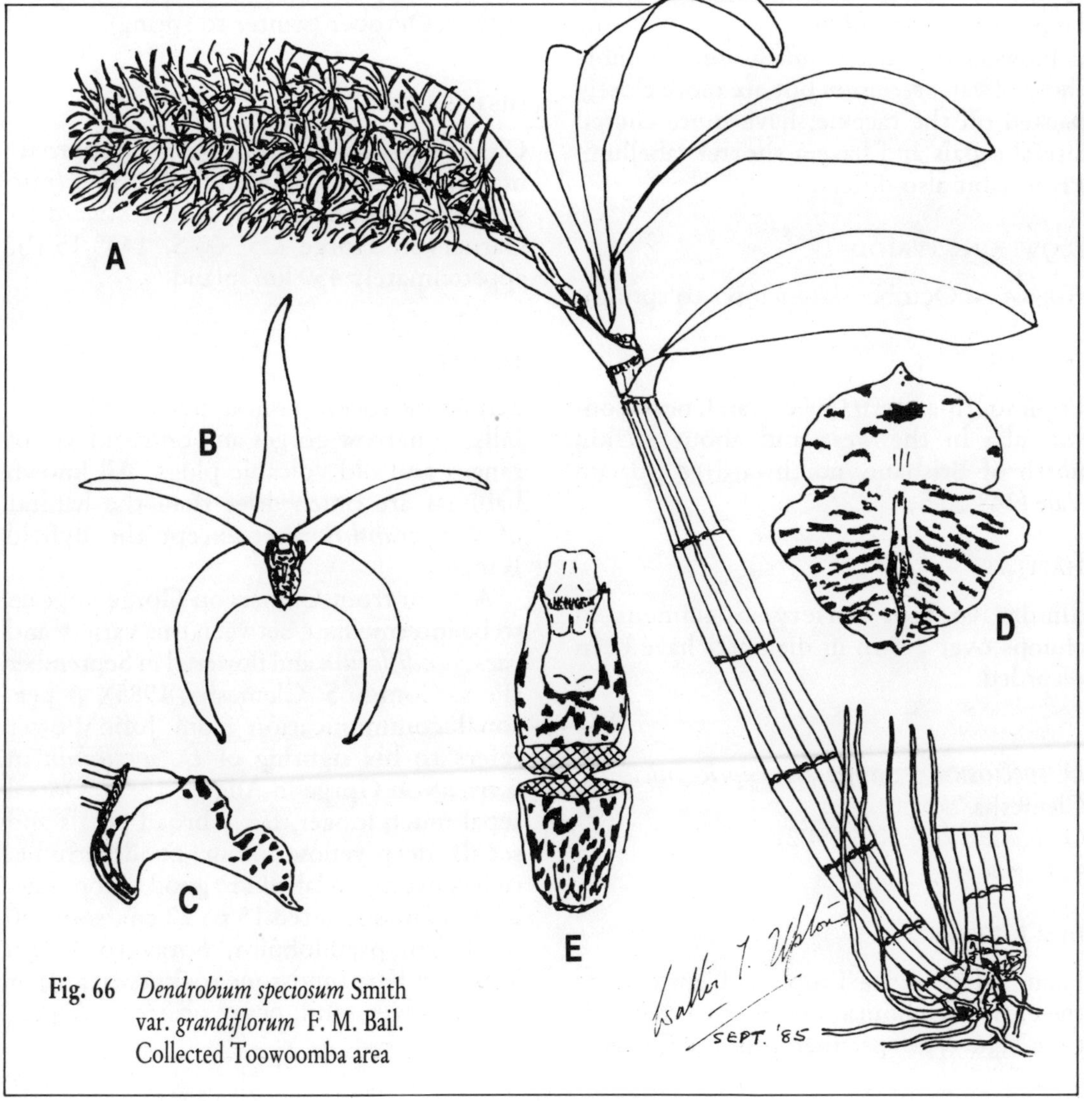

Fig. 66 *Dendrobium speciosum* Smith var. *grandiflorum* F. M. Bail. Collected Toowoomba area

A Plant with only part of raceme 1/5X
B Flower, front-view 1X
C Column and labellum, side-view 2X
D Labellum, pressed flat 3X
E Column, front-view 4X

Racemes, about 40 cm long, not as crowded as in var. *hillii* but more so than in var. *speciosum.* I counted a considerable number of racemes with 60 to 70 flowers, although up to 100 flowers is not uncommon. Flowers from 3 to 7 cm in diameter (occasionally longer), lower sepals generally a little shorter than the dorsal sepal and mostly more in-curved than in var. *hillii* or var. *speciosum.* Colour of flowers from cream-yellow to quite a strong vibrant yellow. A specimen growing in the front garden of a house in the Glasshouse Mountains area of Queensland in 1966, could be seen from a long way off as it was such a deep yellow. (It apparently came from the local bush.) Labellum much shorter than petals and sepals and usually a pale yellow and more heavily marked than in the var. *hillii.* The labellum is much larger than in var. *hillii.*

Flowers of var. *grandiflorum* resemble those of var. *speciosum* but are more closely packed on the raceme; have more curved lateral sepals and have a shorter labellum. Plant habit also differs.

FLOWERING PERIOD

August to October (late winter to spring).

DISTRIBUTION

From south-eastern Queensland, near Toowoomba in the west and about 80 km north of Brisbane, northward to Miriam Vale (24° 20′S).

HABITAT

Similar to type variety. Specimens in clumps over 2·5 m in diameter have been recorded.

D. speciosum Smith var. *capricornicum* Clemesha

Orchadian 7(5):103 (1982)

Fig. 67, Plate 68

DESCRIPTION

Name refers to the Tropic of Capricorn as the entire distribution is close to that latitude. The type specimen is in the Queensland Herbarium and a specimen taken from the same plant is in the National Herbarium of NSW. It was collected by S. Clemesha near a disused quarry on an old volcanic plug about 8 km inland from Yeppoon, Queensland, in May 1978.

This variety is similar to var. *grandiflorum* but is smaller in all its parts. Pseudobulbous stems, 7 to 25 cm tall, slightly tapered to the apex. Racemes about 36 cm long carrying up to 35 or more flowers, white with purple markings on the labellum. Flowers of plants from the inland part of its range (Carnarvon and Blackdown Tableland) have cream to yellow flowers, frequently heavily marked and of heavy substance.

FLOWERING PERIOD

July to October (winter to spring).

DISTRIBUTION

Close to Tropic of Capricorn from approximately 22° 50′S (Byfield) to 23° 30′S (just south of Rockhampton) and westward to Carnarvon Gorge (25° 00′S, 148° 15′E), approximately 450 km inland.

HABITAT

Sandstone rockfaces near creeks and waterfalls, in narrow gorges and on rockfaces of ranges and old volcanic plugs. All known habitats are much drier than the habitat of var. *grandiflorum*, except the Byfield Range.

A form from Carnarvon Gorge appears to be intermediate between this variety and var. *grandiflorum* and flowered in September (Pers. Comm. S. Clemesha, 1985). A personal communication from John Power refers to his sighting of *D. speciosum* in Carnarvon Gorge in August 1973. 'Dorsal sepal much longer, fairly broad petals and sepals, deep yellow colour, good perfume, rich spotting on labellum, good bloom size. Most plants stunted 15 to 22 cm, some 30 to 37 cm pseudobulbs. Spray to 30 cm long. On bare sandstone rocks (vertical) or in fine silver sand, never on trees. Aspect S.E.'.

Fig. 67 *Dendrobium speciosum* Smith var. *capricornicum* Clemesha
From Bowinda Gorge, north Queensland

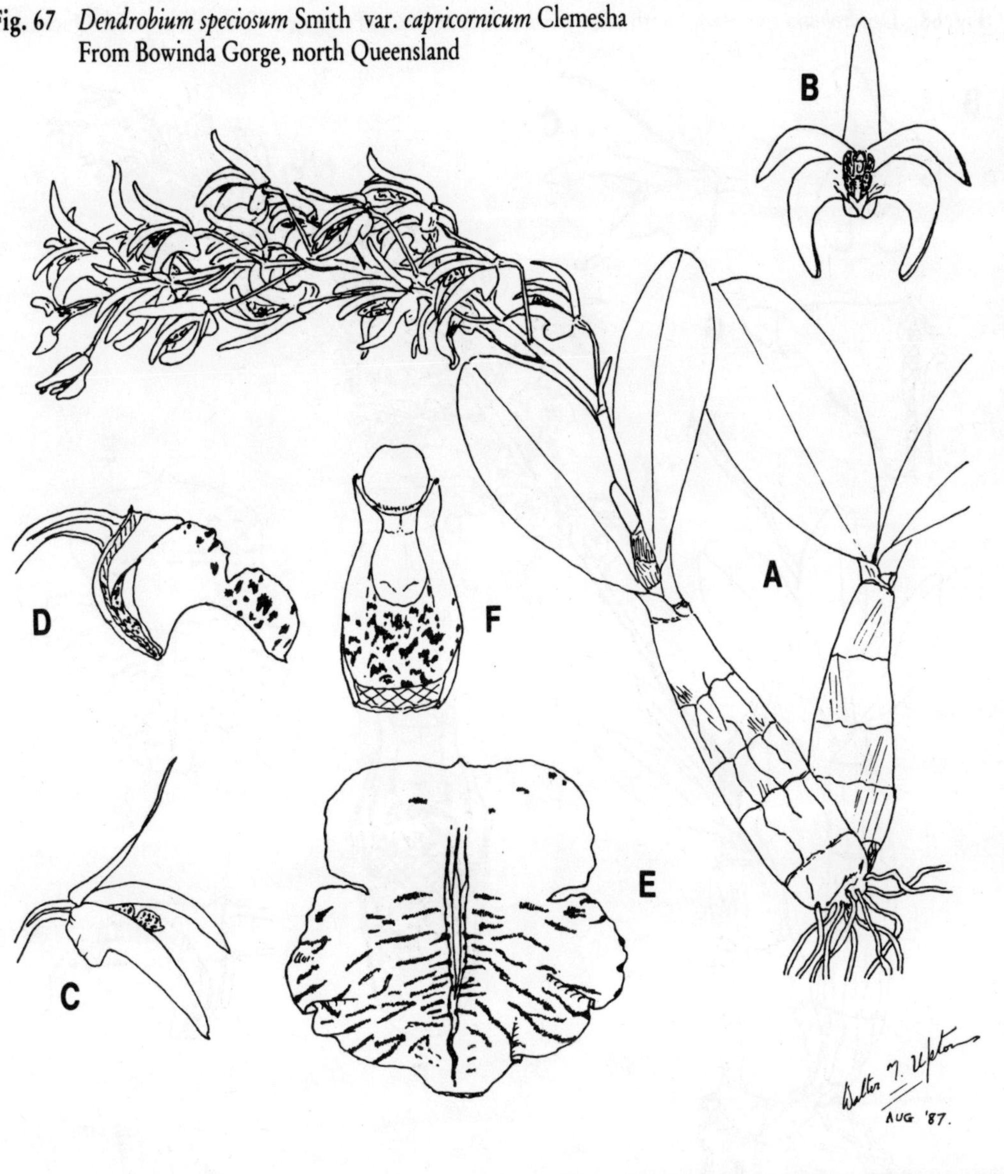

A Plant ½X
B Flower, front-view 1X
C Flower, side-view 1X
D Column and labellum, side-view 2X
E Labellum, pressed flat 4X
F Column, front-view 4X

D. speciosum Smith var. *curvicaule* F. M. Bail.
Bot. Bull. Dept. Agric. Q/ld. 14:12 (1896)
Fig. 68, Plate 70

DESCRIPTION

Name comes from *curvi*, meaning bowed or bent and *caulis*, meaning stem — referring to the curved pseudobulbous stems in some forms. In the Queensland Herbarium there is a specimen sheet of this variety with F. M. Bailey's handwriting on it; this is probably the type. E. W. Bick, Curator of the Brisbane Botanic Gardens, stated '*D. speciosum* var. *curvicaule* was named many years ago from a specimen growing at Bowen Park, which had been collected at Lady Elliott Island'. (*Aust. Orchid Rev.*, December 1938).

Pseudobulbous stems from 7 to about 100 cm long, varying in shape from narrow at base and widest at top, to plants that are broader than thick. Curved pseudobulbous stems to squat straight stems.

Raceme about 20 to 35 cm long. Flowers, lasting about 15 days, are from cream to white, with short, broad sepals. Labellum

A Plant $^1/_3$X
B Flower 1X
C Flower, side-view 1X
D Column and labellum, side-view 2X
E Labellum, pressed flat 3X
F Column, front-view, cut at centre 4X
G Top of column, side-view 4X

Fig. 68 *Dendrobium speciosum* Smith var. *curvicaule* F. M. Bail. From near Tully Falls, north Queensland

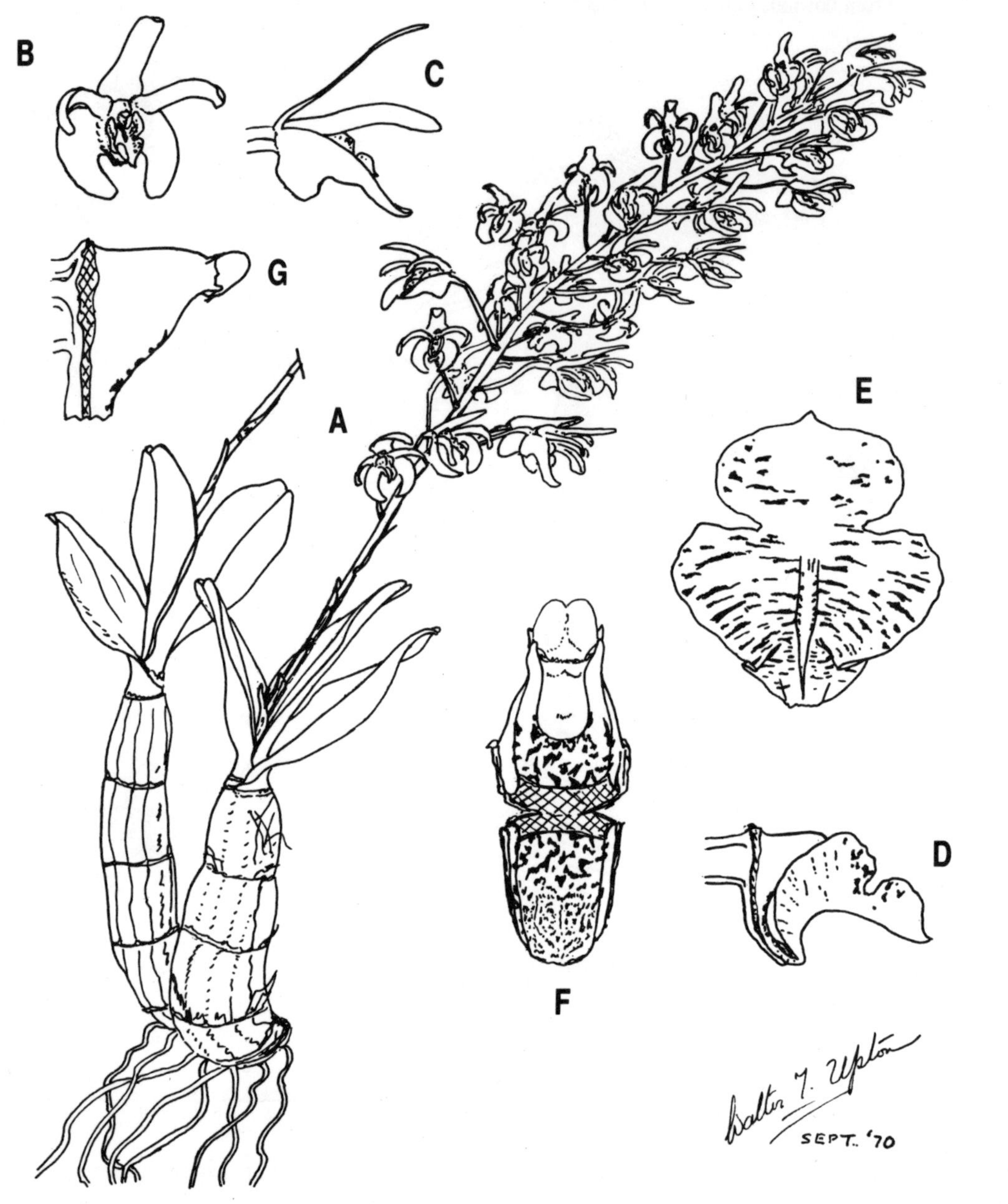

markings purple and vary considerably; mid-lobe separated from lateral lobes by a short neck (in the northern part of its range).

FLOWERING PERIOD

August to October (winter to spring).

DISTRIBUTION

From near Mackay (21° 10′S) to at least as far north as Thornton's Peak (16° 10′S) (south of Cooktown), but possibly much further north.

HABITAT

On trees and rocks from coast to highlands but most common at higher elevations. Plant in Fig. 68 was collected from near Tully Falls (18° 00′S) and was growing on rock in open forest and almost in full sun.

D. speciosum Smith var. *pendunculatum* Clemesha
Orchadian 6(11):261 (1981)
Fig. 69, Plates 69 and 71

DESCRIPTION

Name refers to the long and very rigid flower stalk. Type specimen is in the Queensland Herbarium and was collected by Bruce Gray at Bones Knob near Atherton.

Pseudobulbous stems 5 to 16 cm long but mostly about 7 cm; conical and dwarf. Racemes erect and held up by a stiff and very rigid peduncle which is as long, to twice as long, as the rhachis. Flowers crowded at the top of the raceme. Stems, raceme, buds and occasionally sepals stained with dark purple. Seed capsules dark purple. Flowers vary in shape but generally have shorter and broader segments than var. *curvicaule.* Labellum large in proportion to other segments. Flowers cream to a light dull-yellow and last about 15 days.

FLOWERING PERIOD

July to September (winter to spring).

DISTRIBUTION

From the Palmer River (16° 00′S) in the north to the Herbert River (18° 15′S) at least in the south; and mostly on the western side of the dividing range. Also recorded in Papua New Guinea, from the Owen Stanley Ranges.

HABITAT

Exposed rockfaces in full sun.

CULTURE

For ease of culture few species could surpass *D. speciosum*. It has been in cultivation ever since it was first collected and from the size of many clumps in home gardens, some plants have been grown for a very long time.

Is an ideal subject for well-drained rockeries, preferring the top position. Grows well in the forks of large garden trees or on the top of large tree stumps. In these positions it likes lots of light; full sun does not harm it. Always attach it very firmly to its host to ensure it establishes itself quickly.

Pot culture also suits, but in a very open mix. As growth is extremely robust, it will require frequent repotting. Do not start off with a small piece in a large pot. This does not save repotting as the plant frequently dies. The best time to repot is soon after flowering has finished as this is when the new growth appears. The new growth should develop rapidly, this can be helped by regular watering and feeding at this time. A further growth may develop in autumn. It can tolerate lots of water in the hotter months but prefers drier winter conditions.

Minimum winter temperature of about 2°C for the southern forms, while those from the lower elevations of north Queensland benefit from a minimum temperature of about 10°C. The varieties *grandiflora* and *hillii* prefer more shade than the rest.

IN HYBRIDIZATION

Not obviously dominant in any particular characteristic, but at the same time has played a very important part in both hybrids within the section *Dendrocoryne* and with species outside this section, particularly with the section *Spatulata.* If there are any dominant characteristics, they are the lip shape, number of flowers and the habit of raceme.

Some hybrids with *D. speciosum* in their background seem, in many instances, to take longer to reach maturity.

HYBRIDS

D. speciosum	X *D. adae*	= *D.* Yellow Venture
D. speciosum	X *D. agrostophyllum*	= *D.* Aussie Ochre
D. speciosum	X *D.* Aussie Gypsy	= *D.* Aussie Frontier
D. speciosum	X *D.* Aussie Ira	= *D.* Aussie Starlight
D. speciosum	X *D.* Bardo Rose	= *D.* Dilgry
D. speciosum	X *D. bigibbum*	= *D.* Warringah
D. speciosum	X *D.* Blushing Rose	= *D.* Regal Rose
D. speciosum	X *D. canaliculatum*	= *D.* Memoria Lloyd Bradford
D. speciosum	X *D. x delicatum*	= *D.* Jane Leaney
D. speciosum	X *D.* Double Two	= *D.* Ronald Banks
D. speciosum	X *D.* Ellen	= *D.* Graham Hewitt
D. speciosum	X *D.* Emmy	= *D.* Craig Hewitt
D. speciosum	X *D.* Eureka	= *D.* Lynette Banks

HYBRIDS (con't)

D. speciosum	X *D. falcorostrum*	= *D.* Andrew Persson
D. speciosum	X *D. fleckeri*	= *D.* Sunglow
D. speciosum	X *D. x gracillimum*	= *D.* Gracious Cascades
D. speciosum	X *D.* Gracious Falcon	= *D.* Karloo
D. speciosum	X *D.* Hastings	= *D.* Wonga
D. speciosum	X *D.* Hilda Poxon	= *D.* Essie Banks
D. speciosum	X *D.* John Upton	= *D.* Ingleside
D. speciosum	X *D.* Jamie Upton	= *D.* Lemon Glow
D. speciosum	X *D. johannis*	= *D.* Our Native
D. speciosum	X *D. jonesii* (as *D. ruppianum*)	= *D. x ruppiosum*
D. speciosum	X *D. kingianum*	= *D.* Specio-kingianum
D. speciosum	X *D. kingianum*	= *D. x delicatum*
D. speciosum	X *D.* Ku-Ring-Gai	= *D.* Northmead
D. speciosum	X *D.* Lowana Nioka	= *D.* Justin Wall
D. speciosum	X *D. macropus* subsp. *gracilicaule* (as *D. gracilicaule*)	= *D. x gracillimum*
D. speciosum	X *D. monophyllum*	= *D.* Aussie Choice
D. speciosum	X *D.* Peewee	= *D.* Elegant Heart
D. speciosum	X *D.* Peter	= *D.* Aussie Spectrum
D. speciosum	X *D.* Pink Pencil	= *D.* Wesley Rose
D. speciosum	X *D.* Star of Gold	= *D.* Star of Riverdene
D. speciosum	X *D. x suffusum*	= *D.* Harold Hirsch
D. speciosum	X *D. tetragonum*	= *D.* Hilda Poxon
D. speciosum	X *D.* Willowie Gold	= *D.* Manning
D. speciosum	X *D.* Wonga	= *D.* Woko
D. speciosum	X *D.* Wyuna	= *D.* Nerang
D. speciosum	X *D.* Yondi	= *D.* Jesmond Gem
D. Allyn Star	X *D. tetragonum*	= *D.* Tarantula
D. Andrew Persson	X *D.* Bardo Rose	= *D.* Ultravox
D. Andrew Persson	X *D. kingianum*	= *D.* Alan Printer
D. Aussie Starlight	X *D.* Eureka	= *D.* Aussie Zest
D. Aussie Starlight	X *D.* Kingrose	= *D.* Aussie Utmost
D. x delicatum	X *D. adae*	= *D.* Kevin Banks
D. x delicatum	X *D.* Andrew Persson	= *D.* Tom Jupp
D. x delicatum	X *D.* Aussie Mist	= *D.* Aussie Sandy
D. x delicatum	X *D.* Aussie Starlight	= *D.* Aussie Reward
D. x delicatum	X *D.* Ellen	= *D.* Delicate Ellen
D. x delicatum	X *D.* Emma	= *D.* Delemma
D. x delicatum	X *D.* Eureka	= *D.* Aussie Love
D. x delicatum	X *D. falcorostrum*	= *D.* Delicate Falcon
D. x delicatum	X *D. x gracillimum*	= *D.* Sun Sprite
D. x delicatum	X *D. kingianum*	= *D.* Gillian Leaney
D. x delicatum	X *D.* Kingrose	= *D.* Debra Wray
D. x delicatum	X *D.* Peter	= *D.* Memoria Eric Gordon
D. x delicatum	X *D.* Star of Gold	= *D.* Star One
D. x delicatum	X *D. x superbiens*	= *D.* Elsie Cox
D. x delicatum	X *D. tetragonum*	= *D.* Ku-Ring-Gai
D. Double Two	X *D. kingianum*	= *D.* Nathan Wall

Fig. 69 *Dendrobium speciosum* Smith var. *pedunculatum* Clemesha
Collected on the Atherton Tableland, north Queensland

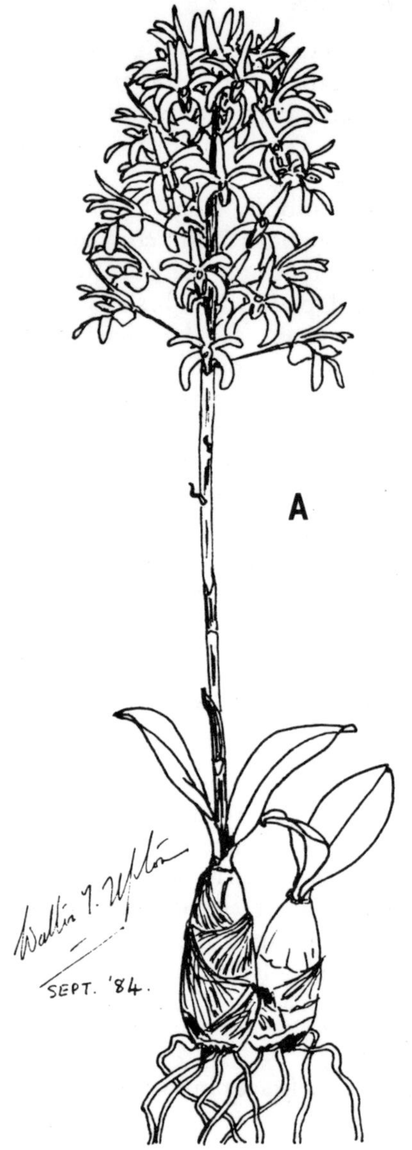

A Plant 1/3X
B Flower, front-view 1X
C Flower, side-view 1X
D Column and labellum, side-view 2X
E Labellum, pressed flat 3X
F Column, front-view 4X

82 *D. tozerensis*

83 *D. trilamellatum*

84 *D. wassellii*

85 *D. bigibbum* var. *venosum* (natural hybrid)

86 *D. cucumerinum* x *D. bowmanii* (as *D. mortii*)

87 *D.* x *foederatum*

88 *D. pugioniforme* x *D. mortii*

89 *D.* x *ruppiosum* 'Baron Falls'

90 *D.* x *delicatum* (*D. kestevenii* form)

91 *D.* x *delicatum* 'Harold's Pink'

92 *D.* x *gracillimum* (84 flowers)

93 *D.* x *delicatum* 'Jill'

94 *D.* x *gracillimum*

95 *D.* x *delicatum* 'Magenta Stripe'

96 *D.* x *suffusum*

B
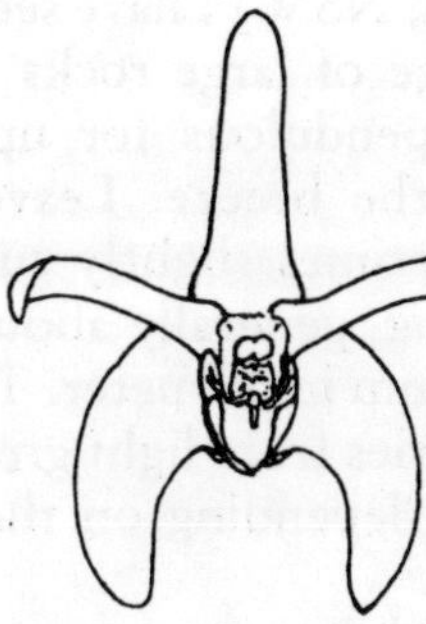

C
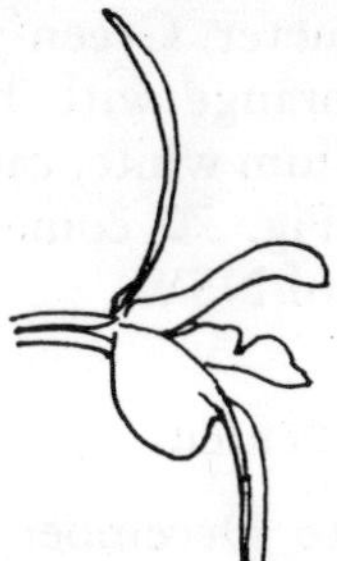

D
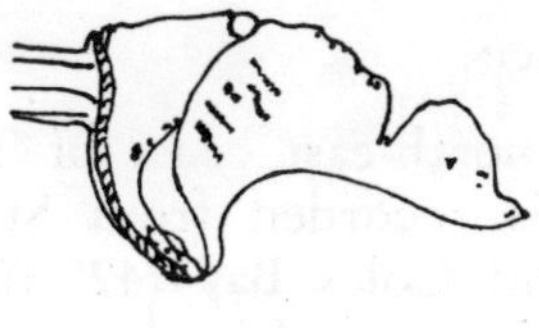

E
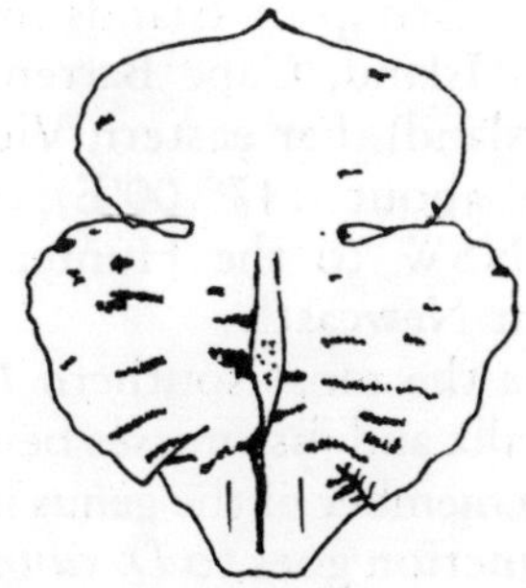

F

HYBRIDS (con't)

D. Gillian Leaney	X *D.* Aussie Starlight	= *D.* Aussie Springtime
D. Gillian Leaney	X *D.* Colin	= *D.* Narraweena
D. Gillian Leaney	X *D.* Ellen	= *D.* Aussie Charm
D. Gillian Leaney	X *D.* Peter	= *D.* Gilpet
D. Gillian Leaney	X *D. tetragonum*	= *D.* Emma
D. x gracillimum	X *D. adae*	= *D.* Double Two
D. x gracillimum	X *D. falcorostrum*	= *D.* Gracious Falcon
D. x gracillimum	X *D. kingianum*	= *D.* Penny Ann
D. x gracillimum	X *D. macropus* subsp. *gracilicaule* (as *D. gracilicaule*)	= *D.* Shan Leaney
D. x gracillimum	X *D. tetragonum*	= *D.* Allyn Star
D. Gracious Cascades	X *D. tetragonum*	= *D.* Rutherford Sunstar
D. Gracious Falcon	X *D.* Eureka	= *D.* Aussie Delight
D. Gracious Falcon	X *D. tetragonum*	= *D.* Gracious Star
D. Graham Hewitt	X *D.* Gillian Leaney	= *D.* Aussie Quest
D. Graham Hewitt	X *D.* Ellen	= *D.* Pinterry
D. Graham Hewitt	X *D. kingianum*	= *D.* Julatten
D. Graham Hewitt	X *D. tetragonum*	= *D.* Elinjaa
D. Gulginni	X *D. tetragonum*	= *D.* Coral Sun
D. Hilda Poxon	X *D.* Aussie Spectrum	= *D.* Aussie Child
D. Hilda Poxon	X *D. bigibbum*	= *D.* Brinawa
D. Hilda Poxon	X *D.* Blushing Star	= *D.* Kim Heinze
D. Hilda Poxon	X *D.* Eureka	= *D.* Aussie Magic
D. Hilda Poxon	X *D. falcorostrum*	= *D.* Miara
D. Hilda Poxon	X *D. fleckeri*	= *D.* Wirruna
D. Hilda Poxon	X *D.* Golden Fleck	= *D.* Hot Shot
D. Hilda Poxon	X *D.* Hastings	= *D.* Peter Hewitt
D. Hilda Poxon	X *D. jonesii* (as *D. ruppianum*)	= *D.* Ira Butler
D. Hilda Poxon	X *D. kingianum*	= *D.* Telekon
D. Hilda Poxon	X *D.* Ku-Ring-Gai	= *D.* Warrior
D. Hilda Poxon	X *D.* Peter	= *D.* Berserker
D. Hilda Poxon	X *D.* Star of Gold	= *D.* Tarean
D. Hilda Poxon	X *D.* Sunglow	= *D.* Graeme Banks
D. Hilda Poxon	X *D. tetragonum*	= *D.* Pauline Rankin
D. Hilda Poxon	X *D.* Wonga	= *D.* Vida Banks
D. Kevin Banks	X *D. tetragonum*	= *D.* Winston Hills
D. Ku-Ring-Gai	X *D.* John Upton	= *D.* Ring of Fire
D. Ku-Ring-Gai	X *D. kingianum*	= *D.* Zip
D. Ku-Ring-Gai	X *D.* Star of Gold	= *D.* Star Child
D. Penny Ann	X *D.* Bardo Rose	= *D.* Penny Rose
D. Penny Ann	X *D.* Ellen	= *D.* Ferg Martin
D. Penny Ann	X *D. falcorostrum*	= *D.* Twinkle
D. Penny Ann	X *D.* Goblin	= *D.* Minuet
D. Penny Ann	X *D. tetragonum*	= *D.* Fairy Floss
D. Sunglow	X *D.* Blushing Star	= *D.* Blushing Sun
D. Sunglow	X *D.* Gracious Falcon	= *D.* Kalang
D. Sunglow	X *D. kingianum*	= *D.* Gulginni
D. Sunglow	X *D.* Peter	= *D.* Tweed

HYBRIDS (con't)

D. Sunglow	X *D.* Star of Gold	= *D.* Carawah
D. Sunglow	X *D. tetragonum*	= *D.* Sun Star
D. Sunglow	X *D.* Wyuna	= *D.* Tully
D. Tully	X *D.* Carawah	= *D.* Omega
D. Wonga	X *D.* Hastings	= *D.* Rutherford Supreme
D. Wonga	X *D.* Jamie Upton	= *D.* Karen Nair
D. Wonga	X *D. kingianum*	= *D.* King-Wong

ETHNOBOTANY

It is known to the NSW Aborigines, who are said to have eaten the pseudobulbs, as 'Wer-gal-derra' and to the Queensland Aborigines as 'Taraberi'. Pademelons, bandicoots and kangaroo rats are reported to eat this orchid in eastern Australia (Lawler 1984).

Dendrobium striolatum H. G. Reichb.

Hamburger Garten-Blumenzeitung 13:313 (1857)

Fig. 70, Plate 63

SYNONYMS

Dendrobium teretifolium Lindley, *Edwards's Bot. Reg.* 25:misc. 32 (1839) non. R.Br. (1810)

Dendrobium milliganii F. Muell., *Fragm. Phyt. Aust.* 1:88, t.6 (1859)

Callista striolata (H. G. Reichb.) Kuntze, *Rev. Gen. Pl.* 2:655 (1891)

Dockrillia striolata (H. G. Reichb.) Rauschert, *Feddes Repert.* 94:447 (1983)

DESCRIPTION

Name comes from *striolatum*, meaning with fine striae — referring to the fine dark lines on the petals and sepals. Common name is streaked rock orchid. Described by Reichenbach from a specimen grown by F. Strange in the greenhouses of Consul Schiller.

One of the smallest of the Australian terete-leaved members of the genus *Dendrobium*. Frequently grows into large mats covering the contours of the rocks giving the appearance of succulent grass. Has a creeping rhizome with branching wiry stems, and produces roots along these stems quite freely. Plant habit generally erect although in the Fitzroy Falls area (near Nowra, NSW) I have seen it growing over the edge of large rocks and hanging completely pendulous for up to 90 cm, blowing in the breeze. Leaves obscurely ribbed, sometimes slightly curved and up to 13 cm long, generally about 5 to 8 cm, and up to 5 mm in diameter. The colour of the leaves varies from light green to almost dark purple depending on the amount of sun received.

Often widely-opening and fragrant flowers, produced in ones and twos, about 2 cm in diameter. Green-yellow to yellow, almost to orange with brown to purple striae. Labellum white, calli creamish.

Plant in Fig. 70 comes from the Katoomba area of NSW.

FLOWERING PERIOD

September to December (spring to summer). I have seen it in flower in early December in Tasmania.

DISTRIBUTION

From the north-east coast of Tasmania, where it is recorded from St Helens, Bicheno and Coles Bay (42° 10′S). The Furneaux Group of Islands in Bass Strait (Flinders Island, Cape Barren Island and Clarke Island). Far eastern Victoria (most westerly about 147° 00′E), and south-eastern NSW to the Hunter River (32° 40′S) near Newcastle.

This is the most southern *Dendrobium* in Australia and just misses being the most southern member of the genus in the world; this distinction goes to *D. cunninghamii* of New Zealand, the only *Dendrobium* which grows as far south as Stewart Island (47° 10′S).

HABITAT

On rocks, very rarely on trees, from the coast to about 900 m, frequently in very exposed places. It is mainly a species of the ranges, seldom growing right on the coast except in Tasmania where it grows on large granite boulders near the sea. The habitat description by Nicholls (1969) is superb. 'The habit of this small lithophyte is up-

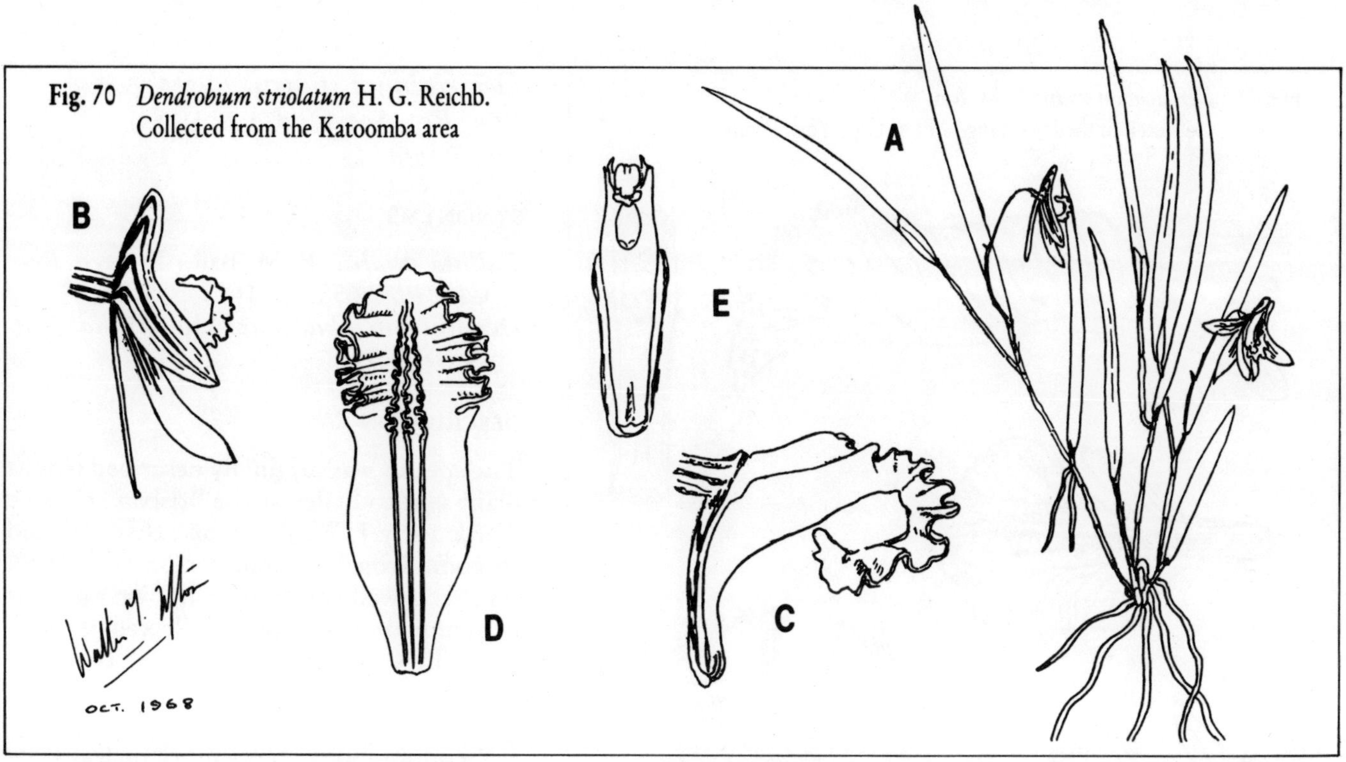

Fig. 70 *Dendrobium striolatum* H. G. Reichb.
Collected from the Katoomba area

A Plant ²/₃X
B Flower, side-view 2X
C Column and labellum, side-view 4X
D Labellum (partially), pressed flat 4X
E Column, front-view 4X

right on the crest of the boulders, while on the sloping faces and sides, where the plants are thickly massed, it hangs in mats. On the under side, the foliage and succulent root-growths descend from the ceilings of caves, the latter like strings of shining beads, to be reflected from the pools of clear water beneath. In the spring the abundant golden flowers stand out conspicuously from the dark-hued background of foliage'.

I have seen it in great masses on granite rocks on a hill overlooking the sea and village of Bicheno in Tasmania. On the rocks that stood out above the trees it was in full sun, the terete leaves being a dark purple all over due to the exposure. Plants growing deep down in the crevices and out of the sun, were a dark green. On 6 December 1985 two or three flowers were still to be seen, a good yellow-orange in colour with a white lip. In the Illawarra region south of Sydney, it is very common, growing in extremely large mats on rocks.

CULTURE

Cork slabs or hardwood blocks are most satisfactory. As with other *Dendrobium* species it must be tied tightly to its host. Some spaghnum moss placed over the roots when tying helps the plant become established. Requires plenty of water during the hottest months, two or more times on some days, and a misting at dusk also beneficial. During the winter on slabs it still needs to be just moist; it must never be allowed to dry out completely. Care must, however, be taken when the new leaf is forming as it will rot off if water is allowed to stay in or near the bract. Frequent weak feedings are also beneficial when in full growth. Minimum winter temperature of 1°C.

IN HYBRIDIZATION

Only six hybrids have been registered and each is about midway between the parents, with perhaps the yellow colour and purple striations coming through.

HYBRIDS

D. striolatum	X *D. mortii* (as *D. tenuissimum*)	= *D.* Julie Skillicorn
D. striolatum	X *D. pugioniforme*	= *D.* Duffy
D. striolatum	X *D. schoeninum* (as *D. beckleri*)	= *D.* Stephen
D. striolatum	X *D. teretifolium*	= *D.* Rosemary Jupp
D. Rosemary Jupp	X *D. mortii* (as *D. tenuissimum*)	= *D.* Super Ruffles
D. Rosemary Jupp	X *D. teretifolium*	= *D.* Fay Moore

Fig. 71 *Dendrobium stuartii* F. M. Bail.
Collected in the Iron Range area of Cape York Peninsula

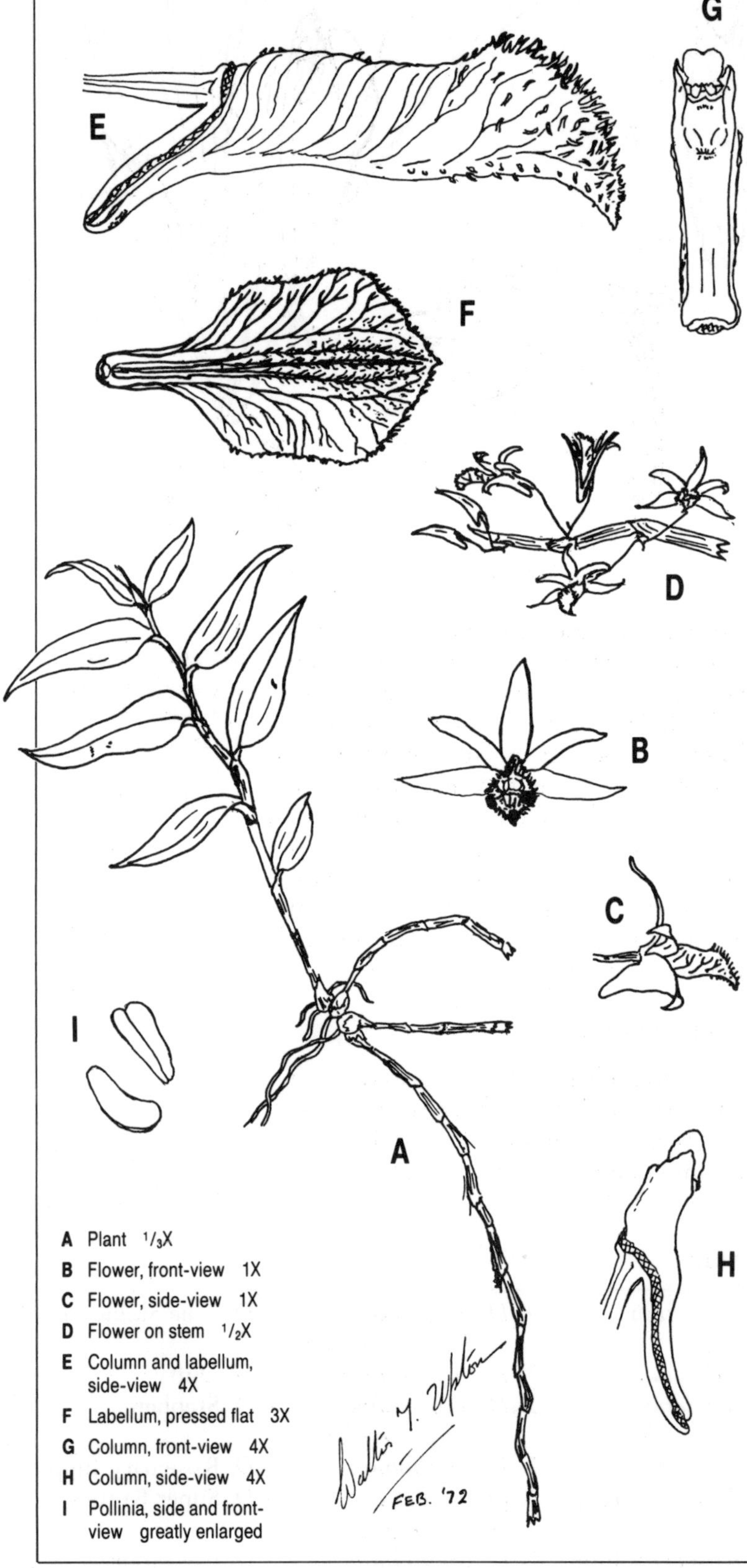

A Plant $^1/_3$X
B Flower, front-view 1X
C Flower, side-view 1X
D Flower on stem $^1/_2$X
E Column and labellum, side-view 4X
F Labellum, pressed flat 3X
G Column, front-view 4X
H Column, side-view 4X
I Pollinia, side and front-view greatly enlarged

Dendrobium stuartii F. M. Bail.

Proc. Roy. Soc. Q/ld. 1:12 (1884)
Fig. 71, Plate 64

SYNONYMS

Callista stuartii (F. M. Bail.) Kuntze, *Rev. Gen. Pl.* 2:655 (1891)
Dendrobium whiteanum Hunt, *Q/ld. Nat.* 14:60 (1951)

DESCRIPTION

The species was originally described from a plant sent to Bailey at the Brisbane Botanic Gardens by J. W. R. Stuart, that was said to come from near Herberton (17° 23′S). As this is well south of its known range, it is assumed that Stuart collected it from much further north and sent it to Bailey from Herberton. It is named after J. W. R. Stuart.

Pseudobulbous stems more or less pendulous, up to 45 cm long. Leafy throughout, leaves seldom lasting more than one season. Old leafless stems produce from one to three fragrant flowers, commonly two. Widely-opening flowers with petals and sepals usually curved back are about 2 cm in diameter. Flowers a dull yellowish green, with a pale yellowish labellum beautifully marked with reddish-mauve veins. Segments almost translucent. (I have never seen flowers with the ovary and spur coloured a darkish red as shown in Fitzgerald's *Australian Orchids.*) Flowers frequently self-pollinating and short lived.

FLOWERING PERIOD

January to February (summer).

DISTRIBUTION

Uncommon but widespread in the Cape York Peninsula from about Coen (14° 00′S) northward to the Iron Range area (12° 40′S) and the Lockhart River and to Bamaga (10° 50′S).

HABITAT

In lowland rainforest, dry gullies and vine scrub, on the ridge tops and plateaus, and in dry scrub bordering mangrove swamps from the coast up to approximately 350 m.

I have seen it in the Iron Range high up in rainforest trees, quite a few plants over a wide area. New stems start growing upright and then, as they mature, become pendulous.

CULTURE

Slab culture suits best. A warm, well-shaded position is satisfactory as foliage is soft and burns easily. Lots of water when in growth. Give a dry rest when the buds appear and withhold water until the new growth is about 4 cm long. Weak fertilizing when in growth is beneficial. Propagate by cutting old stems in lengths of several nodes and placing them on moist sphagnum moss to induce aerial growths. Minimum winter temperature of 15°C (60°F).

IN HYBRIDIZATION

No hybrids registered to date.

Dendrobium teretifolium R.Br.

Prodr. Fl. Novae Holl. 333 (1810)

Figs 72, 73, Plates 73 and 75

SYNONYMS

Callista teretifolia (R.Br.) Kuntze, *Rev. Gen. Pl.* 2:655 (1891)

Dockrillia teretifolia (R.Br.) Brieger, in Schlechter *Die Orchideen* ed. 3 (11/12): 745 (1981)

DESCRIPTION

First described by Robert Brown. In *Prod. Fl. Novae Holl.*, Brown indicated the regions where he collected plants by a single Roman capital, in this case J, which denotes 'the neighbourhood of the colony of Port Jackson, including the banks of the estuary named Hunter's River or Coal River'. Name comes from *teretis*, meaning rounded or cylindrical, and *folium*, meaning leaf—referring to the cylindrical-shaped leaves.

Fig. 72 *Dendrobium teretifolium* R.Br. Collected at Failford near Forster, NSW in August, 1961

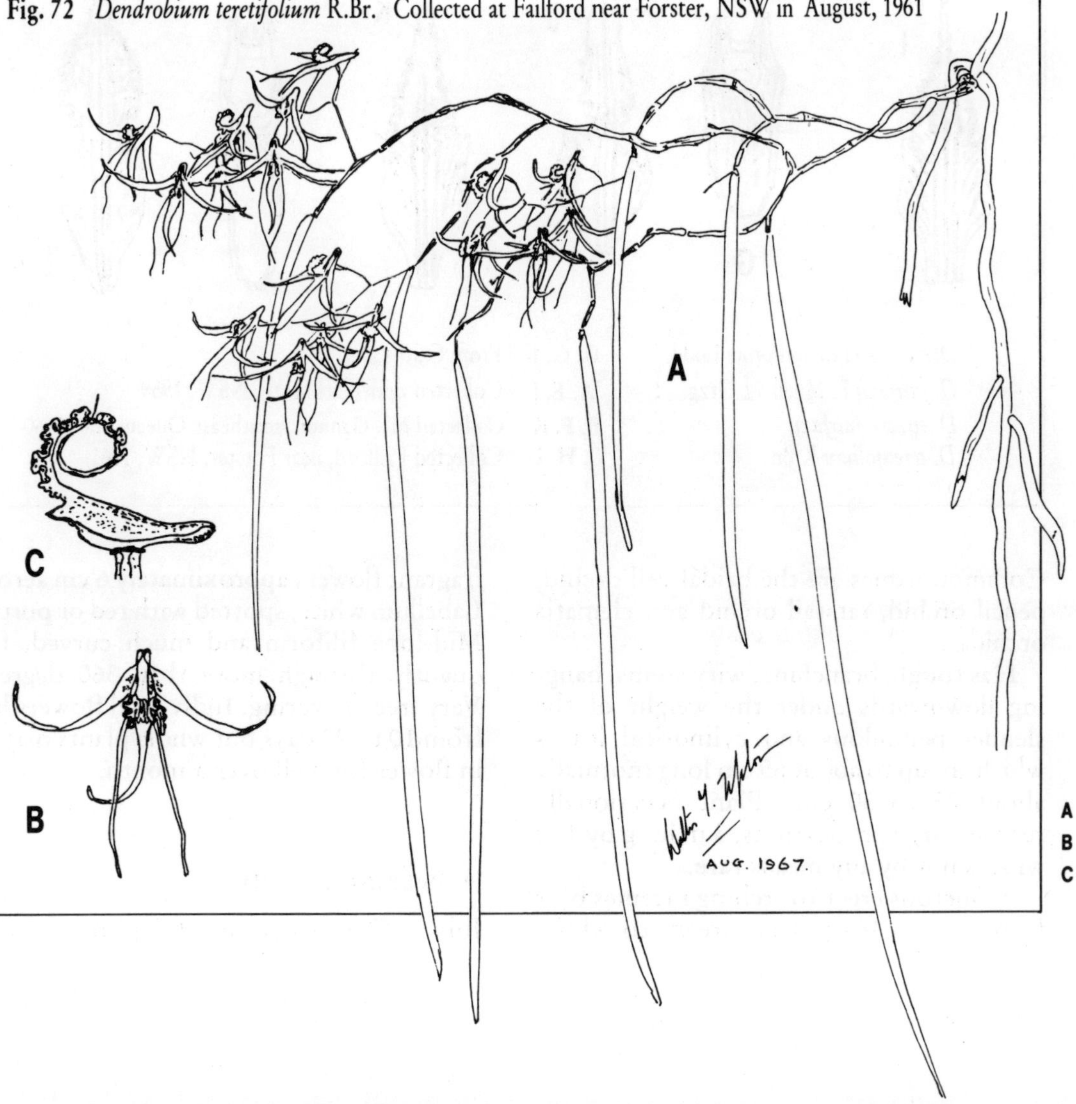

A Plant $^1/_3$X
B Flower, front-view $^2/_3$X
C Column and labellum, side-view 2X

A to D Flower, side-view 1X
E to H Column, front-view 4X
I to L Labellum pressed flat 2X

Fig. 73 Comparison drawings of the floral parts of *D. teretifolium* R.Br. and its closely allied species.

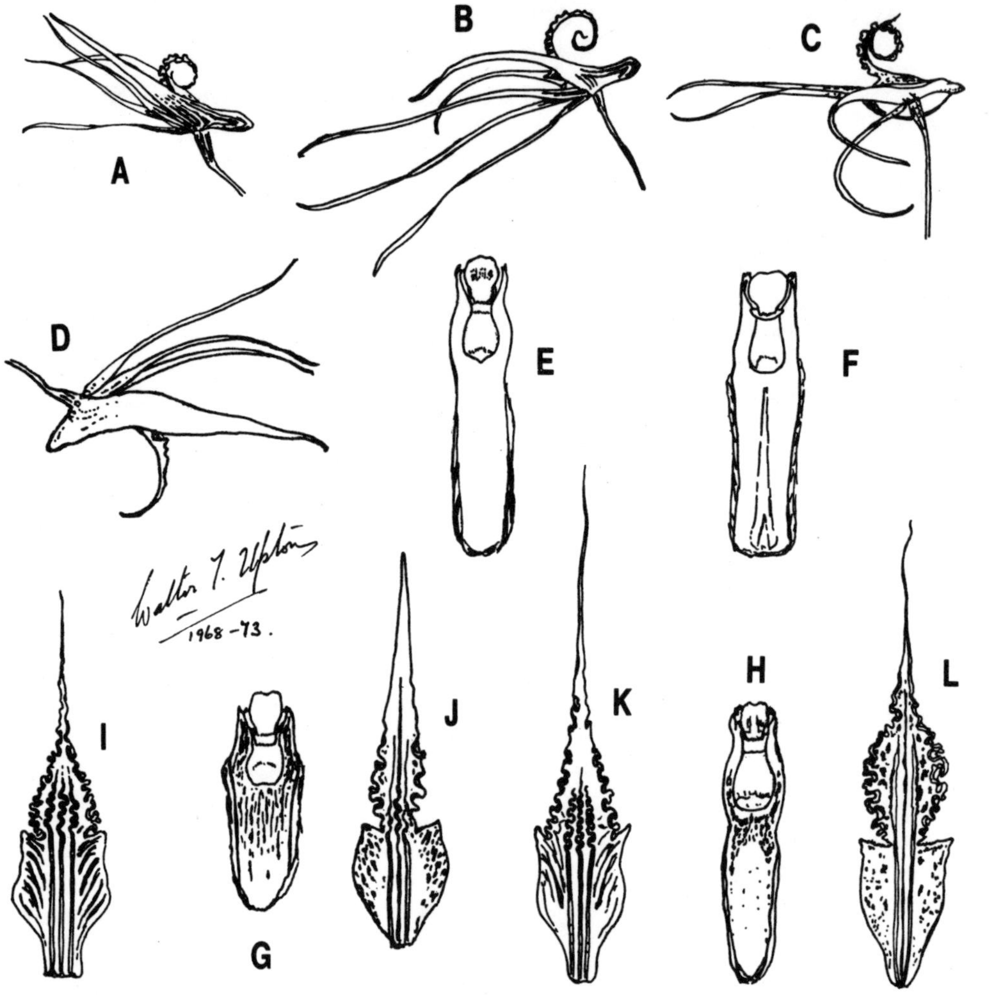

Dendrobium calamiforme Lodd.	D, G, J	From Cairns area
D. fairfaxii F. Muell. et Fitzg.	A, E, I	Collected near Blackheath, NSW, 1959
D. sp. aff. *fairfaxii*	B, F, K	Collected Mt. Gannon, southeast Queensland, 1960
D. teretifolium R.Br.	C, H, L	Collected Failford, near Forster, NSW, 1961

Common names are the bridal veil orchid, pencil orchid, ratstail orchid and clematis orchid.

Has tough, branching, wiry stems, hanging downwards under the weight of the slender pendulous and cylindrical leaves which are up to about 65 cm long (normally about 25 to 30 cm). Plant occasionally reaches large proportions, 3 m long by 1 m wide is not by any means rare.

Numerous erect to arching racemes bear from few to 16 spidery, cream to white, fragrant flowers approximately 6 cm across. Labellum white, spotted with red or purple. Mid-lobe filiform and much curved, frequently through more than 360 degrees. Very free-flowering. Individual flowers last from 10 to 20 days but whole plants may be in flower for well over a month.

FLOWERING PERIOD

July to October (winter to spring).

DISTRIBUTION

From Bega (36° 40′S) in south-eastern NSW to about the Fitzroy River, near Rockhampton in Queensland (23° 20′S).

HABITAT

At low altitudes mostly on trees but occasionally on rocks. Seldom seen in rainforests, preferring a fair amount of light. Favours trees overhanging creeks, lakes, etc. or in swamps, particularly the swamp oak (*Casuarina glauca*). Mostly confined to the coastal gullies and river flats. Also grows on the river oak, *C. cunninghamiana*, and some species of paperbark (*Melaleuca*).

POLLINATION

Pollination by the native bee (*Trigonia carbonaria*) has been observed by Jones (1983).

CULTURE

Requires a sheltered, shady position preferably in a shade-house where it can receive ample air movement, or tied to a limb of a tree where it receives full early morning sun and filtered sunlight for the rest of the day.

Slab-type culture is satisfactory. Cork, hardwood, tree-fern fibre blocks, etc. (I have a plant on a large block of tree fern, and it has been on it for over 30 years; it is healthy and flowers well each year.) Regular watering and feed during the growing season and keep just moist during winter. Minimum winter temperature of 2°C (34°F).

IN HYBRIDIZATION

Of the hybrids registered to date, seven have been within its own section *Rhizobium*, two with species of the section *Dendrocoryne* and one with the section *Spatulata* of the subgenus *Dendrobium.* Within its own section the progeny have been approximately midway between the parents. Outside of its own section it has mostly produced leaves that are partially cylindrical with the result that the primary flowering node in these leaves is at the bottom of a tube, so of necessity it flowers from nodes lower down.

The hybrid *D.* Wesley Rose (*D.* Pink Pencil X *D. speciosum*) is a cross between two species from the section *Dendrocoryne* with *D. teretifolium*, with the result that only a few of these hybrids have tubular-type leaves. Fortunately it does have the lovely lip of *D. teretifolium* and is therefore a very worthwhile hybrid, flowering quite freely.

HYBRIDS

D. teretifolium	X *D. canaliculatum*	= *D.* Ayr Queensland
D. teretifolium	X *D. cucumerinum*	= *D.* Goose Bumps
D. teretifolium	X *D. kingianum*	= *D.* Pink Pencil
D. teretifolium	X *D. linguiforme*	= *D.* Virginia Jupp
D. teretifolium	X *D.* Rosemary Jupp	= *D.* Fay Moore
D. teretifolium	X *D. schoeninum* (as *D. beckleri*)	= *D.* Jiggi
D. teretifolium	X *D. striolatum*	= *D.* Rosemary Jupp
D. teretifolium	X *D.* Virginia Jupp	= *D.* Wesley Pink
D. Pink Pencil	X *D. speciosum*	= *D.* Wesley Rose
D. Rosemary Jupp	X *D. mortii* (as *D. tenuissimum*)	= *D.* Super Ruffles

Dendrobium tetragonum A. Cunn.

Edwards's Bot. Reg. 25: misc. 33 (1839)

DESCRIPTION

Veitch's *A Manual of Orchidaceous Plants* Pt. 3, p. 80, 1888 states 'Discovered by Allan Cunningham about the year 1820, in the dry shady woods around Moreton Bay, Qld., and subsequently detected by other explorers in the neighbourhood of Rockingham Bay, Qld., on the banks of the Hastings or Macleay Rivers, N.S.W., and in other localities near the east coast of Australia'.

From the Greek *tetragon*, meaning figure with four angles — referring to the pseudobulbous stems which are four-sided at the apex. Common name is Spider Orchid.

Gilbert (1939) describes how the yellow-throated scrub wren (*Sericornis lathami*) decorates its nest with orchid plants, including seedlings of *D. tetragonum*, its main purpose appears to be one of camouflage. Also the Satin Bower-bird (*Ptilonorhynchus violaceus*) displays plants and flowers of *D. tetragonum* and the flowers of *D. speciosum* in front of its bower.

Fig. 74 *Dendrobium tetragonum* A. Cunn. var. *tetragonum*
Collected in the Wyong area, NSW

A Plant 1X
B Flower, front-view 1X
C Column and labellum, side-view 4X
D Labellum, pressed flat 4X
E Column, front-view 4X

An extremely variable species with three named varieties, each described separately.

D. tetragonum A. Cunn. var. *tetragonum*

Fig. 74, Plates 78 and 79

SYNONYMS

Callista tetragona (A. Cunn.) Kuntze, *Rev. Gen. Pl.* 2:655 (1891)

Dendrobium tetragonum A. Cunn. var. *variabilis* Gilbert, *Aust. Orchid Rev.* 2(4):20 (1937)

Dendrocoryne tetragonum (A. Cunn.) Brieger in Schlechter, *Die Orchideen*, ed. 3 (11/12):724 (1981)

Tropilis tetragona (A. Cunn.) Rauschert, *Feddes Repert.* 94:471 (1983)

DESCRIPTION

Pseudobulbous stems nearly always semi-pendulous to pendulous in this variety, particularly at the southern end of its distribution. Some Queensland forms are sometimes quite erect. Pseudobulbous stems from about 5 to 10 cm long in some areas and up to 50 cm in others. Very thin and wiry at the base and thicker and distinctly quadrangular near the apex. The proportion of the slender (wiry) part of the stem is extremely variable. Up to five thin but leathery leaves produced at the top of the stems, frequently quite twisted.

Racemes quite short and produce up to nine fragrant (vanilla) flowers, each stem produces flowers for a number of years. A form from the lower north coast area of NSW produces up to three racemes per stem each with about nine flowers. Flowers sometimes as large as the var. *giganteum*, mostly yellow-green, edged in red-brown with a broad, white, lightly-spotted lip—a most magnificent form. Flowers of the var. *tetragonum* most varied in size, shape and colour. In size they vary from 3 cm to about 10 cm long, longer in some clones. In shape they vary from long narrow flowers to those with the lateral sepals spread wide apart, all widely opening. Colour ranges from green to yellow to cream with varying degrees of markings. Sepals completely outlined in red-brown with petals not marked at all. Fine red-brown spotting all over, red-brown blotches or sometimes striations. Generally the southern forms have dark red-brown stripes and the northern forms of this variety have mostly red-brown blotches rather than stripes. The small southern form seldom carries more than one or two flowers per raceme in the south part of its distribution. Flowers last 10 to 20 days.

FLOWERING PERIOD

September to November (spring).

DISTRIBUTION

From about 180 km south of Sydney (35° 00′S) northward to at least south-eastern Queensland. Chiefly coastal but occasionally extends to the eastern slopes of the dividing range 75 km inland. There have been recordings of this variety as far north as south-eastern Cape York Peninsula, but these are more likely to be the var. *giganteum.*

HABITAT

From the coast to the ranges up to at least 1000 m, but generally much lower than this, and up to 100 km inland. Mostly growing on rainforest trees but also on *Melaleuca* and *Casuarina* in swamps. Along creeks in the ranges it is often found on scrub myrtle (*Backhousia myrtifolia*). The swamp mahogany (*Tristaniopsis laurina*) is another of its hosts. I have mostly found it growing fairly low down and on the main trunk of the tree. It prefers very shaded areas like gullies where there is always continuous air movement. Very rarely on rocks. In the northern part of its distribution it appears to seldom grow above 500 m.

Referring to the form from the lower north coast of NSW, Lorrie Friar of South Grafton, a great bushman and lover of native orchids, in a personal communication said 'We have seen many in the bush and the only two trees they occur on are a *Casuarina* (Swamp Oak), and a *Melaleuca* (what we call prickly tea tree or Paper Bark). Those on the Oak are always miserable little plants with a lot of dead ones and yet, within reach of these, plants growing on a paper bark could be up to 2′6″ long, magnificent specimens, and yet the reverse happens with *D. teretifolium* huge plants on the Oak and practically nothing on the "tea tree". Another thing of interest is that growing together in this swampy country is another paper bark which we call broad leafed tea tree, these could be rubbing against the prickly leafed variety but never have we seen a *D. tetragonum* on this fellow'.

D. tetragonum A. Cunn. var. *hayesianum* Gilbert

Austral. Orchid Rev. 2(4):20 (1937)

Fig. 75

DESCRIPTION

Named after Mr F. L. Hayes. Originally described from plants collected in the Illawarra area, south of Sydney, NSW.

A very pretty pale-yellow and green variety with an all-white to creamy-white labellum. (A colour form of the type variety.) Stems and leaves pale green. One or two flowers approximately 50 mm long. Same flowering period and distribution as the type variety.

Appear as individual plants within normal colonies. There is also a similar colour form of the variety *giganteum* on Mt Lewis and adjacent highlands, but in this case a whole race, not an isolated plant in a normal colony. (These plants are not *D. tetragonum* var. *hayesianum.*)

D. tetragonum A. Cunn. var. *giganteum* Gilbert

Aust. Orchid Rev. 2(4):20 (1937)

Fig. 76, Plates 77, 80 and 81

SYNONYM

Dendrobium tetragonum A. Cunn. var. *tomentosum* Nicholls, *Aust. Orchid Rev.* 7:40 (1942)

DESCRIPTION

This is commonly referred to as the large-flowered, north-Queensland form, although flowers equally as big, on the average, are frequently found in the type variety in its mid NSW to northern distribution. Generally the differences between this variety

A Plant 1X
B Flower, front-view 1X
C Flower, side-view 1X
D Column and labellum, side-view 4X
E Labellum, pressed flat 4X
F Column, front-view 4X
G Pollinia greatly enlarged

Fig. 75 *Dendrobium tetragonum* A. Cunn. var. *hayesianum* Gilbert
From the Cambewarra Range near Nowra, NSW

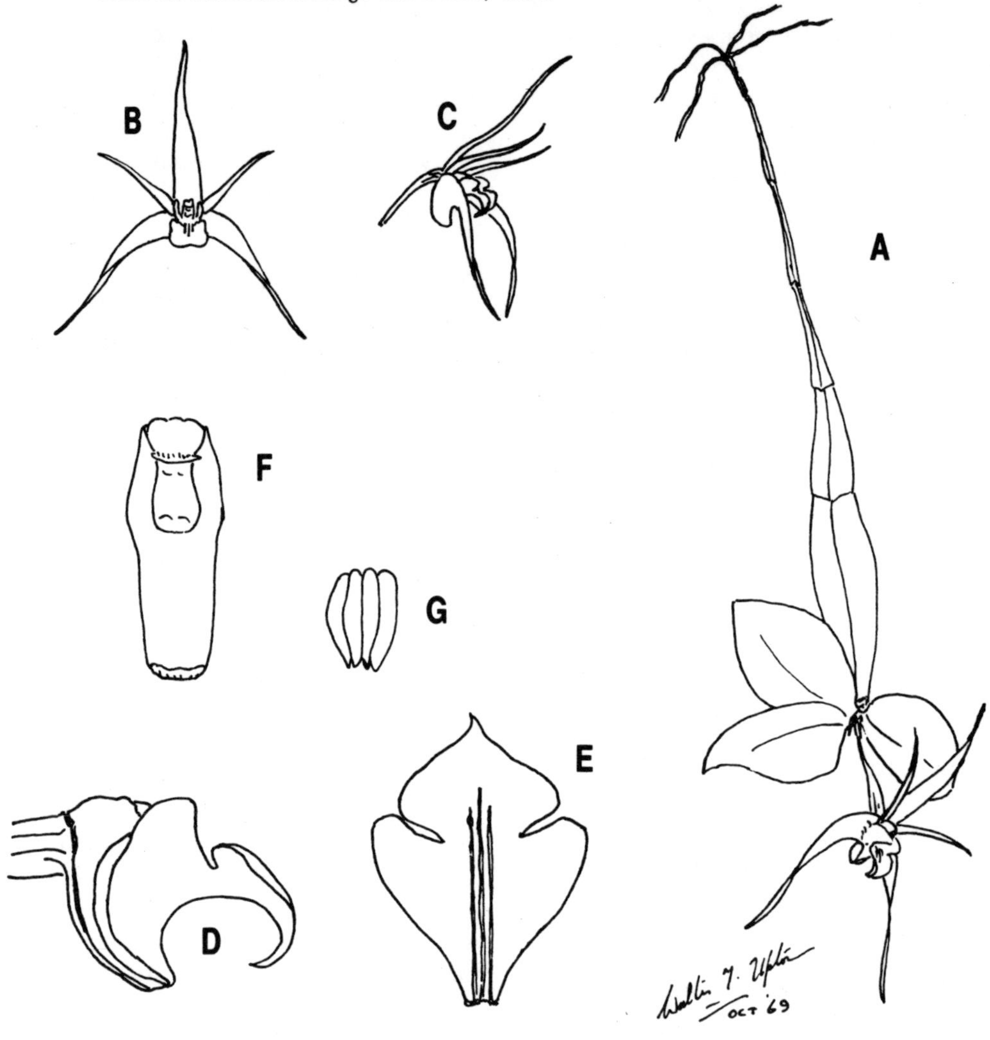

and the type variety are that stems are longer and more upright in their growth, and flowers are often paler and more heavily blotched with red (whereas the type variety more often has striped markings). Flowers consistently larger. Side lobes of the labellum frequently much larger than the pointed mid-lobe; often the mid-lobe in this variety is more angular. Green-yellow coloured form usually has a white lip and is devoid of markings. But it is not generally a complete albino and red pigment is frequently seen in varying degrees. I have seen many flowers up to 20 cm long, but flowers to 25 cm have been reported (see Rupp and White, 1937). Mid-lobe of labellum and often the anterior portion of ridges of disc are tomentose.

It is interesting to note that the seed capsule of the var. *giganteum* take approximately 10 days longer to mature (at least in Sydney, NSW) than the type variety.

CHROMOSOME NUMBER

2n = 38 (K. Y. Lim, 1985).

FLOWERING PERIOD

April to October (autumn, winter to spring). Can flower at other times. Some forms flower almost continuously over the flowering period.

DISTRIBUTION

Eastern tropical variety. Between the Fitzroy River (23° 30′S) and at least as far north as Iron Range (12° 45′S).

HABITAT

In highland and lowland rainforests from sea level to approximately 1250 m. In some localities it is quite common.

CULTURE

Requires high humidity, ample air movement and about 60 per cent shade. Slab culture is satisfactory, provided the plants are not allowed to dry out completely for any length of time or too frequently. A moisture-retentive host, such as paperbark poles, is most satisfactory. It is necessary to tie the plant on tightly, so that it can quickly establish itself on to its host. Minimum winter temperature of about 2°C is suitable for the majority of plants, but some of the low level forms of var. *giganteum* need extra warmth in winter.

IN HYBRIDIZATION

D. tetragonum is very dominant in shape and size, and mostly dominant in colour. (When crossed with *D. kingianum* and *D. bigibbum* those species are then colour dominant.) The habit of raceme is mostly arching to almost pendulous.

The lip shape is very dominant in some forms, particularly in the lower north-coast NSW form. The mid-lobe is wide and flat and this excellent characteristic is being passed on to its progeny.

The var. *giganteum* passes on its habit of flowering more than once over the autumn to spring period.

HYBRIDS

D. tetragonum	X *D. adae*	= *D.* Star Imp
D. tetragonum	X *D. affine* (as *D. dicuphum*)	= *D.* Maron
D. tetragonum	X *D.* Allyn Star	= *D.* Tarantula
D. tetragonum	X *D.* Bardo Rose	= *D.* Silvester
D. tetragonum	X *D. bigibbum*	= *D.* Peewee
D. tetragonum	X *D. x delicatum*	= *D.* Ku-Ring-Gai
D. tetragonum	X *D. discolor* (as *D. undulatum*)	= *D.* Willowie Gold
D. tetragonum	X *D.* Ella Victoria Leaney	= *D.* Phil Deane
D. tetragonum	X *D.* Ellen	= *D.* Aussie Ira
D. tetragonum	X *D.* Eureka	= *D.* Nunkumbil
D. tetragonum	X *D. falcorostrum*	= *D.* Star of Gold
D. tetragonum	X *D. fleckeri*	= *D.* Golden Glory
D. tetragonum	X *D.* Gillian Leaney	= *D.* Emma
D. tetragonum	X *D. x gracillimum*	= *D.* Allyn Star
D. tetragonum	X *D.* Golden Fleck	= *D.* Jamie Upton
D. tetragonum	X *D.* Gracious Cascades	= *D.* Rutherford Sunstar
D. tetragonum	X *D.* Gracious Falcon	= *D.* Gracious Star
D. tetragonum	X *D.* Graham Hewitt	= *D.* Elinjaa
D. tetragonum	X *D.* Gulginni	= *D.* Coral Sun
D. tetragonum	X *D.* Hastings	= *D.* Peach Star
D. tetragonum	X *D.* Hilda Poxon	= *D.* Pauline Rankin
D. tetragonum	X *D.* Judy Leroy	= *D.* Duke
D. tetragonum	X *D. johannis*	= *D.* Star Twist
D. tetragonum	X *D.* John Upton	= *D.* Bright Spark
D. tetragonum	X *D. jonesii* (as *D. ruppianum*)	= *D.* Kungara
D. tetragonum	X *D.* Kevin Banks	= *D.* Winston Hills
D. tetragonum	X *D. kingianum*	= *D.* Ellen
D. tetragonum	X *D.* Peewee	= *D.* Rosella
D. tetragonum	X *D.* Penny Ann	= *D.* Fairy Floss
D. tetragonum	X *D.* Peter	= *D.* Amber
D. tetragonum	X *D. speciosum*	= *D.* Hilda Poxon
D. tetragonum	X *D.* Star Imp	= *D.* Rutherford Star
D. tetragonum	X *D. x suffusum*	= *D.* Blushing Star
D. tetragonum	X *D.* Sunglow	= *D.* Sun Star
D. Aussie Bonanza	X *D.* Aussie Freckles	= *D.* Aussie Flame
D. Aussie Bonanza	X *D.* Aussie Mist	= *D.* Aussie Jewel
D. Aussie Bonanza	X *D.* Golden Fleck	= *D.* Margery Angeleri
D. Aussie Bonanza	X *D. kingianum*	= *D.* Aussie Sequins
D. Aussie Glow	X *D.* Golden Fleck	= *D.* Tambar
D. Aussie Ira	X *D. bigibbum*	= *D.* Aussie Angel
D. Aussie Ira	X *D. falcorostrum*	= *D.* Aussie Bonanza
D. Aussie Ira	X *D. fleckeri*	= *D.* Aussie Glow
D. Aussie Ira	X *D.* Hastings	= *D.* Aussie Gift
D. Aussie Ira	X *D. jonesii* (as *D. ruppianum*)	= *D.* Aussie Nave

Fig. 76 *Dendrobium tetragonum* A. Cunn. var. *giganteum* Gilbert Collected near Cairns

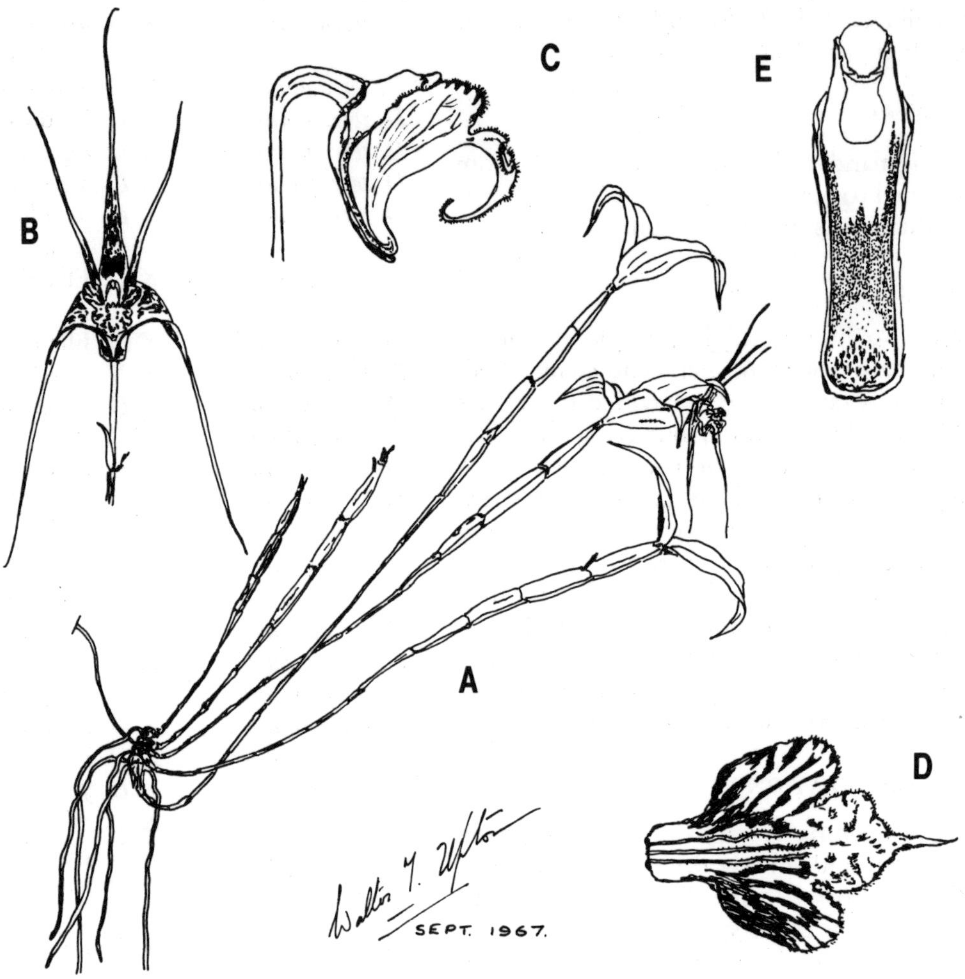

HYBRIDS (con't)

D. Aussie Ira	X *D.* Kingrose	= *D.* Aussie Lust
D. Aussie Ira	X *D.* Peter	= *D.* Aussie Ember
D. Aussie Ira	X *D.* *speciosum*	= *D.* Aussie Starlight
D. Aussie Ira	X *D.* Star of Gold	= *D.* Aussie Parade
D. Aussie Starlight	X *D.* *x delicatum*	= *D.* Aussie Reward
D. Aussie Starlight	X *D.* Eureka	= *D.* Aussie Zest
D. Aussie Starlight	X *D.* Gillian Leaney	= *D.* Aussie Springtime
D. Aussie Starlight	X *D.* Kingrose	= *D.* Aussie Utmost
D. Blushing Star	X *D.* Sunglow	= *D.* Blushing Sun
D. Carawah	X *D.* Tully	= *D.* Omega
D. Colin	X *D.* Eureka	= *D.* Colreka
D. Colin	X *D.* Gillian Leaney	= *D.* Narraweena
D. Colin	X *D.* John Upton	= *D.* Coljohn
D. Colin	X *D.* Judy Leroy	= *D.* Jombock

A Plant 1/5X
B Flower, front-view 2/3X
C Column and labellum, side-view 2X
D Labellum, pressed flat 2X
E Column, front-view 4X

HYBRIDS (con't)

D. Colin	X *D.* Kim	= *D.* Fiesta
D. Ellen	X *D.* Aussie Bonanza	= *D.* Dot Sheen
D. Ellen	X *D. x delicatum*	= *D.* Delicate Ellen
D. Ellen	X *D.* Ella Victoria Leaney	= *D.* Aussie Wonder
D. Ellen	X *D.* Emmy	= *D.* Goblin
D. Ellen	X *D.* Eureka	= *D.* Ocker's Dream
D. Ellen	X *D. falcorostrum*	= *D.* Warrambool
D. Ellen	X *D. fleckeri*	= *D.* Aussie Dream
D. Ellen	X *D.* Gillian Leaney	= *D.* Aussie Charm
D. Ellen	X *D.* Graham Hewitt	= *D.* Pinterry
D. Ellen	X *D. jonesii* (as *D. ruppianum*)	= *D.* Colin
D. Ellen	X *D.* Kim	= *D.* John Upton
D. Ellen	X *D. kingianum*	= *D.* Kathryn Banks
D. Ellen	X *D. macropus* (as *D. gracilicaule*)	= *D.* Ellen Grace
D. Ellen	X *D.* Penny Ann	= *D.* Ferg Martin
D. Ellen	X *D.* Peter	= *D.* Aussie Treat
D. Ellen	X *D. speciosum*	= *D.* Graham Hewitt
D. Ellen	X *D.* Star of Gold	= *D.* Kookaburra
D. Ellen	X *D. x suffusum*	= *D.* Pixie
D. Ellen	X *D.* Susan	= *D.* Sweet Ellen
D. Emma	X *D. x delicatum*	= *D.* Delemma
D. Goblin	X *D. kingianum*	= *D.* King Goblin
D. Goblin	X *D.* Penny Ann	= *D.* Minuet
D. Graham Hewitt	X *D.* Gillian Leaney	= *D.* Aussie Quest
D. Graham Hewitt	X *D. kingianum*	= *D.* Julatten
D. Hilda Poxon	X *D.* Aussie Spectrum	= *D.* Aussie Child
D. Hilda Poxon	X *D. bigibbum*	= *D.* Brinawa
D. Hilda Poxon	X *D.* Blushing Star	= *D.* Kim Heinze
D. Hilda Poxon	X *D.* Eureka	= *D.* Aussie Magic
D. Hilda Poxon	X *D. falcorostrum*	= *D.* Miara
D. Hilda Poxon	X *D. fleckeri*	= *D.* Wirruna
D. Hilda Poxon	X *D.* Golden Fleck	= *D.* Hot Shot
D. Hilda Poxon	X *D.* Hastings	= *D.* Peter Hewitt
D. Hilda Poxon	X *D. jonesii* (as *D. ruppianum*)	= *D.* Ira Butler
D. Hilda Poxon	X *D. kingianum*	= *D.* Telekon
D. Hilda Poxon	X *D.* Ku-Ring-Gai	= *D.* Warrior
D. Hilda Poxon	X *D.* Peter	= *D.* Berserker
D. Hilda Poxon	X *D. speciosum*	= *D.* Essie Banks
D. Hilda Poxon	X *D.* Sunglow	= *D.* Graeme Banks
D. Hilda Poxon	X *D.* Wonga	= *D.* Vida Banks
D. Jamie Upton	X *D.* Golden Fleck	= *D.* Green Star
D. Jamie Upton	X *D.* John Upton	= *D.* Colsboys
D. Jamie Upton	X *D. speciosum*	= *D.* Lemon Glow
D. Jamie Upton	X *D.* Wonga	= *D.* Karen Nair
D. John Upton	X *D.* Bardo Rose	= *D.* John Rose
D. John Upton	X *D.* Kungara	= *D.* Wakehurst
D. John Upton	X *D. speciosum*	= *D.* Ingleside
D. Ku-Ring-Gai	X *D.* John Upton	= *D.* Ring of Fire
D. Ku-Ring-Gai	X *D. kingianum*	= *D.* Zip

HYBRIDS (con't)

D. Ku-Ring-Gai	X *D. speciosum*	= *D.* Northmead
D. Ku-Ring-Gai	X *D.* Star of Gold	= *D.* Star Child
D. Nunkumbil	X *D. adae*	= *D.* Whitehart
D. Peewee	X *D.* Ellen	= *D.* Lorikeet
D. Peewee	X *D. speciosum*	= *D.* Elegant Heart
D. Star of Gold	X *D.* Aussie Mist	= *D.* Aussie Luck
D. Star of Gold	X *D.* Bardo Rose	= *D.* Aquarius
D. Star of Gold	X *D. x delicatum*	= *D.* Star One
D. Star of Gold	X *D. falcorostrum*	= *D.* Star Falcon
D. Star of Gold	X *D.* Hilda Poxon	= *D.* Tarean
D. Star of Gold	X *D. jonesii* (as *D. ruppianum*)	= *D.* Star Delight
D. Star of Gold	X *D.* Judy Leroy	= *D.* Andrew Upton
D. Star of Gold	X *D. kingianum*	= *D.* Star of Dawn
D. Star of Gold	X *D.* Lowana Nioka	= *D.* Natalie Wall
D. Star of Gold	X *D.* Peter	= *D.* Aussie Aurora
D. Star of Gold	X *D. speciosum*	= *D.* Star of Riverdene
D. Star of Gold	X *D.* Sunglow	= *D.* Carawah
D. Willowie Gold	X *D. speciosum*	= *D.* Manning

Dendrobium toressae (F. M. Bail.) Dockr.

Orchadian 1:64 (1964)
Fig. 77, Plate 76

BASIONYM

Bulbophyllum toressae F. M. Bail. *Syn. Q/ld. Fl.* Suppl. 3:72 (1890)

SYNONYM

Dockrillia toressae (F. M. Bail.) Brieger, in Schlechter, *Die Orchideen* ed. 3 (11/12): 745 (1981)

DESCRIPTION

Discovered by Archibald Meston and F. M. Bailey in June 1889, at Whelanian Pools in the Bellenden Ker Range. Named after Toressa Meston.

A small plant, which appears almost rhizomeless, made up of numerous branches of wheat-like, green leaves, frequently growing into quite sizeable masses. The comparatively large flower nestles into the base of the leaf and, as A. W. Dockrill (1964) stated 'The leaves have the appearance of being sessile on the rhizome (sympodium) but actually the base of the leaf is a vestigial pseudobulb and the peduncle arises from the apex of this and through the cavity of the leaf and does not arise from the rhizome'.

The cream to dirty-yellowish flower lies on its back in the cavity of the leaf. The most conspicuous part is the underside of the labellum as this segment is the uppermost part of the flower. Mid-lobe of the labellum slightly more yellow than the rest of the flower, and basal half of the labellum has tiny red spots on it. Tips of sepals faintly mauve and rear of column foot usually green. Flowers last for 2 to 3 weeks.

CHROMOSOME NUMBER

2n = about 36 (K. Y. Lim, 1985).

FLOWERING PERIOD

Almost any time of the year. Some plants seem to be continuously in flower with a flush in the winter months.

DISTRIBUTION

South-eastern Cape York Peninsula from the Tully River (18° 00′S) to south of Cooktown (15° 40′S).

HABITAT

On trees and rocks from the coast to up to approximately 1000 m. Grows in both rainforests and in open forest and frequently in exposed positions. Although frequently carpets a whole rockface it is difficult to detect as its green leaves, with reddish suffusions, blend into the host rock.

CULTURE

Does well on cork slabs but requires frequent watering in the hotter months. Although it does not require as much water during the winter months, it should not be allowed to dry out completely for any length of time. If grown on tree-fern fibre blocks, it does not need as frequent watering as on cork slabs. Tree fern retains the water longer.

Requires about 50 per cent shade, plenty of air movement and fairly high humidity. Prefers shade-house conditions provided the winter temperature does not go below 6°C.

Fig. 77 *Dendrobium toressae* (F. M. Bail.) Dockr.
From Atherton Tableland

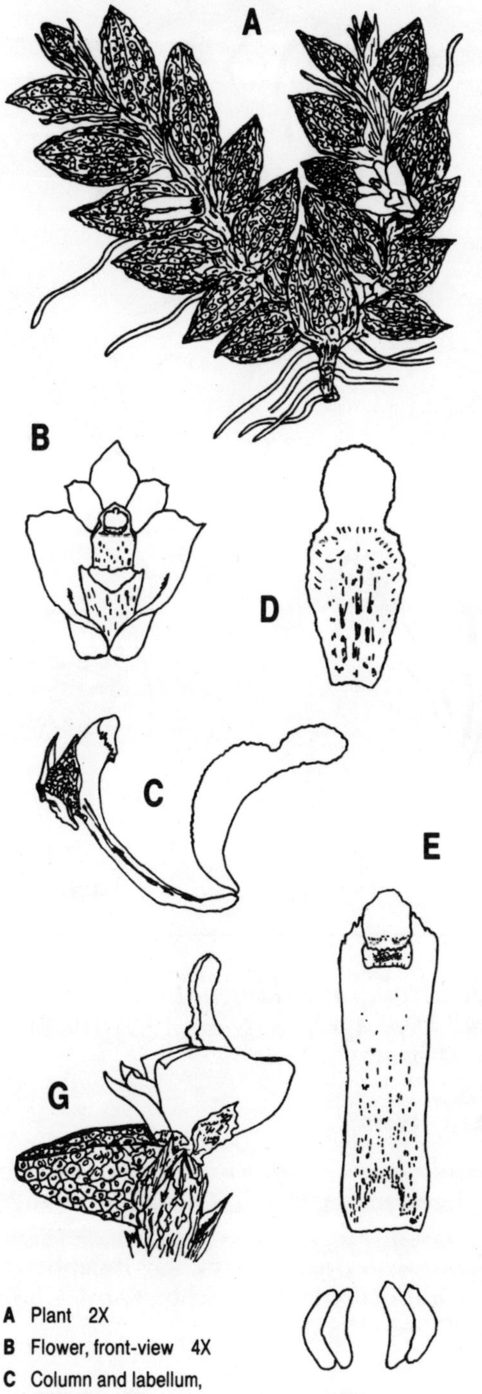

A Plant 2X
B Flower, front-view 4X
C Column and labellum, side-view 6X
D Labellum, pressed flat 6X
E Column, front-view 8X
F Pollinia greatly enlarged
G Flower and leaf from side 4X

JAN. 1967

IN HYBRIDIZATION

No hybrids registered to date.

Dendrobium tozerensis Lavarack

Austrobaileya 1:64 (1977)
Fig. 78, Plate 82

DESCRIPTION

Originally discovered by Lavarack and Gray at Tozer's Gap, north Queensland in 1975. Named after the area in which it was originally found.

Very slender, wiry stems up to about 60 cm tall, growing into large clumps. Stem bears grass-like leaves; in the mature stem the leaves are only on the top half to one-third of the stem. Young growth leafed throughout its entire length.

Almost pure white flowers, which only last for 1 day, usually borne in pairs on a short peduncle. Apical half of the sepals and petals cream. Labellum mid-lobe, usually folded inwards towards the tip. Dr P. Lavarack (1978) states 'In all the specimens examined (about 15 plants collected from localities as much as 5 or 6 km apart) three anthers were present; the two lateral anthers being in all regards similar to the medium anther'.

In the flowers in Fig. 78 there are two vestigial structures on the column that could be the remnants of the additional two anthers. In all the flowers I inspected (20), on two separate plants which were stated to come from Tozer's Gap, the clinandrium was absent, although the rostellum was quite developed. Six of the flowers were pollinated and developed mature capsules.

FLOWERING PERIOD

Several flowerings throughout October to January (spring to summer), but can flower spasmodically throughout the year.

DISTRIBUTION

McIlwraith (13° 50′S), Tozer and Janet Ranges (12° 40′S), Cape York Peninsula.

A Plant 1/3X
B Flower, front-view 1X
C Flower, side-view 1X
D Flower buds 2X
E Column and labellum, side-view 4X
F Labellum, pressed flat 4X
G Column, front-view 8X
H Column, side-view 6X

Fig. 78 *Dendrobium tozerensis* Lavarack From the Tozer Range, north Queensland

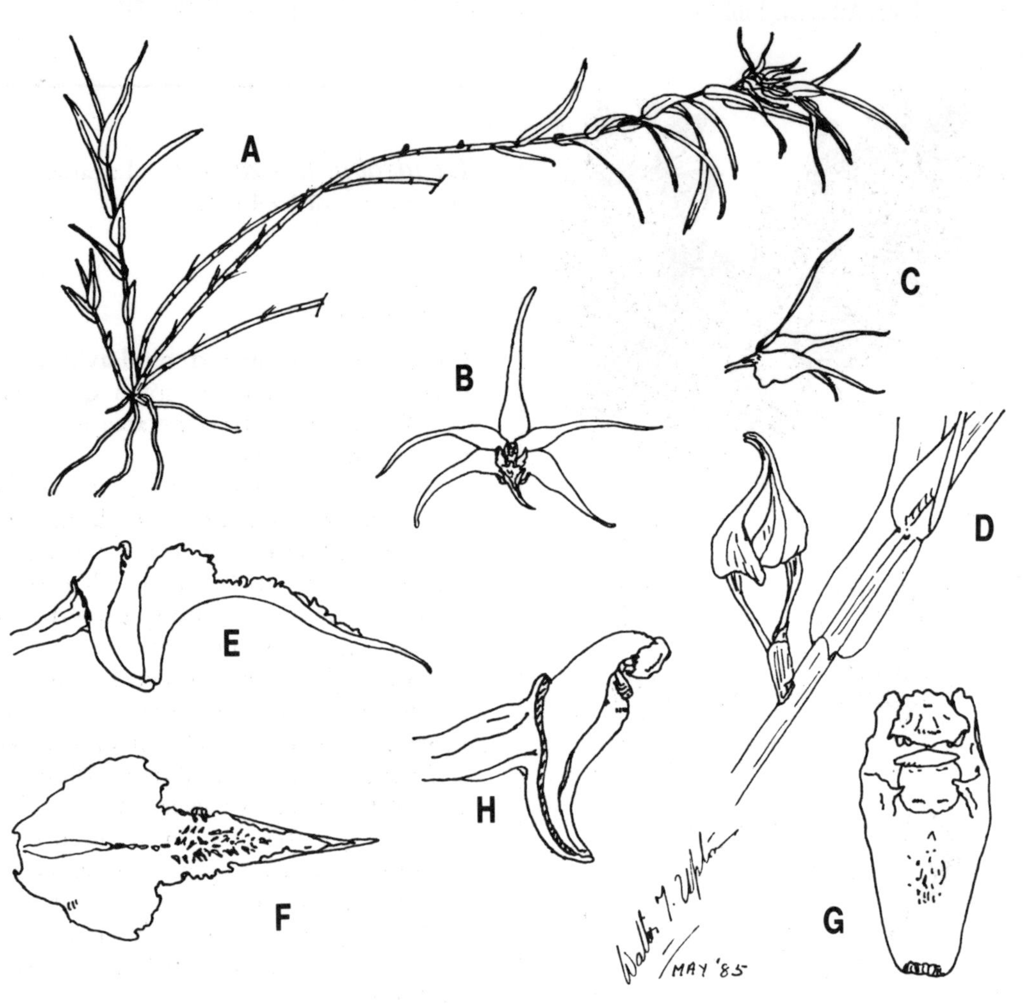

HABITAT

Common on trees on open rocky areas in the rainforests. Occasionally on rocks. Usually in areas of bright light and frequently near waterfalls.

CULTURE

Either pot or slab culture seem satisfactory. Requires high humidity, medium light and frequent watering in the hottest summer months, but never allowed to completely dry out in the winter months. Do not water over the new growths until they are about 5 to 9 cm high. Minimum winter temperature of 12°C.

IN HYBRIDIZATION

No hybrids have been registered to date.

Dendrobium trilamellatum J. J. Sm.

Bull Dep. Agr. Ind. N'eerl. 19:21 (1908)
Fig. 79, Plate 83

SYNONYMS

Dendrobium semifuscum (H. G. Reichb.) Lavarack et Cribb, *Orchadian* 6(9):209 (1980)

Dendrobium johannis H. G. Reichb. var. *semifuscum* H. G. Reichb., *Gard. Chron.* 20:368 (1883)

DESCRIPTION

In recent research Cribb has determined that *D. semifuscum* is a synonym of *D. trilamellatum* from New Guinea (Pers. Comm. M. Clements, 1986). This species in Australia was named by Reichenbach, in 1883,

97 *D.* x *superbiens*

98 *D.* x *superbiens* 'Goldiei'

99 *D.* Andrew Persson

100 *D.* Aussie Glow

101 *D.* Andrew Upton 'Limelight'

102 *D.* Andrew Upton 'Pink Blush'

103 *D.* Bardo Rose 'Berowra' HCC/AOC-NSW

104 *D.* Bardo Rose 'Candy Pink'

105 *D.* Duke

106 *D.* Ellen

107 *D.* Ellen, Schmidt's original cross

108 *D.* Blushing Rose 'Jill'

109 *D.* Craig Hewitt

110 *D.* Big Fleck 'Serenade'

111 *D.* Bright Spark

as *D. johannis* var. *semifuscum* from plants imported from Australia and cultivated by Mr W. Bull of Chelsea, England. Name comes from *semi*, meaning half, and *fuscus*, meaning dark—referring to the partially dark colour of the flowers.

Pseudobulbous stems up to 60 cm long, leaves up to nine, usually about six or seven confined to the top part of the stem. Racemes about 30 cm long, occasionally up to 50 cm carrying up to 15 fragrant flowers. Flowers approximately 4·5 cm in diameter, opening widely. Slightly twisted sepals and petals from yellowish to dark brown, occasionally pure yellow flowers encountered. Labellum yellowish with purplish to red markings on side lobes. Flowers long-lasting and emit quite a pleasant sweet odour. The notch in the anther cap is an aid to identification of this species. Refer to *D. johannis* for the difference between this species and *D. johannis*.

FLOWERING PERIOD

August to October (late winter to spring).

DISTRIBUTION

Cape York Peninsula from about Archer Point (15° 30′S) northward. In the northern part of the Northern Territory including Melville and Bathurst Islands (11° 30′S), and the islands in the Torres Straits and in New Guinea. (A pure yellow colour from Melville Island and a yellow-brown form from Bathurst Island.)

Fig. 79 *Dendrobium trilamellatum* J. J. Sm. Collected Wenlock River area, north Queensland

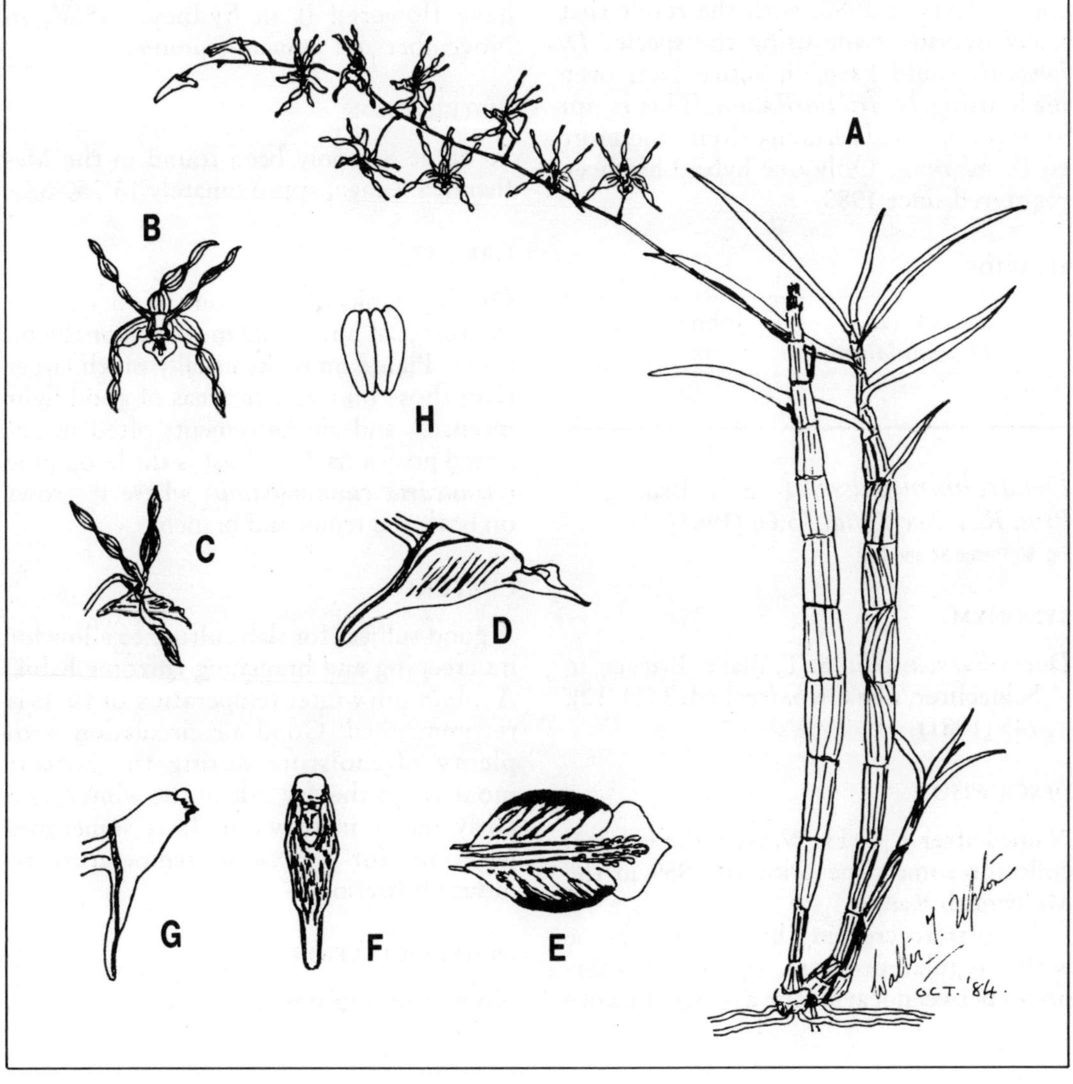

A Plant 1/3X
B Flower, front-view 2/3X
C Flower, side-view 2/3X
D Column and labellum, side-view 2X
E Labellum, pressed flat 2X
F Column, front-view 2X
G Column, side-view 2X
H Pollinia greatly enlarged

HABITAT

In *Melaleuca* (paperbark) woodland where it is usually dry over winter and spring (May to December) and very wet the rest of the year. Very strong light and high temperature even in winter.

CULTURE

Good drainage is essential. Slab culture suits well. If in pots a very open mix is necessary. Requirements are good light, good air movement, keep dryish in cold winter months and plenty of water when in active growth. Minimum winter temperature 15°C.

IN HYBRIDIZATION

This species has been grown and used under the name *D. johannis* until it was raised to species level in 1980, with the result that many hybrids made using the species *D. johannis* could have, in actual fact, been made using *D. trilamellatum*. This is not known with total certainty. Refer therefore to *D. johannis.* Only one hybrid has been registered since 1980.

HYBRIDS

D. trilamellatum (as *D. semifuscum*)	X *D. discolor* (as *D. undulatum*)	= *D.* John's Charm

Dendrobium wassellii S. T. Blake

Proc. Roy. Soc. Q/ld. 73:66 (1963)

Fig. 80, Plates 84 and 222

SYNONYM

Dockrillia wassellii (S. T. Blake) Brieger, in Schlechter, *Die Orchideen*, ed. 3 (11/12): 745 (1981)

DESCRIPTION

Named after J. L. H. Wassell, the original collector some time prior to 1959 in the McIlwraith Range.

A prostrate, creeping, branching rhizome with irregular progress, up to six internodes between leaves, the average distance about 20 mm. Leaves erect, subterete and generally conspicuously, but irregularly, grooved. Leaves mostly a lightish green and up to 120 mm long, generally about 50 to 90 mm.

Racemes up to 20 cm long, averaging about 9 to 12 cm and carrying approximately 40 to 50 flowers. In cultivation about 30 flowers seems to be a good number. Flowers, mostly with labellum uppermost, white to creamy-white with a yellowish labellum that has purple markings and stripes — these mostly on the basal half. Individual flowers last from 10 to 20 days.

FLOWERING PERIOD

The main flowering period seems to be from May to June (early winter), but in culture it flowers at other times as well. I have flowered it in Sydney, NSW, in November and January (summer).

DISTRIBUTION

So far it has only been found in the McIlwraith Range, approximately 13° 50′S.

HABITAT

On the trunks of trees, on the edge of the rainforest at about 300 m. Occasionally on rocks. Plants on rocks usually much larger than those on trees. In areas of good light intensity and air movement; often in exposed positions. One host is the hoop pine (*Araucaria cunninghamii*) where it grows on both the trunk and branches.

CULTURE

A good subject for slab culture to allow for its creeping and branching rhizome habit. A minimum winter temperature of 10°C is recommended. Good air circulation with plenty of moisture during the hottest months, on the dry side in the winter, and fairly moist in between. It is sometimes necessary for a drop in temperature to induce flowering.

IN HYBRIDIZATION

No hybrids registered to date.

Fig. 80 *Dendrobium wassellii* S. T. Blake From the McIlwraith Range, north Queensland

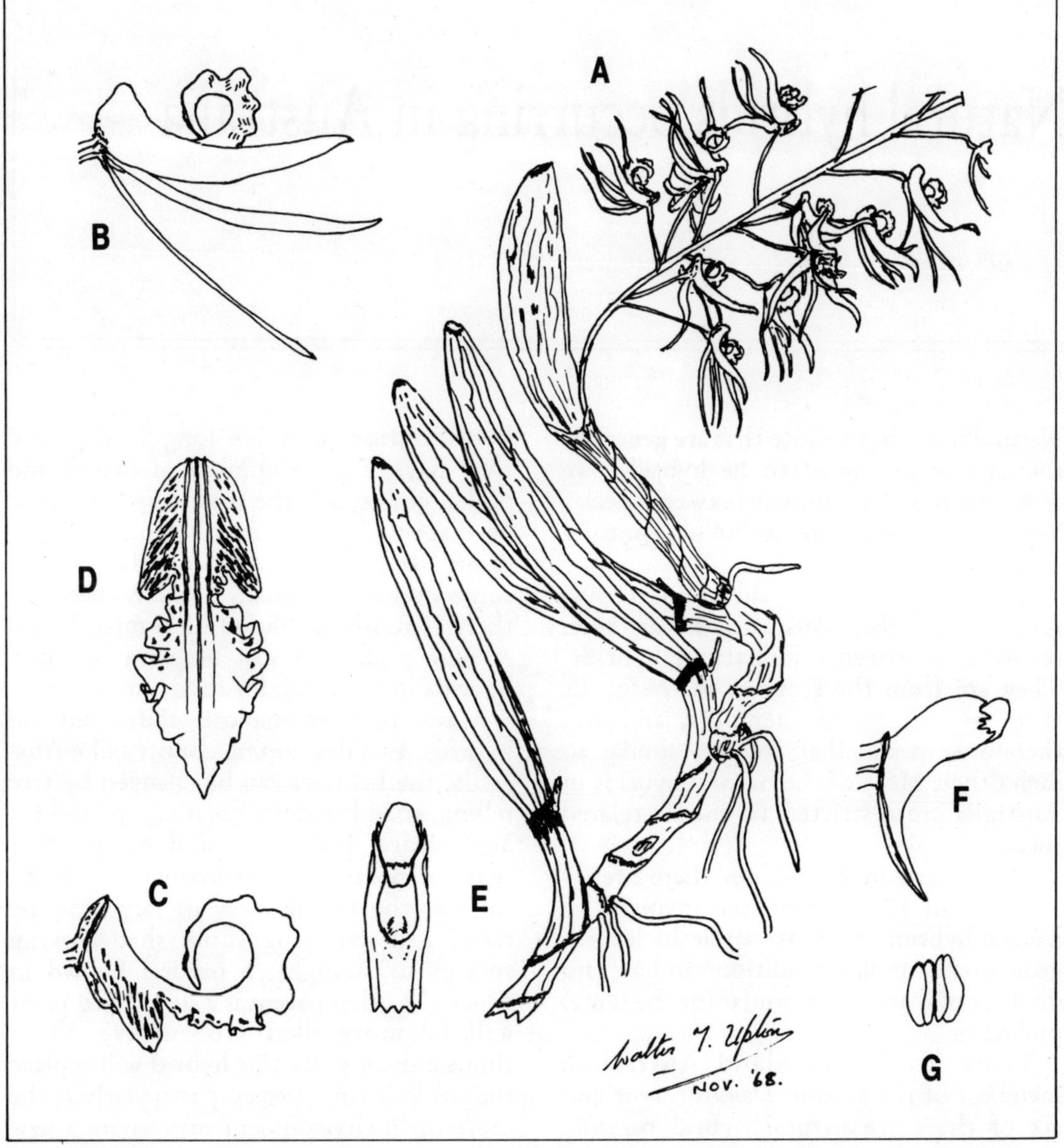

A Plant 1X
B Flower, side-view 3X
C Column and labellum, side-view 4X
D Labellum, pressed flat 4X
E Column, front-view 8X
F Column, side-view 8X
G Pollinia greatly enlarged

4

Natural hybrids occurring in Australia

Natural hybrids, or those that are generally thought or presumed to be hybrids, are common in orchids, mostly between species belonging to the one genus but occasionally between different genera.

Of the 54 species in the genus *Dendrobium* in mainland Australia, 25 have been recorded as parents in natural hybrids. They are from the sections in the genus that have the most members, and are, therefore, species that are very similar to each other. Most of the natural hybrids in Australia are restricted to closely-related species.

In the section *Rhizobium* there are 17 species and 12 of these are involved in natural hybrids. They are all fleshy leaved, grow under similar conditions and are, no doubt, pollinated by a similar (or the same) pollinator.

There are 11 mainland Australian members of the section *Dendrocoryne* and six of these are natural hybrid parents. Once again these species all grow under similar conditions making hybrids possible. Two species, *D. adae* and *D. fleckeri*, have *never* been recorded as natural hybrid parents yet they live together and are quite similar in many respects; although they have entirely different scents and no doubt different pollinators.

Sections *Phalaenanthe* and *Spatulata* have ten species between them and seven of these are natural hybrid parents—the sections crossing freely with each other. The growing conditions of the species in these two sections vary greatly.

Orchids are good subjects for natural hybridization, they are long-lived plants, they have a great number of ovules and pollen grains, and the seed is light and can be dispersed over great distances. In the genus *Dendrobium* in Australia, the chromosome number is usually similar, therefore there is more likelihood of compatibility. A new pollinator can help initiate new hybrids in its search for food, etc.

Many phenomena can cause natural hybrids. In a developing country like Australia, the habitats can be changed by tree felling, road building, opening up land for agriculture, bush fires and many other reasons. Because of disturbance of habitat, such as the taking away of large canopy trees and exposing some shade-loving species to sunlight, a natural hybrid in which the other parent is a sun-loving plant will be more likely to survive. Over thousands of years this hybrid will replace the shade-loving species, particularly as the more open environment may invite a new pollinator.

Up to now, the effect of natural hybridization on the parental species seems very limited, if at all. Perhaps confusion to the taxonimist is the result, particularly with much back crossing as witnessed with such natural hybrids as *D. x superbiens*, *D. x delicatum* and *D. x gracillimum*.

There are many factors that hinder natural hybridization—different flowering seasons, different habitats, different pollinators, or different position in habitat. Those plants that grow on the top canopy of the host tree are less likely to hybridize with those that grow low down on the trunk. Incompatibility between species, thus preventing fertilization, is a major factor against hybridization.

The following is a list of natural hybrids in the genus *Dendrobium* in mainland Australia by Section. Descriptions and figures are given for some of these natural hybrids.

RHIZOBIUM

D. cucumerinum	X *D. bowmanii*
D. cucumerinum	X *D. linguiforme*
D. linguiforme	X *D. calamiforme (D. x grimesii)*
D. linguiforme	X *D. racemosum*
D. pugioniforme	X *D. mortii*
D. pugioniforme	X *D. striolatum*
D. rigidum	X *D. calamiforme (D. x foederatum)*
D. schoeninum (as *D. beckleri*)	X *D. mortii*
D. schoeninum (as *D. beckleri*)	X *D. pugioniforme*
D. schoeninum (as *D. beckleri*)	X *D.* sp. aff. *fairfaxii*
D. striolatum	X *D. teretifolium*

DENDROCORYNE

D. aemulum	X *D. kingianum*
D. falcorostrum	X *D. kingianum*
D. falcorostrum	X *D. macropus* subsp. *gracilicaule*
D. falcorostrum	X *D. speciosum*
D. macropus subsp. *gracilicaule*	X *D. kingianum (D. x suffusum)*
D. speciosum	X *D. jonesii (D. x ruppiosum)*
D. speciosum	X *D. kingianum (D. x delicatum)*
D. speciosum	X *D. macropus* subsp. *gracilicaule (D. x gracillimum)*

PHALAENANTHE/SPATULATA

D. affine	X *D. canaliculatum* (*D.* Judy Leroy)
D. bigibbum	X *D. discolor (D. x superbiens)*

SPATULATA

D. antennatum	X *D. discolor*
D. canaliculatum	X *D. discolor*
D. canaliculatum	X *D. trilamellatum*
D. discolor	X *D. nindii*
D. trilamellatum	X *D. discolor*

Dendrobium aemulum R.Br. *X D. kingianum* Bidw.

Fig. 81

DESCRIPTION

The plant in Fig. 81 was given to me well over 10 years ago by Harold Hirsch who collected it near Gloucester, NSW.

Plant form is more like *D. kingianum*, but pseudobulbous stems do not taper towards the apical portion as much. Stems up to 24 cm long and packed closely together, almost radiating from the centre, as in the growth habit of some *D. aemulum*. Some stems therefore growing upright and some completely pendulous. Three to five leaves.

Flowers about 25 mm across the lateral sepals and a creamy-white with pale-purple suffusions on the rear of the segments. Labellum lightly spotted with purple. Up to 12 flowers.

FLOWERING PERIOD

August to September (late winter to early spring).

DISTRIBUTION

So far it has only been recorded from the Gloucester area.

HABITAT

In open forest on trees.

Fig. 81 *Dendrobium aemulum* R.Br. X *D. kingianum* Bidw.
Collected near Gloucester, NSW by H. Hirsch

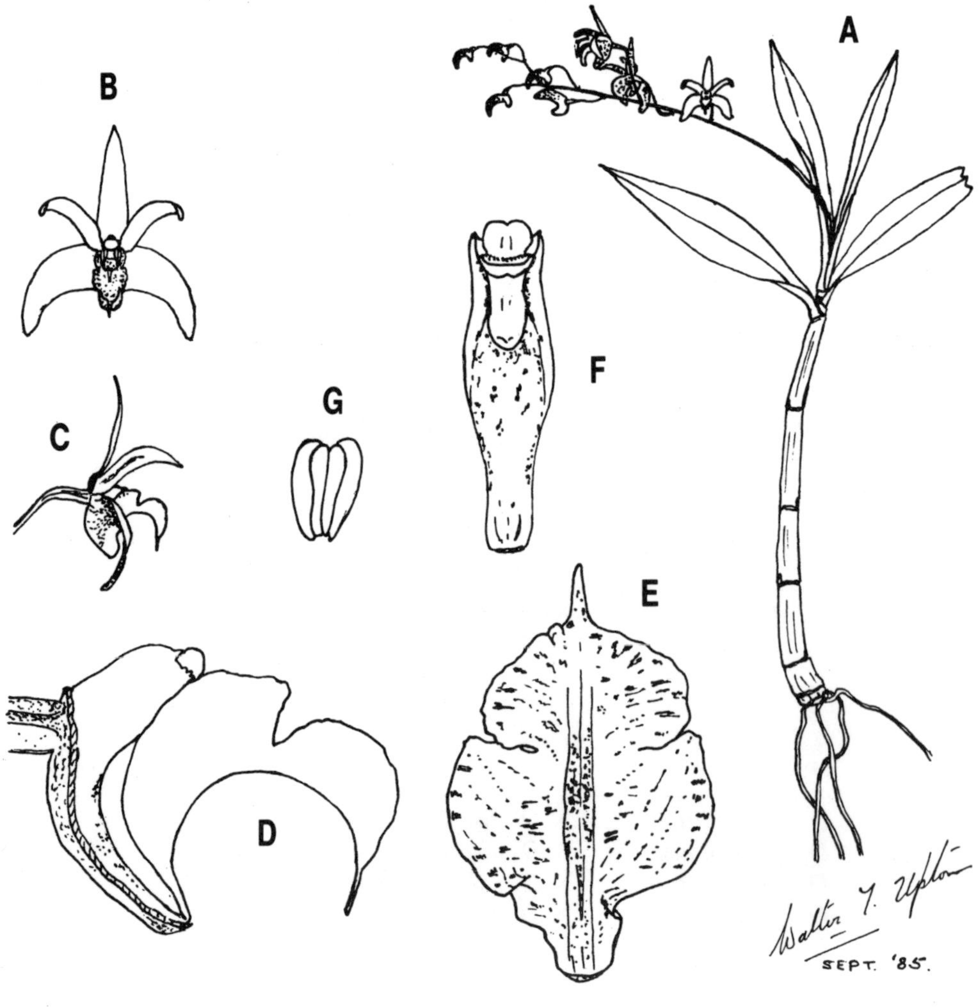

A Plant $^1/_3$X
B Flower, front-view 1X
C Flower, side-view 1X
D Column and labellum, side-view 4X
E Labellum, pressed flat 4X
F Column, front-view 4X
G Pollinia greatly enlarged

CULTURE

Slab culture is necessary because of its habit of growth. Must be allowed to dry out between waterings. In the hottest summer months requires frequent daily mistings. Minimum winter temperature of 2°C.

IN HYBRIDIZATION

One hybrid made using *D. aemulum* and *D. kingianum* was registered by I. Butler in 1966 and called *D.* Emmy.

Dendrobium cucumerinum Macleay ex Lindley *X D. bowmanii* Benth.

Fig. 82, Plate 86

DESCRIPTION

Originally found by Mrs Gladys Walters of Toowoomba and recorded in *Orchadian*, February 1970.

Stems and leaves approximately midway between *D. cucumerinum* and *D. bowmanii*, although habit pendulous like that of *D. bowmanii.* Slightly curved leaves up to 10 cm long, distinctly grooved and rough to the touch.

Flowers pale green to pale yellow streaked with red-purple, on a short raceme of two to six flowers. Labellum white and keels green with mauve markings. Some have faint mauve striations on the labellum.

FLOWERING PERIOD

November to May (late spring to autumn).

DISTRIBUTION

Only recorded from near Towoomba, Queensland, where it was growing together with numerous plants of *D. cucumerinum* and *D. bowmanii.*

HABITAT

In very dense, dryish scrub on trees, generally about 1·5 to 2 m from the ground.

CULTURE

Slab culture suits well. Needs frequent mistings in the hot summer months. Requires high humidity, ample air movement and 60 per cent shade. Minimum winter temperature of 2°C.

IN HYBRIDIZATION

One hybrid made using *D. bowmanii* (as *D. mortii*) and *D. cucumerinum* was registered by W. Skillicorn of Maitland in 1985 and called *D.* Limestone.

A Plant 1/3X
B Portion of plant with raceme 1X
C Flower, side-view 2X
D Column and labellum, side-view 4X
E Labellum, pressed flat 4X
F Column, front-view 4X
G Column, side-view 4X
H Pollinia greatly enlarged

Fig. 82 *Dendrobium cucumerinum* Mackay ex Lindley X *D. bowmanii* Benth. From near Toowoomba

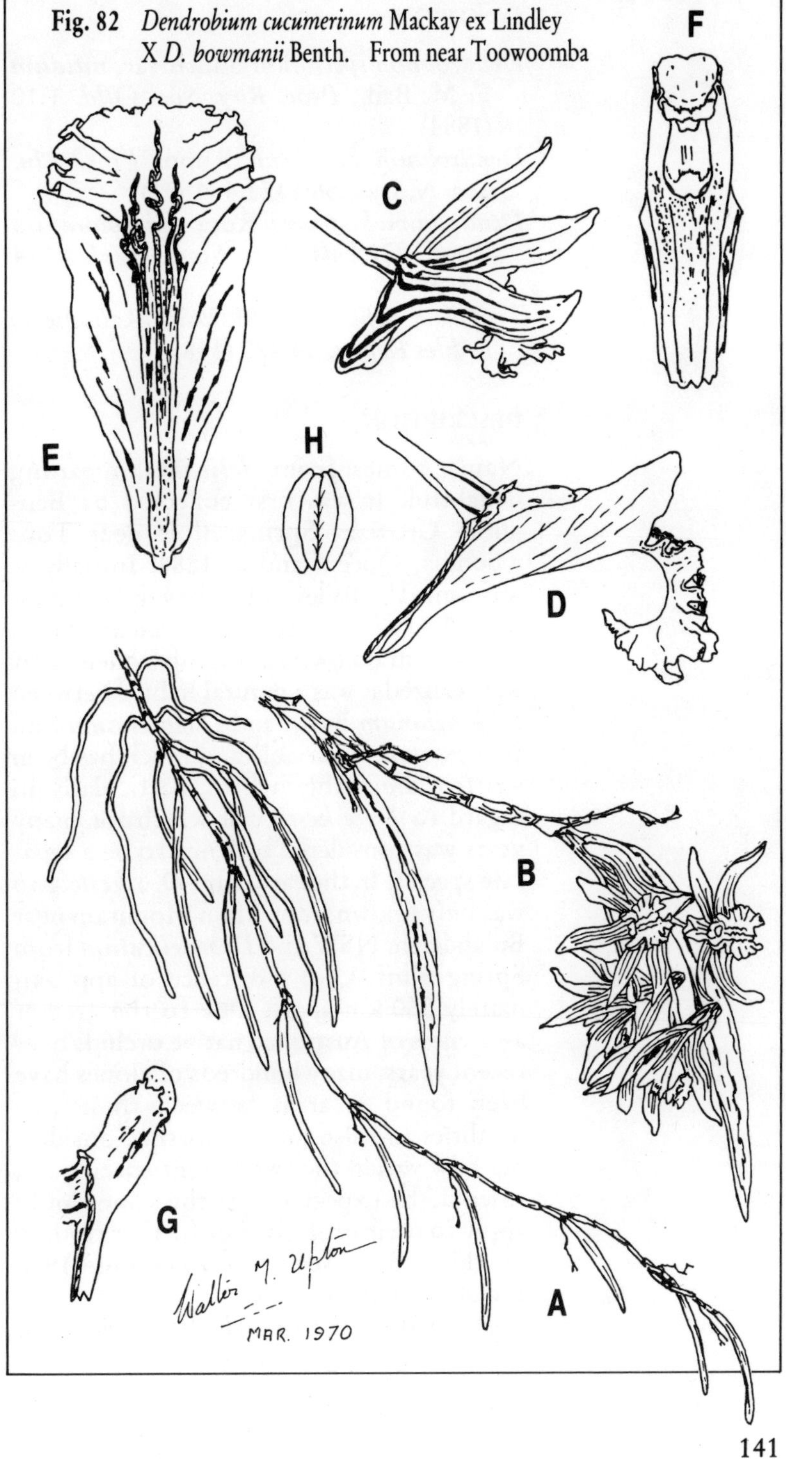

Dendrobium x delicatum (F. M. Bail.) F. M. Bail.

Qld. Fl. 5:1527 (1902)

Fig. 83, Plates 90, 91, 93, 95, 213 and 214

BASIONYM

Dendrobium speciosum Smith var. *delicatum* F. M. Bail., *Proc. Roy. Soc. Qld.* 1:11 (1884)

SYNONYMS

Dendrobium speciosum Smith var. *nitidum* F. M. Bail., *Proc. Roy. Soc. Qld.* 1:10 (1884)

Dendrobium kestevenii Rupp, *Proc. Lin. Soc. N.S.W.* 56:137 f.4 (1931)

Dendrobium kestevenii Rupp var. *coloratum* Rupp, *Orchids New South Wales* 114 (1943)

Tropilis x delicata (F. M. Bail.) Rauschert, *Feddes Repert.* 94:470 (1983)

DESCRIPTION

Name comes from *delicatus*, meaning delightful. It was first collected by Benjamin Crow at Spring Bluff near Toowoomba, Queensland in 1884. Initially it was named by Bailey as a variety of *D. speciosum*, then a few years later he renamed it as a species in its own right. Subsequently, it was realized it was a natural hybrid between *D. kingianum* Bidw. and *D. speciosum* Sm. It is extremely variable and much has been written about this hybrid, particularly in regard to *D. x kestevenii* which for many years was considered by some to be a separate species. In the early days *D. x kestevenii* was only known from Alum Mountain near Bulahdelah, NSW and *D. x delicatum* from Spring Bluff, Qld; a distance of approximately 560 km apart. Due to the greater awareness of Australian native orchids over recent years, many hundreds of clones have been found in areas between these two localities and also further north. Considering how varied the two parent species are, it would be expected that the same would apply to their progeny. For further reading on this subject see *Orchadian*, June 1965 and August 1965.

Original plant from which *D. kestevenii* was described was collected by a Mr Slater in 1930 on Alum Mountain at Bulahdelah, NSW and given to Dr H. Leighton Kesteven, after whom it was named. Photographs of the original plant in Dr Kesteven's garden taken in 1935 by H. J. R. Overall appear in *The Orchadian* Vol. 1, No. 11, April 1965.

The drawing is from a piece of *D. x kestevenii* given to me by Overall. The pseudobulbous stems are generally stouter, shorter and the leaves shorter in the southern than the northern form of this natural hybrid.

Pseudobulbous stems up to 55 cm long, most commonly about 35 cm long. Frequently growing into large clumps. Leaves two to six, commonly three or four.

Racemes arching to erect up to 35 cm long carrying up to 20 flowers. Flowers to 30 mm in diameter, most commonly about 20 mm, from hardly opening at all to widely opening. Slightly fragrant. Sepals and petals white to cream to pink to mauve with varying degrees of pink to mauve spotting, suffusions or tinting. Very rarely a pure yellow form is seen. Labellum generally white to cream spotted or suffused with pink to purple.

FLOWERING PERIOD

August to October (late winter to spring).

DISTRIBUTION

From the Hunter River (32° 30′S) in central-eastern NSW to the Blackall Range in south-eastern Queensland (26° 50′S).

HABITAT

On rocks or cliff faces at altitudes up to 900 m. Generally in open forests and occasionally exposed to the full sun. Sometimes on trees.

CULTURE

Requires abundant air movement and plenty of light. A vigorous grower and pot culture suits well. When potting allow enough room for at least 2 year's growth. Shallow pots best. Does not like 'wet feet'. Minimum winter temperature of 2°C.

IN HYBRIDIZATION

There is nothing really constant in the dominant characteristics of this natural hybrid. It comes from two species that are extremely variable in many respects; and one, *D. kingianum*, is extremely dominant as a parent, while the other is generally not. The resultant progeny are generally nearer to *D. kingianum* shape except where the dominant *D. tetragonum* is used as well, then both shapes come through.

HYBRIDS

D. x delicatum	X *D. adae*	= *D.* Kevin Banks
D. x delicatum	X *D.* Andrew Persson	= *D.* Tom Jupp
D. x delicatum	X *D.* Ellen	= *D.* Delicate Ellen
D. x delicatum	X *D.* Emma	= *D.* Delemma
D. x delicatum	X *D.* Eureka	= *D.* Aussie Love
D. x delicatum	X *D. falcorostrum*	= *D.* Delicate Falcon
D. x delicatum	X *D. x gracillimum*	= *D.* Sun Sprite
D. x delicatum	X *D. kingianum*	= *D.* Gillian Leaney
D. x delicatum	X *D.* Kingrose	= *D.* Debra Wray
D. x delicatum	X *D.* Peter	= *D.* Memoria Eric Gordon
D. x delicatum	X *D. speciosum*	= *D.* Jane Leaney
D. x delicatum	X *D.* Star of Gold	= *D.* Star One
D. x delicatum	X *D. x superbiens*	= *D.* Elsie Cox
D. x delicatum	X *D. tetragonum*	= *D.* Ku-Ring-Gai
D. Gillian Leaney	X *D.* Graham Hewitt	= *D.* Aussie Quest
D. Gillian Leaney	X *D.* Peter	= *D.* Gilpet
D. Gillian Leaney	X *D. tetragonum*	= *D.* Emma
D. Ku-Ring-Gai	X *D.* Hilda Poxon	= *D.* Warrior
D. Ku-Ring-Gai	X *D.* John Upton	= *D.* Ring of Fire
D. Ku-Ring-Gai	X *D. speciosum*	= *D.* Northmead
D. Ku-Ring-Gai	X *D.* Star of Gold	= *D.* Star Child

Fig. 83 *Dendrobium x delicatum* (F. M. Bail.) F. M. Bail.
From Spring Bluff near Toowoomba, Queensland (the plant is drawn from the original plant described by Bailey)

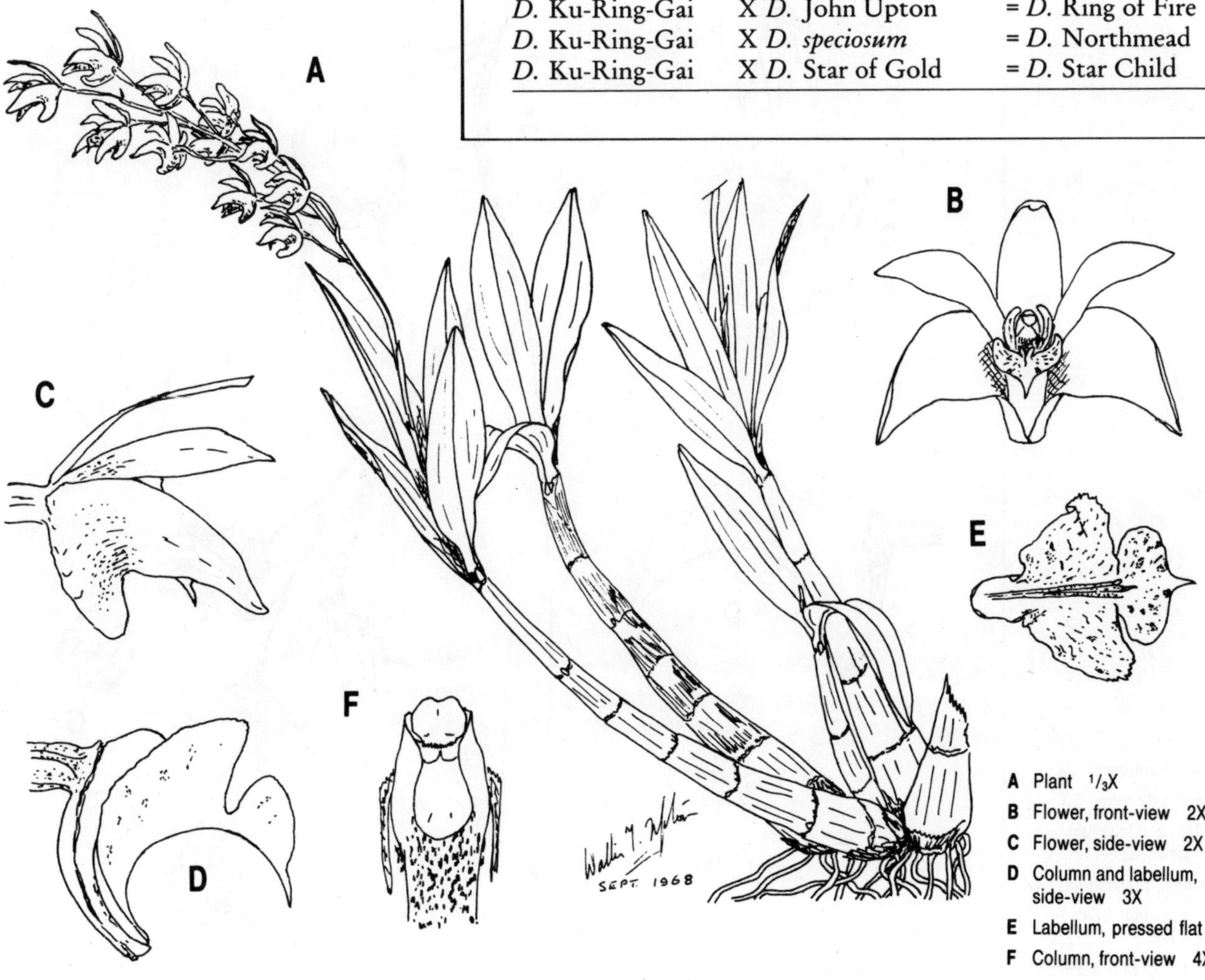

A Plant 1/3X
B Flower, front-view 2X
C Flower, side-view 2X
D Column and labellum, side-view 3X
E Labellum, pressed flat 2X
F Column, front-view 4X

Dendrobium x foederatum St. Cloud

Nth. Q/ld. Nat. 23(111):2 (1955)

Fig. 84, Plate 87

SYNONYM

Dockrillia x foederata (St. Cloud) Rauschert, *Feddes Repert.* 94:446 (1983)

DESCRIPTION

A natural hybrid between *D. calamiforme* Lodd. and *D. rigidum* R.Br. Name comes from *foederatus*, meaning confederated. St Cloud in his original description stated 'The specific name is an allusion to the apparent link between *D. teretifolium* (at that time *D. calamiforme* was known by its synonym *D. teretifolium* var. *fasciculatum*) and *D. rigidum* which is shown in the flowers'.

Type specimen is in the North Queensland Herbarium, Cairns; collected by J. Dyson-Holland at Aeroglen near Cairns in September 1954.

More or less pendulous, rather a small plant. Stems 15 to 20 cm long, sometimes branching, covered with white sheathing

Fig. 84 *Dendrobium x foederatum* St. Cloud From Aeroglen, north Queensland

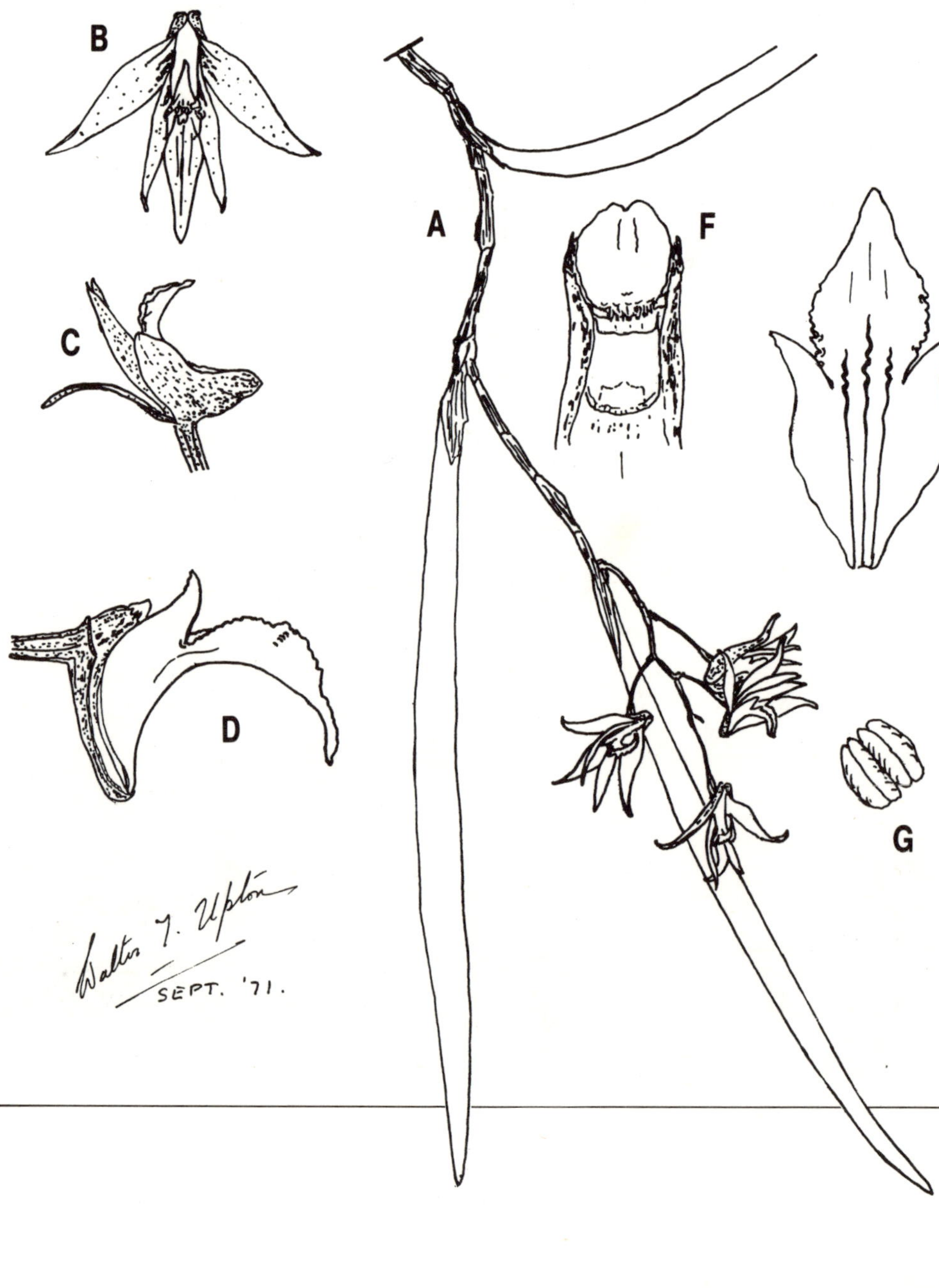

- **A** Plant 1X
- **B** Flower, front-view 2X
- **C** Flower, side-view 2X
- **D** Column and labellum, side-view 4X
- **E** Labellum, pressed flat 4X
- **F** Top of column, front-view 12X
- **G** Pollinia greatly enlarged

bracts. Leaves terete, about 5 to 10 cm long. Green to purplish, depending on amount of sun.

Racemes short. Flowers whitish with pink tinge due to fine purple spotting over the segments; spotting heavier on the rear of the segments. Approximately 15 to 20 mm across lateral sepals. Labellum green at base to white at mid-lobe; sometimes with reddish purple markings on side lobes.

FLOWERING PERIOD

Mainly October to January (spring to summer).

DISTRIBUTION

This species was found in a limited area near the Cairns Aerodrome but unfortunately was destroyed with the expansion of the same, with the exception of two or three specimens collected and cultivated before the expansion. It was found growing with *D. rigidum* and *D. calamiforme.*

HABITAT

Found growing on *Heritiera littoralis* in a mangrove swamp at Aeroglen, Cairns near the mouth of the Barron River.

CULTURE

Slab culture suits best. Plenty of fresh air and humidity. Minimum winter temperature of 8°C.

IN HYBRIDIZATION

No hybrids registered to date.

Dendrobium x gracillimum (Rupp) Rupp

Victorian Nat. 61:200 (1945)

Fig. 85, Plates 92 and 94

BASIONYM

Dendrobium speciosum Smith var. *gracillimum* Rupp, *Proc. Linn. Soc. N.S.W. 54:550 (1929)*

SYNONYMS

Dendrobium speciosum Smith var. *hillii* F. M. Bail. forma *bancroftianum* (H. G. Reichb.) F. M. Bail., *Q/ld. Fl.* 5:1526 (1902)

Dendrobium x gracilosum Clemesha, *Orchadian* 6(11):257 (1981) nom. illeg.

Tropilis x gracillima (Rupp) Rauschert, *Feddes Repert.* 94:470 (1983)

DESCRIPTION

A natural hybrid between *D. speciosum* Smith and *D. macropus* subsp. *gracilicaule* (F. Muell.) P. S. Green. In most cases, the *D. speciosum* parent is *D. speciosum* var. *hillii.* In hybrids made using other varieties the plant form is generally quite different.

In 1925, Mr and Mrs John Tucker of Paterson, NSW, collected several plants from the neighbouring forests and brought them to Rupp's attention. Originally named by Rupp as a variety of *D. speciosum* and subsequently raised to specific rank by him. Name comes from *gracillimum*, meaning slender.

Plant looks like a robust *D. macropus* subsp. *gracilicaule* with fleshier leaves. Pseudobulbous stems up to 80 cm long, commonly about 40 cm.

Racemes up to 25 cm long, peduncle generally shorter than the rhachis. Up to 65 or more flowers (I have counted 84 on one raceme) usually quite densely set on the raceme. Do not generally open widely, cream through cream-yellow to quite a strong yellow. Occasionally a white form seen and very rarely flowers with red-brown spotting on the inside of the perianth segments. Labellum faintly marked with red-brown. Stems produce flowers for many successive years, particularly well-grown specimens, where four or five racemes can be produced from flowering nodes located down the pseudobulbous stem. Large specimens with numerous racemes frequently seen at Australian orchid shows, and are a magnificent sight often winning 'Best Specimen Native Orchid of the Show'.

FLOWERING PERIOD

August to October (late winter to spring).

A Plant 1/5X
B Flower, front-view 2X
C Flower, side-view 2X
D Column and labellum, side-view 2X
E Labellum, pressed flat 4X
F Column, front-view 4X

Fig. 85 *Dendrobium x gracillimum* (Rupp) Rupp Collected near Gloucester in 1963

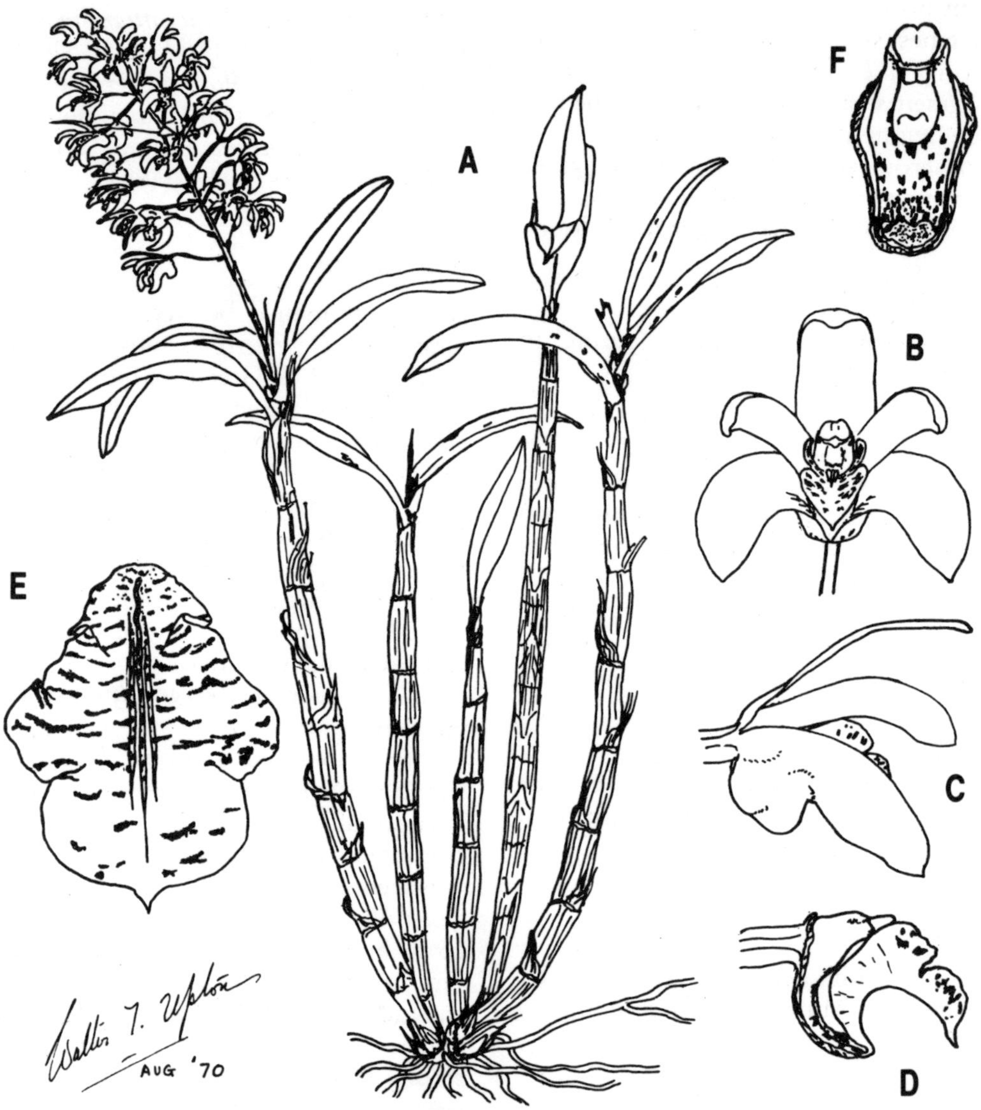

DISTRIBUTION

From about 50 km north of Sydney (33° 20′S) to south-eastern Queensland (approximately 25° 00′S). 'In recent times isolated clones have been found along the ranges—from the N.S.W. border to at least as far north as the Leichhardt Range (19° 20′S)' (Pers. Comm. Lawler, 1986).

HABITAT

On both trees and rocks, mostly on rainforest trees, generally in fairly strong light. I have seen a large specimen growing on a large rock on the top of a hill in a very open forest position where it received lots of sunlight. Always found where both *D. macropus* subsp. *gracilicaule* and *D. speciosum* grow together. Although mostly rare, in some areas it is relatively common. The myrtle tree is one of its favourite hosts; also grows frequently on the stinging tree.

CULTURE

An excellent subject for a specimen plant. Grows vigorously and regularly makes two new growths in the one season. Requires a largish but shallow terracotta pot and a very open mix. When in active growth regular feeding is beneficial. Minimum winter temperature of 2°C.

IN HYBRIDIZATION

No particular dominant characteristics are consistently being passed on to its progeny. *D. macropus* subsp. *gracilicaule*, one of its parents, is a dominant species so this flower shape, particularly the labellum, is frequently seen in progeny. When crossed back to *D. tetragonum*, this shape is fairly dominant.

HYBRIDS

D. x gracillimum	X *D. adae*	= *D.* Double Two
D. x gracillimum	X *D. x delicatum*	= *D.* Sun Sprite
D. x gracillimum	X *D. falcorostrum*	= *D.* Gracious Falcon
D. x gracillimum	X *D. kingianum*	= *D.* Penny Ann
D. x gracillimum	X *D. macropus* subsp. *gracilicaule* (as *D. gracilicaule*)	= *D.* Shan Leaney
D. x gracillimum	X *D. speciosum*	= *D.* Gracious Cascades
D. x gracillimum	X *D. tetragonum*	= *D.* Allyn Star
D. Double Two	X *D. kingianum*	= *D.* Nathan Wall
D. Double Two	X *D. speciosum*	= *D.* Ronald Banks
D. Gracious Falcon	X *D.* Eureka	= *D.* Aussie Delight
D. Gracious Falcon	X *D. speciosum*	= *D.* Karloo
D. Gracious Falcon	X *D.* Sunglow	= *D.* Kalang
D. Penny Ann	X *D.* Bardo Rose	= *D.* Penny Rose
D. Penny Ann	X *D.* Ellen	= *D.* Ferg Martin
D. Penny Ann	X *D. falcorostrum*	= *D.* Twinkle
D. Penny Ann	X *D.* Goblin	= *D.* Minuet
D. Penny Ann	X *D. tetragonum*	= *D.* Fairy Floss

Dendrobium x grimesii C. T. White et Summerh.

Bull. Misc. Inform. Kew 3:106 (1934)
Fig. 86

SYNONYM

Dockrillia x grimesii (C. T. White et Summerh.) Rauschert, *Feddes Repert.* 94:446 (1983)

DESCRIPTION

A natural hybrid between *D. calamiforme* Lodd. and *D. linguiforme* Sw. var. *nugentii* F. M. Bail. Found by Ken Wadsworth on the Atherton Tableland, Queensland. Named after B. D. Grimes.

Stems mostly semi-erect to pendulous, plant generally about 30 cm long. Leaves up to 15 cm long. Racemes to 7 cm long carrying up to 12 flowers; generally about seven flowers. Flowers from barely opening to widely opening; white to cream with pale purple to red-purple markings near the base of sepals and petals. Labellum whitish with mauve markings on the side lobes; sometimes extending on to mid-lobe. Three keels (sometimes as many as five) cream to greenish.

FLOWERING PERIOD

May to October (winter to spring).

DISTRIBUTION

Tablelands to south-west of Cairns, Queensland (between 17°S and 18°S), but should be found in the North Queensland Highlands when both parents are found growing together.

HABITAT

On various trees in exposed positions or in or near rainforests. Although it has not been reported growing on rocks, both its parents are found on rocks as well as trees. Up to 1000 m.

CULTURE

Slab culture under 50 per cent shade. Needs frequent mistings in the hot summer months, and ample air movement. Minimum winter temperature 2°C.

IN HYBRIDIZATION

No hybrids registered to date.

A Plant 1X
B Column and labellum, side-view 3X
C Labellum, pressed flat 4X
D Column, front-view 8X
E Pollinia greatly enlarged

Fig. 86 *Dendrobium x grimesii* C. T. White et Summerh. From Atherton Tableland, Queensland

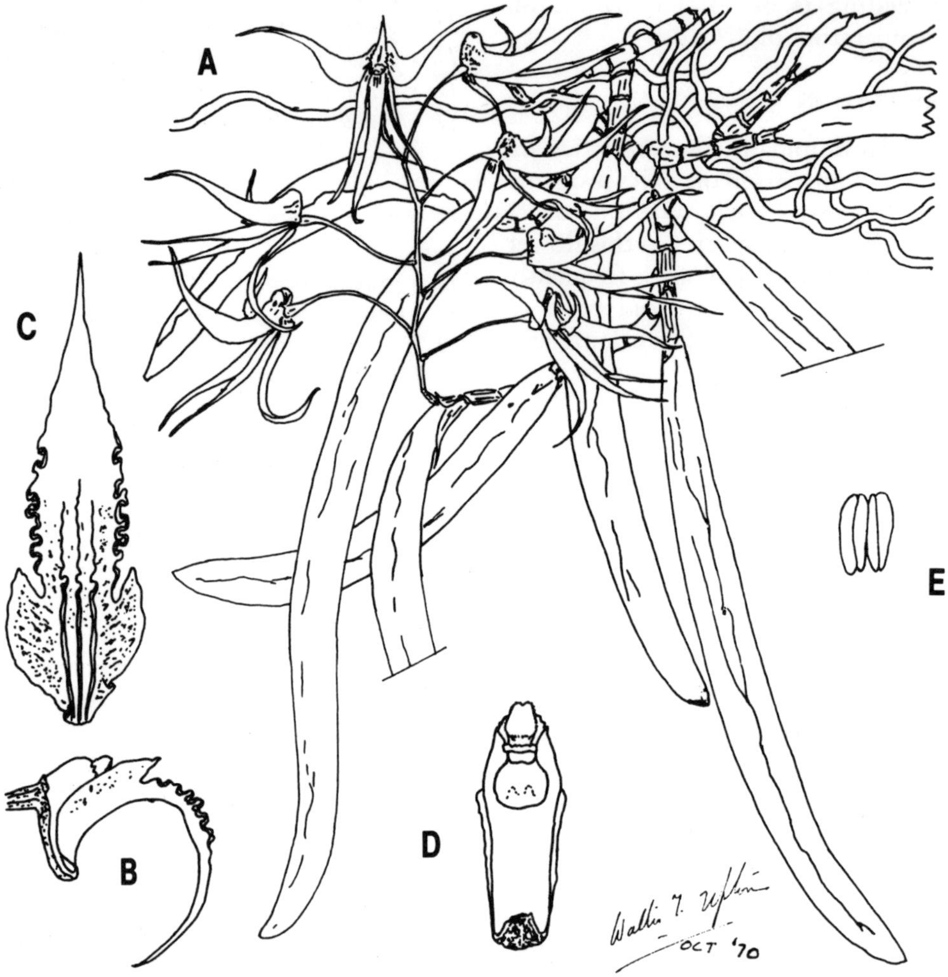

Dendrobium pugioniforme A. Cunn. *X D. striolatum* H. G. Reichb.

Fig. 87

DESCRIPTION

This natural hybrid has been reported on a number of occasions from near Nowra, NSW in the south, to near Newcastle, NSW in the north; its earliest recorded collection being by S. Penman in September 1978 from the Shoalhaven River area (*Orchadian*, December 1983).

Stems and leaves approximately midway between the parents *D. pugioniforme* and *D. striolatum* and the habit favours both parents depending on where it is growing. Some leaves very curved, others quite straight. Leaves of the plants seen up to 140 mm long and up to 4 mm wide, averaging about 80 to 100 mm long and 3 mm wide. In the full sun the leaves turn reddish-purple. Flowers approximately 25 mm across the lateral sepals, and green to yellow with purple striations; the striations being much darker on the back of the segments. Labellum white with two dark purple marks on the mid-lobe, the three keels light green. So far I have only seen small to medium-size plants of this natural hybrid and flowers always in ones, but no doubt large plants will produce some flowers in pairs. I did not detect any perfume.

FLOWERING PERIOD

September (early spring).

DISTRIBUTION

Wherever the two species are found growing together from approximately Nowra, NSW (35° 00′S) to the Hunter River, NSW (32° 40′S).

HABITAT

In trees and rocks in fairly open forest to quite dense rainforest in the coastal ranges. Plant in Fig. 87 came from the Watagan Mountains, near Wyong, NSW.

CULTURE

Slab culture suits well. Needs to be hung up in 50 per cent shade and frequently watered. Minimum winter temperature of 2°C.

IN HYBRIDIZATION

A hybrid made using *D. striolatum* and *D. pugioniforme* was registered by W. Skillicorn of Maitland in 1984 and called *D.* Duffy.

Fig. 87 *Dendrobium pugioniforme* A. Cunn. X *D. striolatum* H. G. Reichb.
From Watagan Mountain, NSW

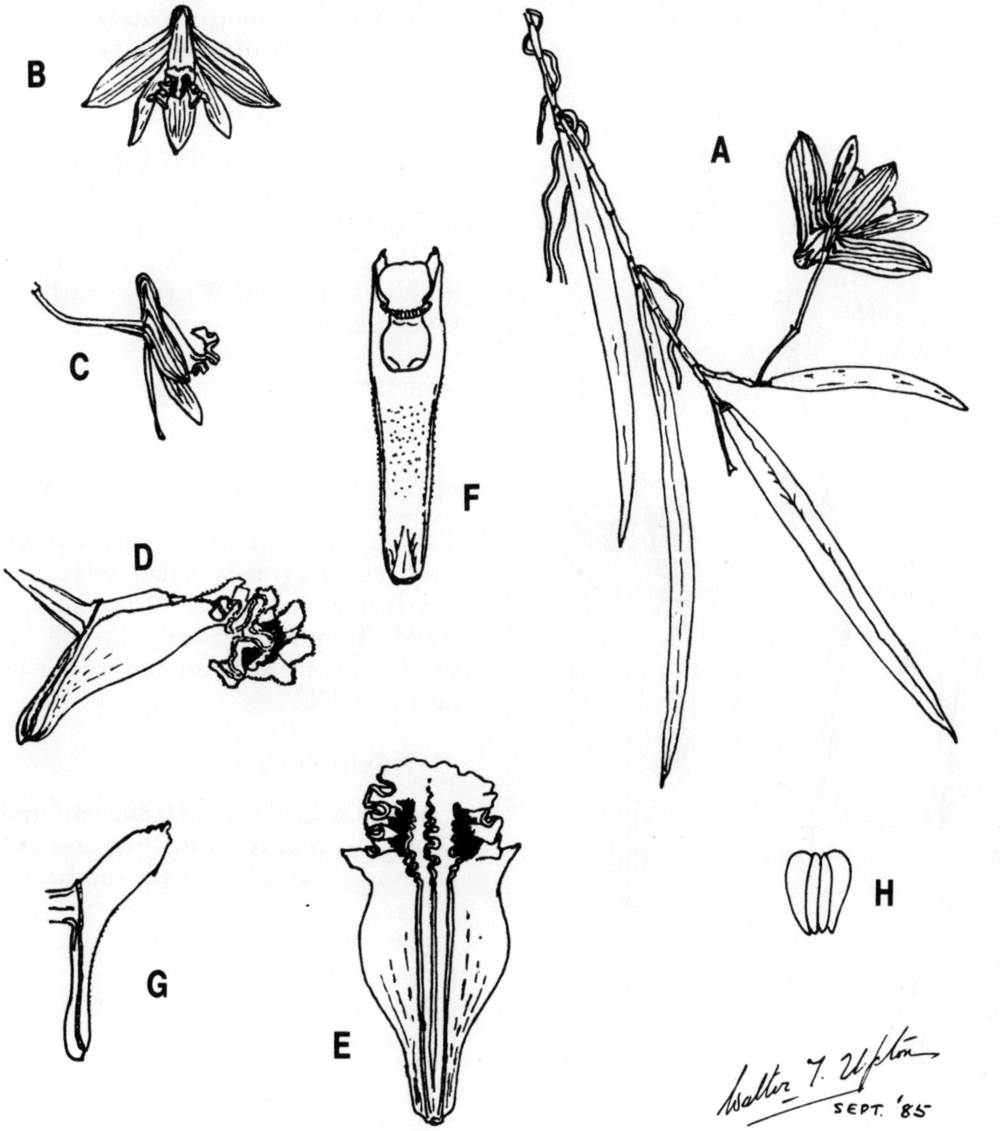

A Plant 1X
B Flower, front-view 1X
C Flower, side-view 1X
D Column and labellum, side-view 3X
E Labellum, pressed flat 3X
F Column, front-view 4X
G Column, side-view 3X
H Pollinia greatly enlarged

Fig. 88 *Dendrobium pugioniforme* A. Cunn. X *D. mortii* F. Muell. From Bellingen River district of north-eastern NSW

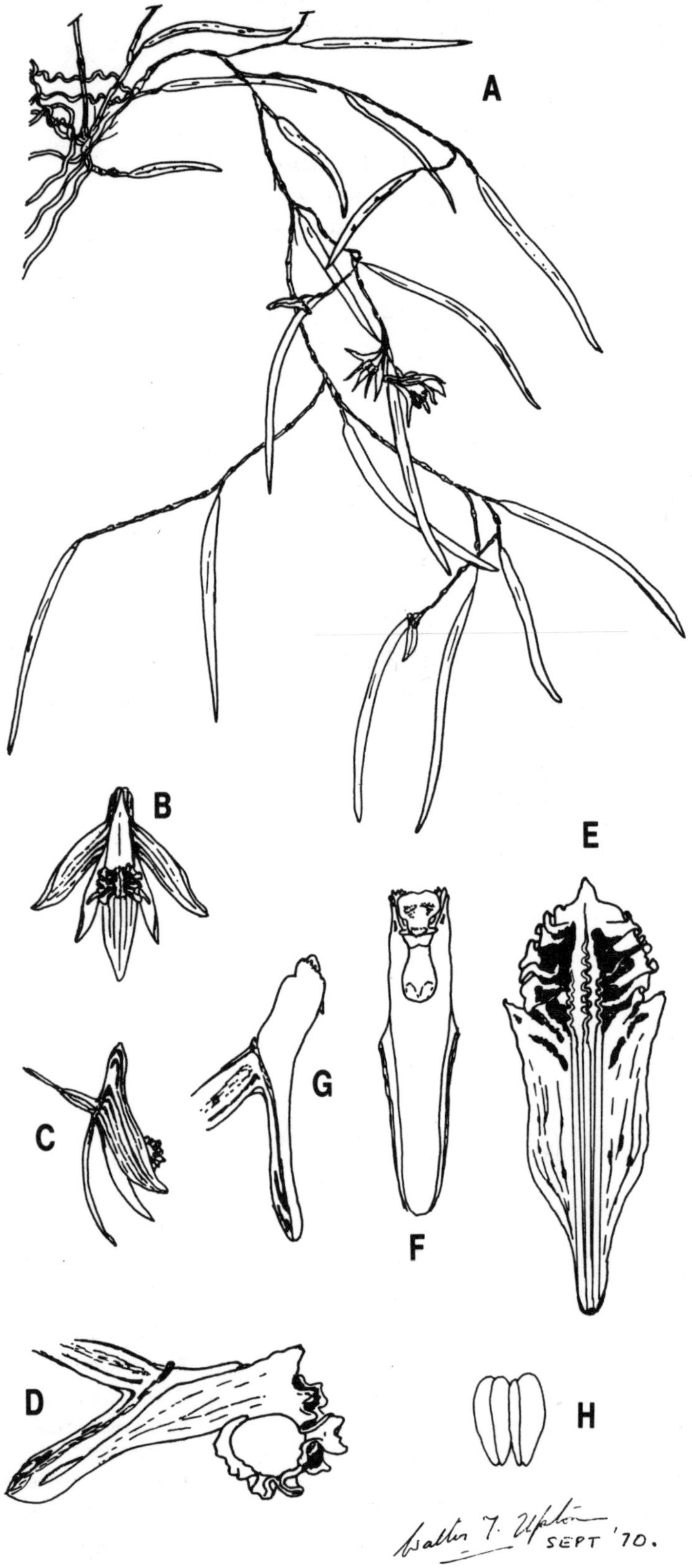

Dendrobium pugioniforme A. Cunn. *X D. mortii* F. Muell.

Fig. 88, Plate 88

DESCRIPTION

This natural hybrid was originally found by L. W. Archer in 1962 in the Bellingen River district of north-eastern NSW. A few recordings have been made since, two from south-eastern Queensland. Fig. 88 is taken from a piece of the original plant, given to M. Corrigan by Archer.

It has the appearance and habit of a very robust *D. mortii.* Leaves intermediate between parents in shape and up to 14 cm long by 5 mm wide; average leaf about 7 cm long. Flowers approximately 3 cm wide and also intermediate in shape and colour.

FLOWERING PERIOD

September to November (spring).

DISTRIBUTION

At the present time it is only known from north-eastern NSW and south-eastern Queensland.

HABITAT

On trees in the ranges.

CULTURE

Slab culture suits well. I have a plant that has been on a tree-fern slab for years and it is growing well under 70 per cent shade. Needs frequent mistings in the hottest summer months. Minimum winter temperature of 2°C.

IN HYBRIDIZATION

I made a hybrid using *D. pugioniforme* and *D. mortii*, registered in 1984 and called it *D.* Dainty Cascades. I made another hybrid,

A Plant 1/3X
B Flower, front-view 1X
C Flower, side-view 1X
D Column and labellum, side-view 3X
E Labellum, pressed flat 3X
F Column, front-view 4X
G Column, side-view 3X
H Pollinia greatly enlarged

112 *D.* Colin 'Pink Blush'

113 *D.* Coljohn 'Supreme'

114 *D.* Colreka

115 *D.* Delemma

116 *D.* Brinawa

117 *D.* Elegant Heart 'Vibrant'

118 *D.* Double Two

119 *D.* Ella Rose

120 *D.* Ella Victoria Leaney 'Red Hot'

121 *D.* Eleebana

122 *D.* Emma 'Newport'

123 *D.* Emmy

124 *D.* Eureka

125 *D.* Gillian Leaney

registered as *D.* Elfin, using *D.* Dainty Cascades and *D. kingianum.*

Dendrobium x ruppiosum Clemesha

Orchadian 7:91 (1982)
Fig. 89, Plate 89

DESCRIPTION

A natural hybrid between *D. jonesii* Rendle and *D. speciosum* Sm. The type specimen is in the Queensland Herbarium and is from a plant collected by S. C. Clemesha in the mountains near Tinaroo Dam, north Queensland in January 1964. It was growing with plants of *D. speciosum* var. *curvicaule* and *D. jonesii.* Fig. 89 is of a plant collected by C. Taylor from the Barron Falls area near Cairns, north Queensland in the early 1960s.

Plant approximately intermediate between *D. jonesii* and *D. speciosum*. Pseudobulbous stems up to 40 cm long and 3 cm broad in the centre portion, narrowing towards the base and apex but not as much as in *D. jonesii.* Two to five leaves, generally four, carried at the top of the pseudobulbous stems.

Racemes up to 55 cm long, carrying about 50 flowers; I have counted as many as 70 flowers. Flowers white with some purple markings on the labellum.

This natural hybrid is uncommon in its natural habitat, occurring mostly as isolated plants. There appears to be no evidence that its two parents, although they are in many respects similar, are merging.

FLOWERING PERIOD

September to November (spring).

CULTURE

Very easy to grow and flower in open shade-house conditions with a minimum winter temperature of 5°C. Needs regular feeding during its very vigorous growth period.

IN HYBRIDIZATION

No hybrids using this natural hybrid have been registered to date. I have used it as a parent, but it has not yet flowered.

Fig. 89 *Dendrobium x ruppiosum* Clemesha
Collected at Baron Falls, north Queensland

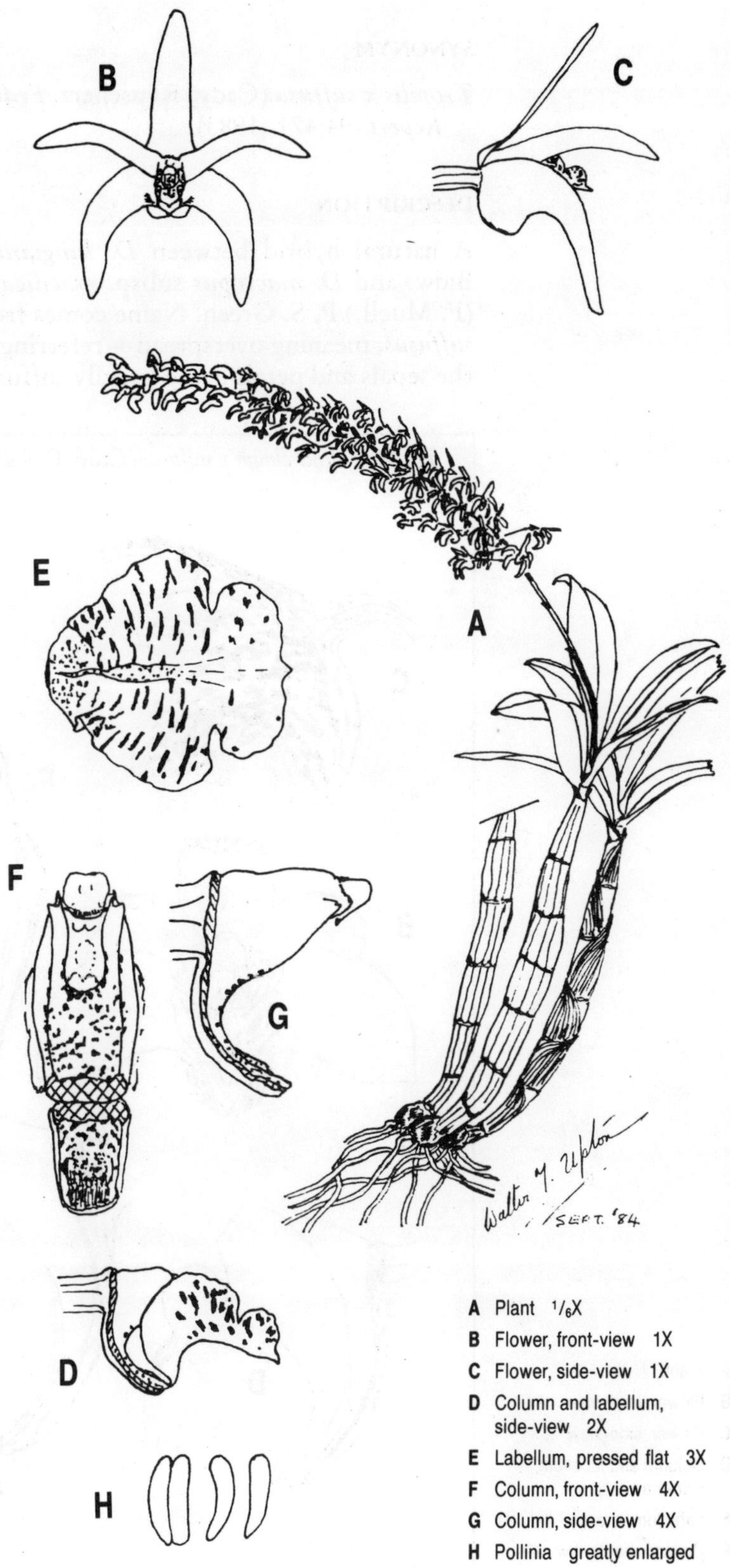

A Plant $^1/_6$X
B Flower, front-view 1X
C Flower, side-view 1X
D Column and labellum, side-view 2X
E Labellum, pressed flat 3X
F Column, front-view 4X
G Column, side-view 4X
H Pollinia greatly enlarged

Dendrobium x suffusum Cady

Orchadian 1:80 (1964)
Fig. 90, Plate 96

SYNONYM

Tropilis x suffusa (Cady) Rauschert, *Feddes Repert.* 94:471 (1983)

DESCRIPTION

A natural hybrid between *D. kingianum* Bidw. and *D. macropus* subsp. *gracilicaule* (F. Muell.) P. S. Green. Name comes from *suffusus*, meaning overspread—referring to the sepals and petals being heavily suffused with deep mauve-pink in the type flowers. The type material is at the NSW National Herbarium, Sydney, and was from a large plant collected by Alan Willows on 9 September 1961 on North Brother Mountain, Laurieton, NSW.

Pseudobulbous stems to 40 cm tall. Most frequently about 20 to 25 cm tall, because most forms that have so far been found are in areas where the shorter-stemmed *D. kingianum* is found. Some of the artificial hybrids made with *D. suffusum* have stems up to 60 cm tall. Up to seven leaves at the top of the stem, most commonly three to five. Racemes erect to semi-pendulous and about 12 cm long, bearing from four to 12

Fig. 90 *Dendrobium x suffusum* Cady Collected at North Brother Mountain, Laurieton, NSW

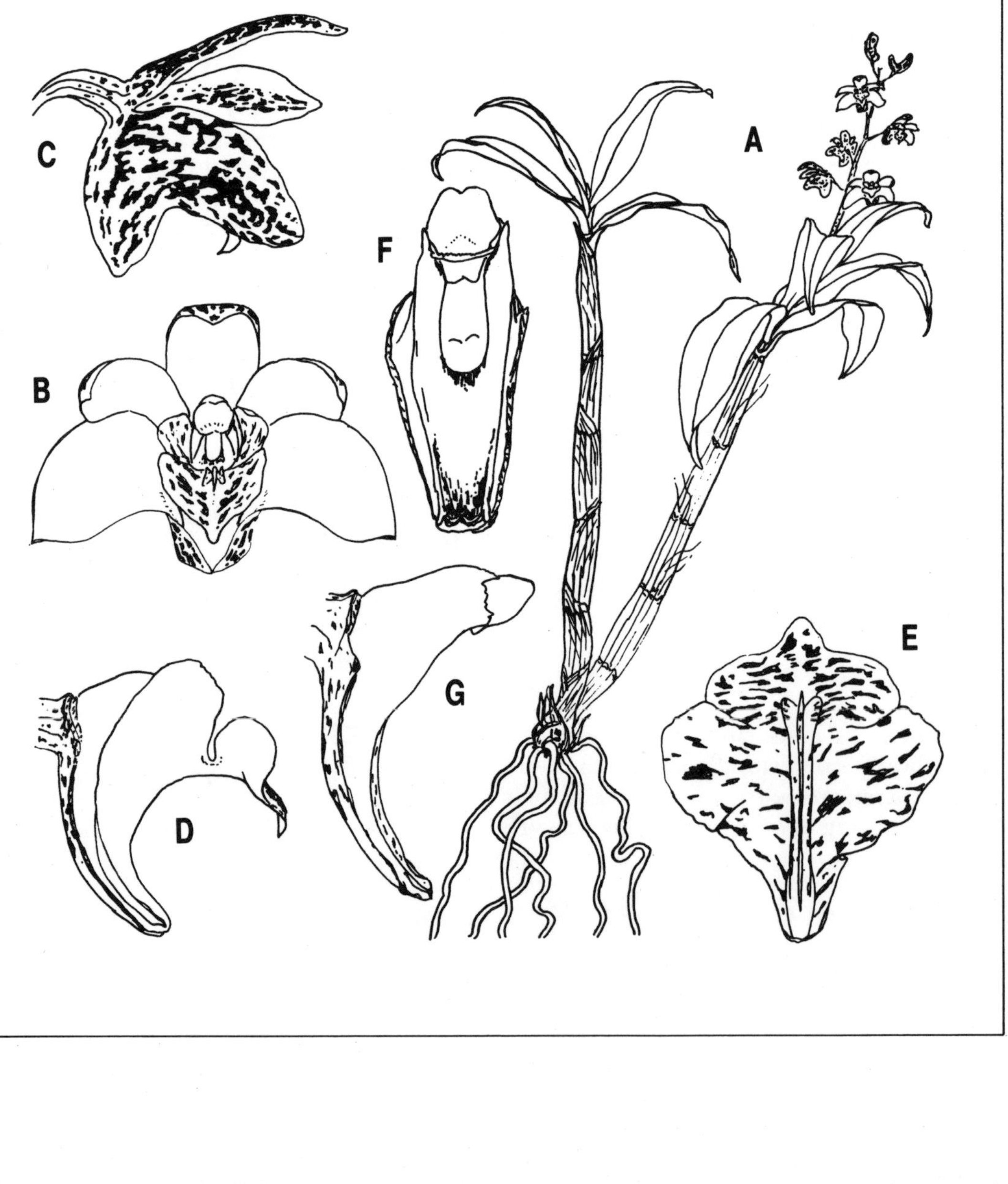

A Plant $^{1}/_{3}$X
B Flower, front-view 3X
C Flower, side-view 3X
D Column and labellum, side-view 4X
E Labellum, pressed flat 4X
F Column, front-view 6X
G Column, side-view 6X

flowers, occasionally more. Flower colour is very variable, from white on the inside of the segments and very heavily marked with a dark purple on the outside, to a yellowy-creamish flower slightly suffused with very pale mauve on the outside.

FLOWERING PERIOD

August to October (late winter to spring).

DISTRIBUTION

Each year more plants of this natural hybrid are being found, thus extending the range. It is quite safe to say it will be found where both parent species are growing in fairly large numbers together and a suitable pollinator is available, say from about Dungog in NSW (32° 30′S) to south-eastern Queensland (26° 00′S).

HABITAT

Many clones have been and continue to be found between Taree, NSW (32° 00′S) and Coffs Harbour, NSW (say 30° 00′S) growing on rocks in open-forest type areas at moderately low altitudes — up to 900 m. With the increasing awareness of Australian native orchids many more are being found each year in areas north of Coffs Harbour, such as Kyogle, Casino and Towoomba.

CULTURE

Pot culture suits well, but also grows well on slabs. Requires plenty of fresh air, a good open medium, fairly high light, a regular feeding programme and ample water when in active growth. It does not like 'wet feet'. Minimum winter temperature of 2°C.

IN HYBRIDIZATION

As both of the parents of this natural hybrid are dominant, so is *D. x suffusum.* The main exception is when used with the dominant *D. tetragonum*; then the results are intermediate between the parents.

HYBRIDS

D. x suffusum	X *D.* Bardo Rose	= *D.* Blushing Rose
D. x suffusum	X *D.* Ellen	= *D.* Pixie
D. x suffusum	X *D. falcorostrum*	= *D.* Aussie Freckles
D. x suffusum	X *D. kingianum*	= *D.* Cawongla
D. x suffusum	X *D. speciosum*	= *D.* Harold Hirsch
D. x suffusum	X *D. tetragonum*	= *D.* Blushing Star
D. Aussie Freckles	X *D. falcorostrum*	= *D.* Aussie Majestic
D. Aussie Freckles	X *D. kingianum*	= *D.* Fiona Banks
D. Blushing Rose	X *D.* Ella Victoria Leaney	= *D.* Ella Rose
D. Blushing Rose	X *D. kingianum*	= *D.* Red Baron
D. Blushing Rose	X *D. speciosum*	= *D.* Regal Rose
D. Blushing Star	X *D.* Hilda Poxon	= *D.* Kim Heinze
D. Blushing Star	X *D.* Sunglow	= *D.* Blushing Sun

Dendrobium x superbiens H. G. Reichb.

Gard. Chron. (new series) 6:516 (1876)

Figs 91 and 92, Plates 97, 98, 204 and 208

SYNONYMS

Dendrobium goldiei H. G. Reichb., *Gard. Chron.* (new series) 9:652 (1878)

Dendrobium bigibbum Lindley var. *superbiens* (H. G. Reichb.) F. M. Bail., *Syn. Q/ld. Fl.* 509 (1883)

Dendrobium fitzgeraldii F. Muell. ex F. M. Bail., *Syn. Q/ld. Fl.* suppl. 1:55 (1886)

Dendrobium bigibbum Lindley var. *albo-marginatum* F. M. Bail., *Bot. Bull. Dept. Agric. Q/ld.* 2:18 (March 1891), non Linden (August 1891)

Callista fitzgeraldii (F. Muell.) Kuntze, *Rev. Gen. Pl.* 2:654 (1891)

Callista superbiens (H. G. Reichb.) Kuntze, *Rev. Gen. Pl.* 2:655 (1891)

Dendrobium x vinicolor St. Cloud, *Nth. Q/ld. Nat.* 25(116):4 (1956)

DESCRIPTION

A natural hybrid between *Dendrobium bigibbum* Lindley and *D. discolor* Lindley. Name comes from *superbus*, meaning magnificent or superb — referring to the magnificent flowers.

In Veitch's *A Manual of Orchidaceous Plants* Vol. 1 (1888), it states 'it is a native of York Peninsula and some of the islands in Torres Straits, whence it was introduced by us, in 1876, through the late Sir William Macarthur of Camden Park, near Sydney, New South Wales'.

Pseudobulbous stems up to 120 cm long. Up to 14 leaves on the upper half of the stem. Racemes usually arching and carrying about 10 to 25 flowers, about 4·5 to 5 cm in diameter. Vary in colour from pale pink to a deep mauve-purple, sometimes the segments have white margins. Sepals and petals twisted to varying degrees.

Wooster (1940) wrote beneath a picture of *D. x superbiens*, from Townsville, north Queensland 'The plant illustrated was grown by me on an open lawn in an easterly aspect. It flowered last year from January to the end of August, producing 21 sprays with up to 20 flowers on a spray. It commenced flowering again at the end of September, carrying 12 sprays, finishing on February 20, 1940, and immediately after flowering spikes again were in evidence. Last month (January) the plant made five new pseudobulbs'.

One of the most beautiful forms of this natural hybrid was originally named as *D. goldiei*, it is a rich shade of purple-red with smallish but well-shaped flowers. The plant form is average *D. x superbiens* shape and size. Fig. 92 is of a plant grown under the name of *D. goldiei* which came from Limberlost Nursery, Cairns, many years ago and was given to me by Bert Cardwell.

FLOWERING PERIOD

Mainly February to June (summer, autumn, early winter) and at other times.

Fig. 91 *Dendrobium x superbiens* H. G. Reichb.
From a plant purchased in a nursery in Cairns and said to be collected on Cape York

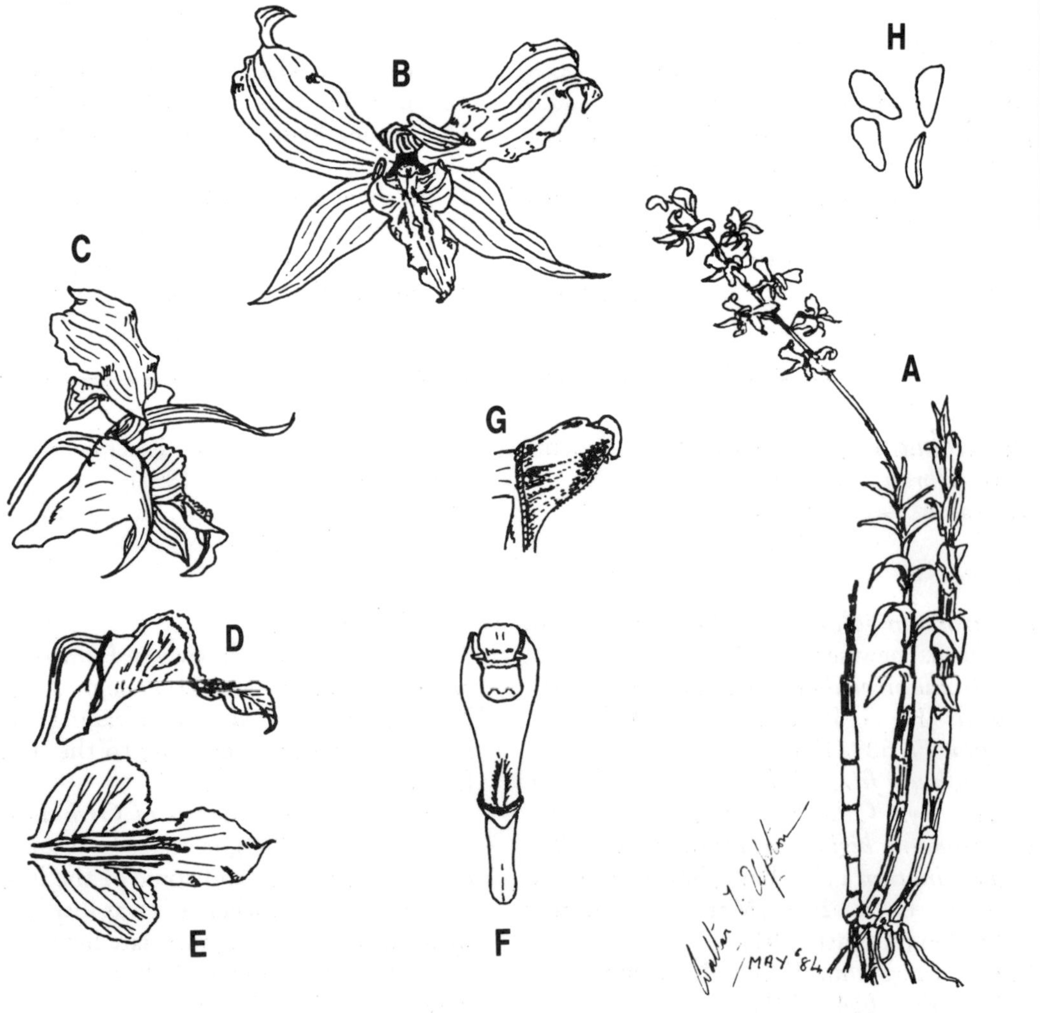

A Plant $^1/_9$X
B Flower, front-view 1X
C Flower, side-view 1X
D Column and labellum, side-view 1X
E Labellum, pressed flat 1X
F Column, front-view 2X
G Top of column, side-view 2X
H Pollinia greatly enlarged

Fig. 92 *Dendrobium x superbiens* as *Dendrobium goldiei* H. G. Reichb. From Limberlost Nursery, Cairns

A Flower, front-view 1X
B Flower, side-view 1X
C Column and labellum, side-view 3X
D Labellum, pressed flat 3X

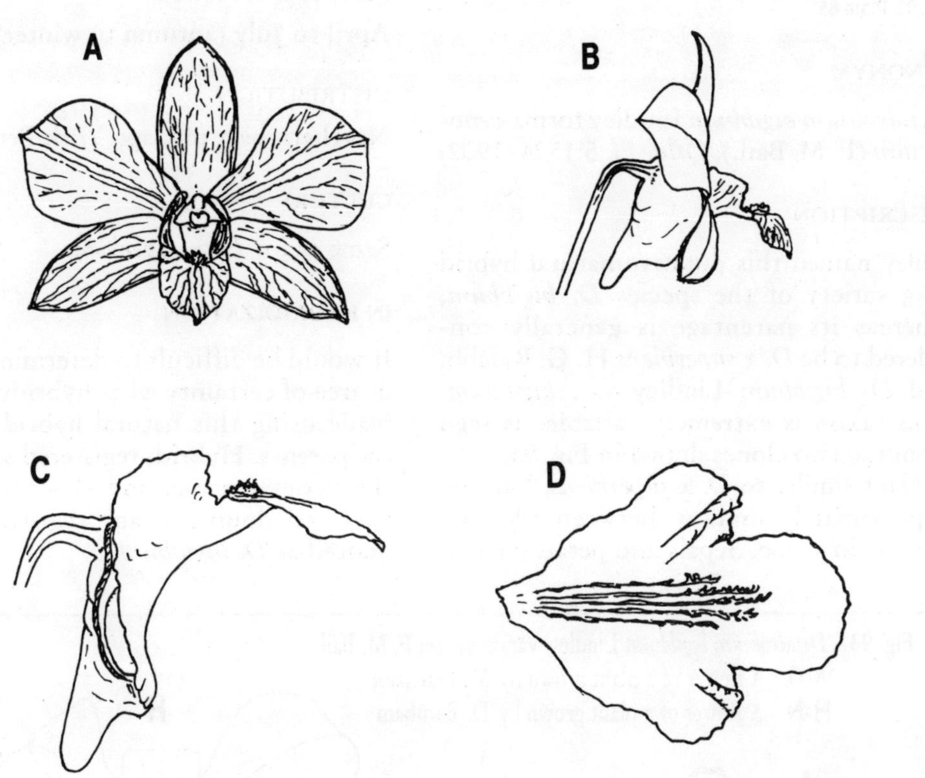

DISTRIBUTION

Northern tip of Cape York Peninsula and extending through the Torres Strait Islands.

HABITAT

On rocks and trees often in almost full sun.

CULTURE

Requires lots of light, ample air movement, plenty of humidity and water in the active growing season and a rest during the winter, when it should be hung up high if grown in a greenhouse and kept dryish. Minimum winter temperature of 12°C.

IN HYBRIDIZATION

The progeny of matings using *D. x superbiens* do not appear to be dominant and are approximately midway between the parents.

HYBRIDS

D. x superbiens	X *D. bigibbum*	= *D.* Saskia
D. x superbiens	X *D. bigibbum*	= *D. bigibbum* var. *venosum*
D. x superbiens	X *D. canaliculatum*	= *D.* Lloyd Stainton
D. x superbiens	X *D. x delicatum*	= *D.* Elsie Cox
D. x superbiens	X *D. discolor* (as *D. undulatum*)	= *D.* Albanense
D. x superbiens (as *D. goldiei*)	X *D. discolor* (as *D. undulatum*)	= *D.* Dang Toi
D. x superbiens	X *D. johannis*	= *D.* D'Bush Chic
D. x superbiens (as *D. goldiei*)	X *D. johannis*	= *D.* Three Star
D. x superbiens	X *D.* Judy Leroy	= *D.* Gold Baby
D. x superbiens (as *D. goldiei*)	X *D. mirbelianum*	= *D.* Moira Stewart
D. x superbiens	X *D. nindii* (as *D. tofftii*)	= *D.* Blue Gloucester
D. x superbiens	X *D.* Peter Petersen	= *D.* Bluey
D. Albanense	X *D. nindii* (as *D. tofftii*)	= *D.* Frances Henton
D. Dang Toi	X *D. canaliculatum*	= *D.* Matilda
D. Dang Toi	X *D.* Champagne	= *D.* Rajah

Dendrobium bigibbum Lindley var. *venosum* F. M. Bail.

Bot. Bull. Dept. Agric. Q/ld. 2:18 (1891)
Fig. 93, Plate 85

SYNONYM

Dendrobium bigibbum Lindley forma *venosum* (F. M. Bail.), *Q/ld. Fl.* 5:1524 (1902)

DESCRIPTION

Bailey named this putative natural hybrid as a variety of the species *D. bigibbum*, whereas its parentage is generally considered to be *D. x superbiens* H. G. Reichb. and *D. bigibbum* Lindley var. *bigibbum*. This taxon is extremely variable, as seen from the two clones drawn in Fig. 93.

Plant similar to *D. x superbiens*. Flowers approximately midway between the two parents in shape. Sepals and petals pink to mauve and frequently outlined in white. Labellum occasionally darker in colour.

FLOWERING PERIOD

April to July (autumn to winter).

DISTRIBUTION

Northern part of Cape York Peninsula.

CULTURE

Same as for *D. x superbiens.*

IN HYBRIDIZATION

It would be difficult to determine with any degree of certainty what hybrids have been made using this natural hybrid as one of the parents. Hybrids registered seldom give the varietal name, and this taxon would have, no doubt, in any registration been quoted as *D. bigibbum*.

Fig. 93 *Dendrobium bigibbum* Lindley var. *venosum* F. M. Bail.
A-G A flower of a plant grown by M. Harrison
H-N A flower of a plant grown by D. Barnham

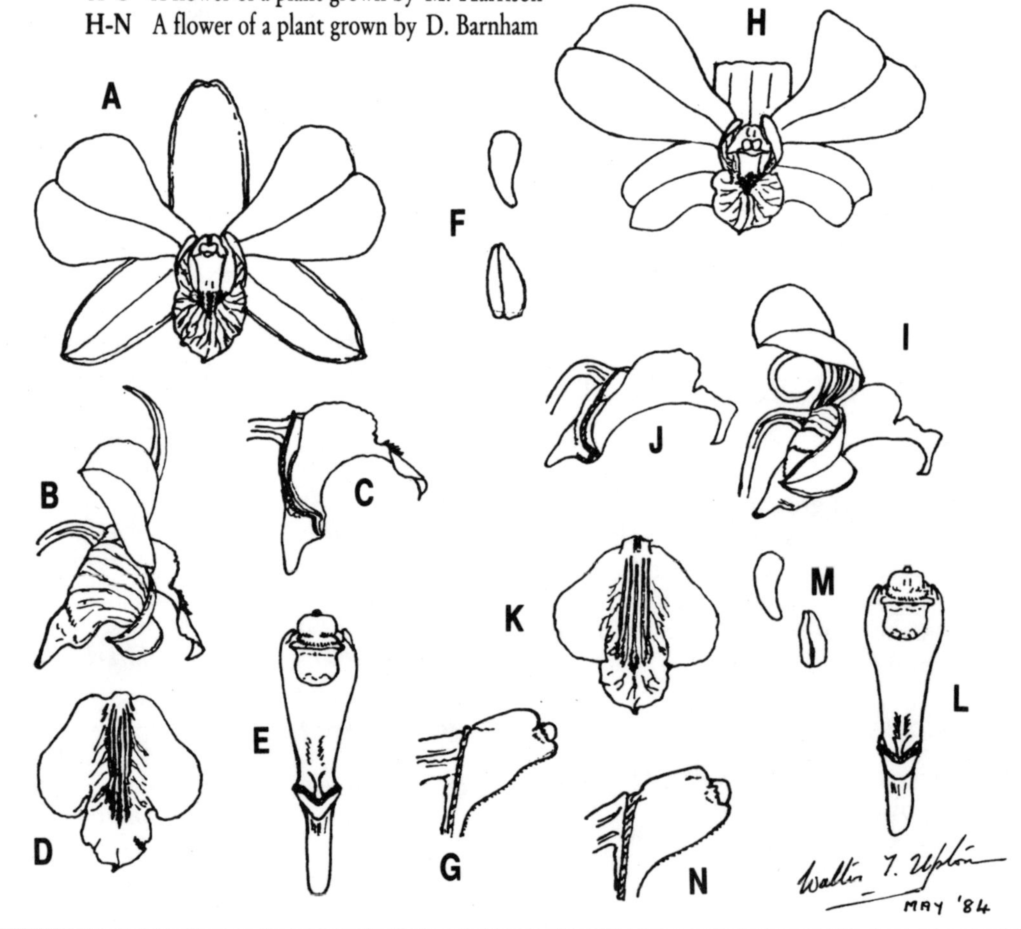

A & H Flower, front-view 1X
B & I Flower, side-view 1X
C & J Column and labellum, side-view 1X
D & K Labellum, pressed flat 1X
E & L Column, front-view 2X
F & M Pollinia greatly enlarged
G & N Top of column 2X

Other natural hybrids

The following are various recordings of isolated specimens of natural *Dendrobium* hybrids recorded in Australia.

D. schoeninum Lindley (as *D. beckleri*) x *D. pugioniforme* A. Cunn.
Found at Dorrigo, NSW in September 1978 by B. J. Wallace, recorded in *Orchadian* Vol. 6, No. 3, March 1979.

D. schoeninum Lindley (as *D. beckleri*) x *D. mortii* F. Muell. (as *D. tenuissimum*)
Reported by R. Ford of Caloundra in *Orchadian* Vol. 3, No. 7, February 1970.

D. schoeninum Lindley (as *D. beckleri*) x *D.* sp. aff. *fairfaxii* R.Br.
Found at Mt Glorious, near Brisbane, Queensland in 1964 by S. Clemesha and W. Mitchell. Recorded in *Orchadian* Vol. 1, No. 10, February 1965. Only one plant was found in association with suspected parents *D. schoeninum* and *D.* sp. aff. *fairfaxii* (previously known as *D. teretifolium* var. *fairfaxii* forma *aureum*).

D. striolatum H. G. Reichb. x *D. teretifolium* R.Br.
Isolated plants. Recorded by Rupp.

D. falcorostrum Fitzg. x *D. macropus* subsp. *gracilicaule* (F. Muell.) P. S. Green
Collected near Taree by D. L. Jones.

D. falcorostrum Fitzg. x *D. kingianum* Bidw.
Isolated specimens collected by L. W. Archer.

D. falcorostrum Fitzg. x *D. speciosum* Smith
Reported by H. J. R. Overall as being collected from the western Dorrigo area of NSW, *Orchadian* Vol. 3, No. 5, September 1969.

D. canaliculatum R.Br. x *D. affine* (Dcne.) Steud. (as *D. dicuphum*)
Collected by F. Woerle on Coburg Peninsula, Northern Territory, prior to 1979, *Orchadian* Vol. 6, No. 7, March 1980.

D. discolor Lindley x *D. antennatum* Lindley
Recorded from the McIlwraith Range, North Queensland in *Tropical Orchids of Australia*, Lavarack and Gray (1985).

D. discolor Lindley x *D. trilamellatum* J. J. Sm. (as *D. semifuscum*)
Collected by D. R. Kitchen near the tip of Cape York Peninsula in 1970, *Orchadian* Vol. 3, No. 11, December 1970.

The following natural hybrids were recorded in the *Proceedings of the 8th Australian Orchid Conference* by D. L. Jones (1983).

Dendrobium canaliculatum R.Br. x *D. trilamellatum* J. J. Sm. (as *D. semifuscum*)
Dendrobium canaliculatum R.Br. x *D. discolor* Lindley
Dendrobium cucumerinum Macleay et Lindley x *D. linguiforme* Sw.
Dendrobium linguiforme var. *nugentii* F. M. Bail. x *D. racemosum* (Nicholls) Clemesha et Dockr.
Dendrobium discolor Lindley x *D. nindii* W. Hill

In *Sanders List of Orchid Hybrids*, *D.* Leeanum is given as a natural hybrid, the parents being *D. schroederianum* and *D. x superbiens.* It was introduced to Europe by F. Sanders & Co. in 1890, along with *D. phalaenopsis* var. *schroederianum.* As *D. x superbiens* is a natural hybrid from Queensland, one must presume, if the parents of this natural hybrid are correct, the imports must have come from Australia, yet *D. phalaenopsis* var. *schroederianum* originally came from Tanimbar Islands. This natural hybrid is therefore suspect.

A number of the hybrids are pictured in plates 99–201.

5

History of hybridization

In Australia there has, back to at least 1950, been classes in annual orchid shows, both winter and spring, for Australian native orchids. The classes generally were for 'Best Australian Native Orchid', 'Best Australian Native Orchid Hybrid', and 'Best Specimen Australian Native Orchid'. Only orchids indigenous to Australia or hybrids using indigenous parents were eligible.

As the greater proportion of Australian native orchids grown are dendrobiums, this has certainly influenced *Dendrobium* hybridizing in Australia. But it was not until the early 1960s that hybridizing really began in earnest as will be seen from the following list of *Dendrobium* hybrids using only Australian indigenous parents.

A quick look at this list makes it obvious that in the early years, except for the first two hybrids, only dendrobiums from the tropical regions were used, and mostly by hybridists in other countries. The species used (except for the first two) were all in the subgenus *Dendrobium* and in the sections *Phalaenanthe* and *Spatulata*.

From 1960 onwards, the momentum grew, slowly at first, but by 1980 there were 83 Australian native *Dendrobium* hybrids registered. The number of hybrids since the late 1960s had more than doubled. In the next 4 years 116 hybrids were registered and the very big majority of these were from parents derived from species in the subgenus *Athecebium*. The total number of hybrids of native *Dendrobium* parents registered to February 1988 is 324. Quite a revolution!

HYBRIDS

D. Specio-kingianum	*D. kingianum* X *D. speciosum*	Lawrence, UK 1892
D. Ellen	*D. kingianum* X *D. tetragonum*	Schmidt, Aust. 1928
D. Albertine	*D. bigibbum* X *D. antennatum*[1]	Nagrok Java about 1940
D. Australia	*D. bigibbum* X *D. nindii*[2]	Nagrok Java about 1940
D. Dang Toi	*D. goldiei* X *D. discolor*[3]	Singapore Bot. Gardens 1941
D. Albanense	*D. x superbiens* X *D. discolor*[3]	Sanders, UK 1948
D. Champagne	*D. discolor*[3] X *D. mirbelianum*	Singapore Bot. Gardens 1951
D. Saskia	*D. x superbiens* X *D. bigibbum*	Barendsen, Holland 1951
D. Red Wings	*D. antennatum*[1] X *D. johannis*	V. L. Deane, USA 1954
D. Three Star	*D. johannis* X *D. goldiei*	Imura, Hawaii 1954
D. Yellow Curls	*D.* Champagne X *D. discolor*[3]	Singapore Bot. Gardens 1956
D. Peter Petersen	*D. discolor*[3] X *D. nindii*[2]	Bourner, Aust. 1956
D. Tusco	*D. nindii*[2] X *D. johannis*	Komatsubara, Hawaii 1957
D. Kaipu	*D. mirbelianum* X *D. johannis*	Kirsch, Hawaii 1958
D. Frances Henton	*D.* Albanense X *D. nindii*[2]	Lee Kim Hong, Singapore 1958
D. Andrew Persson	*D. speciosum* X *D. falcorostrum*	Slade, Aust. 1960
D. Moira Stewart	*D. goldiei* X *D. mirbelianum*	L. B. Stewart, Singapore 1960
D. Bardo Rose	*D. falcorostrum* X *D. kingianum*	Overall, Aust. 1961
D. Elsie Cox	*D. x superbiens* X *D. x delicatum*	Osment, USA 1961

HYBRIDS (con't)

D. Moses Mark	*D. antennatum*[1] X *D. mirbelianum*	Shee Shim, Malaysia 1961 (Voon Orchids)
D. Judy Leroy	*D. canaliculatum* X *D. affine*[4]	Limberlost, Aust. 1961
D. Rajah	*D.* Dang Toi X *D.* Champagne	Khoo Kay Ann, Singapore 1961
D. Verninha	*D. discolor*[3] X *D. antennatum*[1]	McAdam, Singapore 1961
D. Princess Stephanie	*D. mirbelianum* X *D. bigibbum*	Limberlost, Aust. 1962
D. Lowana Nioka	*D. canaliculatum* X *D. antennatum*[1]	Limberlost, Aust. 1962
D. Gloucester Sands	*D. discolor*[3] X *D. canaliculatum*	MacPherson, Aust. 1963
D. Chieno	*D. discolor*[3] X *D. affine*[4]	Aoki, Hawaii 1964
D. Gillian Leaney	*D. x delicatum* X *D. kingianum*	Leaney, Aust. 1964
D. Jane Leaney	*D. x delicatum* X *D. speciosum*	Leaney, Aust. 1964
D. Shan Leaney	*D. x gracillimum* X *D. macropus* subsp. *gracilicaule* (as *D. gracilicaule*)	Leaney, Aust. 1964
D. Blue Gloucester	*D. x superbiens* X *D. nindii*[2]	MacPherson, Aust. 1966
D. Emmy	*D. aemulum* X *D. kingianum*	Butler, Aust. 1966
D. Glenys King	*D.* Lowana Nioka X *D. canaliculatum*	King, USA 1966
D. Wee Cheow Lay	*D.* Verninha X *D.* Yellow Curls	Wee Sian Hee (SBG), Malaysia 1966
D. Gloucester Gem	*D.* Gloucester Sands X *D. bigibbum*	MacPherson, Aust. 1966
D. Gloucester Dawn	*D. mirbelianum* X *D. canaliculatum*	MacPherson, Aust. 1966
D. Kingrose	*D.* Bardo Rose X *D. kingianum*	Butler, Aust. 1966
D. Lowana Tiki	*D.* Judy Leroy X *D. antennatum*	Limberlost, Aust. 1966
D. Rachelle Simpson	*D. canaliculatum* X *D. johannis*	Simpson, Aust. 1967

[1] as *D. d'albertisii* [2] as *D. tofftii* [3] as *D. undulatum* [4] as *D. dicuphum*

Now (1988) it is not unusual for orchid shows in Australia to have many classes for Australian native orchids, the majority in the genus *Dendrobium*, and the number exhibited is increasing every year. In a few shows they predominate.

The above is based on hybrids using only Australian indigenous parents, however there is a problem with hybrids using the species *D. bigibbum*, which is universally known. This is a very complex species, and it has, in the past, been known under many names due to much confusion surrounding it and the species previously known as *D. phalaenopsis* and *D. schroederianum*.

In the above list only registered hybrids using the parent *D. bigibbum* are included, as this is the only parent that can, with any degree of certainty be considered of Australian origin. But due to the use of the names *D. phalaenopsis* and *D. schroederianum* as hybrid parents in the past, the similarity between these parents and *D. bigibbum*, and their doubtful origin,, it would be wise to include a list of hybrids using *D. phalaenopsis* and *D. schroederianum* as parents. This would give a more accurate assessment of hybridizing, particularly as these plants are so popular throughout the world, and the confusion of names is not generally known or understood.

There is little doubt, due to the confusion in the identification of the various subspecies and varieties of *D. bigibbum*, *D. phalaenopsis* and *D. schroederianum*, that many clones grown under the names *D. phalaenopsis* and *D. schroederianum* are of Australian origin and not from the Tanimbar Islands.

The following list covers the same period as the first list and shows how popular *D. phalaenopsis* and *D. schroederianum* were with overseas hybridists. Of the 80 crosses made, only 10 were by hybridists in Australia. I do not feel this has any bearing on the statistics given after the first list, as there is doubt on the origin of some of the species in the second list.

D. Boisseyense	*D. bigibbum* X *D. schroederianum*	Vacherot-Lecoufle 1926
D. Louis Bleriot	*D. schroederianum* X *D. x superbiens*	Vacherot-Lecoufle 1929
D. Guynemer	*D.* Boisseyense X *D.* Pompadour	Vacherot-Lecoufle 1930
D. Pauline	*D. phalaenopsis* X *D. discolor* (as *D. undulatum*)	Stuart Low 1932
D. Pompadour	*D.* Louis Bleriot X *D. schroederianum*	Vacherot-Lecoufle 1934
D. Orchidwood	*D. bigibbum* X *D. phalaenopsis*	Orchidwood 1934
D. Roger Sander	*D. mirbelianum* X *D. schroederianum*	Sanders 1936
D. Colin Potter	*D. phalaenopsis* X *D. nindii* (as *D. tofftii*)	Lyon 1944
D. David Sander	*D. schroederianum* X *D. nindii* (as *D. tofftii*)	de Saram 1944
D. White Gem	*D. affine* (as *D. dicuphum*) X *D. schroederianum*	Sanders 1946
D. Lady Constance	*D.* Orchidwood X *D. phalaenopsis*	Weber 1947
D. Nora Potter	*D.* Colin Potter X *D. schroederianum*	Colin Potter 1947
D. Liliha	*D.* Pauline X *D. phalaenopsis*	Kazamura 1948
D. Joyful	*D. bigibbum* X *D.* Pauline	Stuart Low 1950
D. Anouk	*D.* Pompadour X *D. phalaenopsis*	Vacherot-Lecoufle 1950
D. Margaret Paxton	*D.* Pauline X *D. johannis*	Singapore Bot. Gardens 1950
D. Topsy de Jong	*D. schroederianum* X *D.* Pompadour	de Jong 1951
D. Momi Fujimoto	*D. discolor* (as *D. undulatum*) X *D.* Louis Bleriot	Miyamoto 1951
D. Raimonde de Jong	*D.* Louis Bleriot X *D.* Roger Sander	de Jong 1951
D. Dark Victory	*D.* Pompadour X *D. discolor* (as *D. undulatum*)	Conklin 1951
D. Marianne de Jong	*D.* Pompadour *D.* Roger Sander	de Jong 1951
D. Polynesian	*D.* Pompadour X *D.* Orchidwood	McCoy 1951
D. Helen Fukumura	D. Lady Constance X *D. schroederianum*	Fukumura 1952
D. Naxos	*D.* Guynemer X *D.* Pompadour	Vacherot-Lecoufle 1952

(con't)

D. Garden	*D.* Lady Constance X *D.* Louis Bleriot	Miyamoto 1953
D. Rubicon	*D.* Guynemer X *D. phalaenopsis*	Vacherot-Lecoufle 1953
D. Dorothy Deane	*D.* Red Wings X *D. phalaenopsis*	V. L. Deane 1954
D. Jean	*D.* Lady Constance X *D. x superbiens*	Mrs E. Yamasaki 1954
D. Garnet	*D. bigibbum* X *D.* Roger Sander	Muriel Bridges 1954
D. E. V. Dulin	*D.* Garnet X *D. schroederianum*	E. T. Iwanaga 1954
D. Eugene	*D.* Liliha X *D. schroederianum*	Mujata 1954
D. Gladys Oumae	*D.* Liliha X *D. discolor* (as *D. undulatum*)	Oumae 1954
D. Sarie Marijs	*D. antennatum* X *D. schroederianum*	Tjipeganti
D. Puppchen	*D. schroederianum* X *D. discolor* (as *D. undulatum*)	— —
D. Boey Chun Heng	*D. schroederianum* X *D.* Roger Sander	Kuah 1955
D. Florence Foo	*D.* Albanense X *D. phalaenopsis*	Lee Kim Hong 1955
D. Mary Katherine	*D.* Pompadour X *D.* Lady Constance	Y. Inouye 1955
D. Ora Tatsumara	*D.* Colin Potter X *D.* Pompadour	Kodama 1955
D. Marisa	*D.* Sarie Marijs X *D. phalaenopsis*	H.S.H. Prince Chalerm Phol 1956
D. Top Hat	*D.* Anouk X *D. schroederianum*	McCoy 1956
D. David Baver	*D. johannis* X *D. phalaenopsis* var. *compactum*	Baver 1956
D. Full Moon	*D.* Louis Bleriot X *D.* Helen Fukumura	C. Miwa 1956
D. Lehua	D. Lady Constance X *D.* Helen Fukumura	Fukumura 1957
D. Betty Matsuda	*D. bigibbum* X *D.* Pompadour	Takiguchi 1957
D. Gregor Duruty	*D. phalaenopsis* X *D. goldiei*	Duruty 1957
D. Tau	*D.* Dark Victory X *D.* discolor var. *broomfieldii*	A. C. Chang 1957
D. Agnes Cheok	*D.* Anouk X *D.* Lady Constance	Woodlawn 1958

HYBRIDS (con't)

D. Tim Tam	*D.* Anouk X *D.* Pompadour	McCoy 1958
D. Hale Kulani	*D.* Helen Fukumura X *D.* Top Hat	Kimball 1958
D. Madame Butterfly	*D.* Mary Katherine X *D.* Helen Fukumura	Y. Inouye 1958
D. Lorraine Hanai	*D.* Dark Victory X *D.* Lady Constance	Joseph 1958
D. Top Notch	*D.* Top Hat X *D.* Anouk	Kushima 1958
D. Gigi	*D.* Lady Constance X *D.* Garnet	Y. Inouye 1959
D. Lovely Hawaii	*D.* Orchidwood X *D.* Lady Constance	McCoy 1959
D. Garnouk	*D.* Garnet X *D.* Anouk	H. Iwanaga 1959
D. Bridgjen	*D.* Saskia X *D.* Orchidwood	Moestopo 1959
D. Lim Kit Yee	*D.* Pompapour X *D. discolor* var. *broomfieldii*	Lim Chooi Seng 1959
D. Digger	*D.* Colin Potter X *D. discolor* var. *broomfieldii*	J. H. Miller 1959
D. Maui Queen	*D.* Anouk X *D.* Helen Fukumura	Fukumura 1959
D. Makassar	D. Louis Bleriot X *D.* Orchidwood	Perimpunan 1959
D. Blossom Ross	*D.* E. V. Dulin X *D.* Louis Bleriot	Pickell 1960
D. Rosy Tips	*D. canaliculatum* X *D. phalaenopsis*	Muir 1960
D. Endeavour	*D.* Liliha X *D.* Lady Constance	Limberlost 1961
D. Joan Waddell	*D.* Pompadour X *D.* Louis Bleriot	Waddell 1961
D. Bailys Blue	*D. nindii* (as *D. tofftii*) X *D.* Pauline	Limberlost 1961
D. Taprobane	*D.* x *superbiens* X *D.* Pompadour	Ceylon Bot. Gardens 1961
D. Ruth MacDonald	*D. phalaenopsis* X *D.* Saskia	D. J. Macdonald 1962
D. Harry Wheeler	*D. nindii* (as *D. tofftii*) X *D.* Anouk	Toogood 1962
D. Robyn Jones	*D.* Peter Petersen X *D.* Joan Waddell	Limberlost 1962
D. Lee Siew Sim	*D.* Pauline X *D.* Frances Henton	Singapore Orchids 1963
D. Oriel Mai	*D.* Anouk X *D.* Pauline	M. B. Cook 1963

HYBRIDS (con't)

D. Meng Aun	*D. phalaenopsis* X *D.* Yellow Curls	Lim Swee Aun 1964
D. Nah Hiok Hwee	*D.* Pompadour X *D. antennatum*	Tay Boon Kim 1965
D. Suzanne	*D. phalaenopsis* X *D. tetragonum*	Willersdorf 1965
D. Coneac	*D.* Endeavour X *D. phalaenopsis* subvar. *compactum*	Limberlost 1965
D. Kailua	*D.* Garnet X *D. bigibbum*	Rev. M. Yamada 1965
D. Painted Doll	*D.* Judy Leroy X *D. phalaenopsis*	Limberlost 1966
D. Lady Marie	*D.* Tim Tam X *D.* Lady Constance	George Black 1966
D. Daintree	*D.* Bailys Blue X *D. phalaenopsis*	Limberlost 1967
D. Beauty Queen	*D.* Lee Siew Sim X *D.* Pompadour	Singapore Orchids 1967

The beginnings

Records tell us that the first registered Australian native *Dendrobium* hybrid, made in Australia, was *D.* Ellen and was registered by Wilhelm Schmidt of Warrawee, NSW in 1928. It was named in honour of his daughter. The seeds were sown in April 1914 and the first plants flowered 4½ years later. The capsule parent was *D. kingianum* and the pollen parent was *D. tetragonum* var. *tetragonum*, the southern form.

But, the very first hybrid using Australian indigenous *Dendrobium* parents was made in Dorking, England by Sir Trevor Lawrence. It was *D.* Specio-kingianum, registered in 1892, the parents being *D. kingianum* and *D. speciosum*. Subsequent to this, F. M. Bailey in 1902 named a natural hybrid of the same parents as *D. x delicatum* (F. M. Bail.) F. M. Bail. (he had previously, in 1884, named this same plant as *D. speciosum* var. *delicatum*, not realizing that it was a natural hybrid). The name *D.* Specio-kingianum is seldom, if ever, used.

The International Authority for the Registration of Orchid Hybrids has advised that the name *D. x delicatum* is no longer available for use as a parent in *Dendrobium*. Therefore for hybrid registration purposes only, the name *D.* Specio-kingianum is to be used. All the necessary corrections covering this change to the Sander's List of Orchid Hybrids will appear in the 1981/85 Lists addendum. However, the botanically correct name for this natural hybrid is *Dendrobium x delicatum* and is therefore used throughout this work.

The next hybrid produced in Australia was registered in 1956 by James L. Bourner of Cairns, Queensland; it was *D.* Peter Petersen, the parents were *D. discolor* (as *D. undulatum*) X *D. nindii* (as *D. tofftii*). In 1960 the next Australian hybrid was registered by G. Hermon Slade of Manly, NSW; it was named *D.* Andrew Persson after a well-known orchid identity, the parents were *D. speciosum* and *D. falcorostrum*.

D. Bardo Rose, the next hybrid registered by an Australian was probably one of the most popular of the early hybrids of parents in the subgenus *Athecebium*. It was the first Australian Native *Dendrobium* hybrid to win an award in Australia—*D.* Bardo Rose 'Berowra' HCC/AOC—NSW. The year was 1975. A lovely white, owned by Ron Kerr, a well known native orchid enthusiast and grower and editor of the Australian Orchid Review since 1965. Herbert James Rumley Overall registered *D.* Bardo Rose in 1961 naming it after the road in which he lived in Newport, NSW. Overall was a great naturalist, but ferns and orchids were his greatest love; he produced many hybrids. The cross was made by Overall in September 1952, using a red *D. kingianum* collected by Wilhelm Schmidt (the originator of *D.* Ellen) on the McPherson range as the pollen parent and *D. falcorostrum* as the capsule parent.

In 1961, J. E. Jones of Limberlost Nursery in Cairns, North Queensland, registered the first of their nursery's many excellent hybrids mainly using *Dendrobium* species within the subgenus *Dendrobium.* This was *D.* Judy Leroy and the parents were *D. canaliculatum* var. *pallidum* X *D. affine (as D. dicuphum*). I obtained an unflowered seedling plant of this from Limberlost in March 1960. It was an excellent flower on a long, upright raceme (see plate 147). I have used it in many hybrids both within its own subgenus and with the subgenus *Athecebium.* See plates 101, 102, 105, 121 and 152.

Since 1961 quite a few keen native growers have become active in hybridizing. In Queensland, K. MacPherson of Proserpine (in the subgenus *Dendrobium*) and in NSW Reg Leaney, Ira Butler, Noel Jupp, Roger Bedford, Cannons of Wayside Nursery and, of course, myself. My initial approach to hybridizing in 1960–61 was to improve the existing crosses, such as *D. x delicatum*, *D. x gracillimum*, *D.* Ellen, *D.* Bardo Rose and *D.* Ella Victoria Leaney (at that time an unregistered Reg Leaney cross, subsequently registered in 1980) to give me superior clones on which to base my hybrid programme.

HYDRIDIZING

It is a waste of time to hybridize without thought. To hybridize just because you have two dendrobes out at the same time (or to use inferior parents) is pointless. You should aim to improve on both parents or at least have an aim, even if it may take one, two or more generations before it is achieved.

In hybridizing with Australian dendrobiums, I think many factors should be looked at such as:

- upright raceme;
- more flowers per raceme;
- prolific flowering;
- flowering time and frequency of flowering;
- duration of flowering;
- colour;
- size;
- shape;
- short pseudobulbous stems with flowers well above the leaves;
- compact growth habit;
- cool-growing hybrids using part-tropical species;
- plants that are vigorous and have ease of culture.

Special attention should be paid to the shape and size of the labellum; try to visualize whether it will be well balanced in the resultant progeny. *D. speciosum* for instance has a poor factor in the lip, its mid-lobe folds inwards and unfortunately this appears to be a dominant factor with this species.

Consider substance and texture of flower—it should preferably be fleshy. Flowers that lack substance generally lack lasting qualities and don't present themselves well.

Hybrids registered to February 1988

(This does not include hybrids using non Australian species)

Within sections of the subgenus *Athecebium*

Section *Dendrocoryne*	210 hybrids	
Section *Rhizobium*	19 hybrids	
Section *Monophyllaea*	1 hybrid	230

Between sections within the subgenus *Athecebium*

Section *Rhizobium* X section *Dendrocoryne*	4 hybrids	
Section *Monophyllaea* X section *Dendrocoryne*	3 hybrids	7

Between subgenera		
Subgenus *Athecebium* X subgenus *Dendrobium*	26 hybrids	26
Within sections of subgenus *Dendrobium*		
Section *Spatulata*	29 hybrids	
Section *Phalaenanthe*	1 hybrid	30
Between sections within the subgenus *Dendrobium*		
Section *Phalaenanthe* X section *Spatulata*	30 hybrids	
Section *Phalaenanthe* X section *Trachyrhizum*	1 hybrid	31
		Total 324 hybrids

Section *Dendrocoryne* hybrids

By far the greatest number of hybrids are from species within the section *Dendrocoryne* of the subgenus *Athecebium*. No doubt the main reasons for this are that they are cool-growing and very floriferous, and are the easiest to cultivate. This is particularly so as the greater number of orchid growers in Australia come from the southern half of the continent, and demand is important to hybridists.

There are eleven species and four named natural hybrids within this section and all have been used as parents. The main objective is to produce cool-growing plants that are compact, very florific and have upright racemes. There are, of course, many other considerations such as size of flowers, colour, shape, longevity of flowering, etc.

Although it has only been about 25 years since hybridizing really began in this section, 210 hybrids have been registered, plus natural hybrids. It may be considered too early to come to definite conclusions, but it would seem that three basic shapes are appearing. In a paper I gave at the 8th Australian Orchid Conference in Townsville in 1983 I called these the *D. kingianum* shape, the *D. tetragonum* shape and the *D. falcorostrum*/*D. speciosum*/*D. jonesii* shape. Now, five years later, I see no reason to change. If anything, this situation is being confirmed. No doubt if crossing continues within these three shapes, there will eventually be only one shape.

The main section *Dendrocoryne* hybrid of the not too distant future will have large flowers with wide segments and be of the *D. falcorostrum*/*D. speciosum*/*D. jonesii* shape, with a solid upright raceme of 10 or more flowers in colours of white, cream, yellow, apricot, red and purple or combinations of these colours. The delightful shapes of *D. kingianum* and *D. tetragonum* could be retained by wise back crossing to these species.

D. kingianum shape

Fig. 94 shows the *D. kingianum* shape. The species *D. kingianum*, *D. macropus* subsp. *gracilicaule* and *D. macropus* subsp. *howeanum* are dominant in shape and size and *D. kingianum* is also generally dominant in colour. The habit of raceme is mostly arching to upright. (*D. speciosum* and *D. adae* are helping to strengthen and improve the upright habit of the raceme.) The plant form generally favours *D. kingianum*.

D. kingianum is a most variable species so the clone to be used should be closely considered. For instance, I have a very dark-red clone and this colour characteristic is dominant; I have a clone with extra large side lobes to the labellum and this characteristic is dominant.

D. kingianum crossed with *D. macropus* gives *D. x suffusum*. This is approximately midway between the parents except that in most of the progeny the colour of the *D. kingianum* is dominant in the suffused markings. The only *D. x suffusum* hybrids made to date are with *D. macropus* subsp. *gracilicaule* and *D. macropus* subsp. *howeanum*. The natural hybrid in Australia is between *D. kingianum* and *D. macropus* subsp. *gracilicaule*. When *D. x suffusum* is used as a parent, strangely enough it is not always dominant in shape, particularly when crossed with *D. tetragonum* and *D. speciosum*.

When taking a primary hybrid back to *D. kingianum* the shape is dominant, e.g. *D.* Penny Ann, *D.* Kingrose, *D.* Alan Printer, *D.* Kim.

Although the species *D. fleckeri* is

Section *Dendrocoryne* Hybrids

Fig. 94 *Dendrobium kingianum* shape

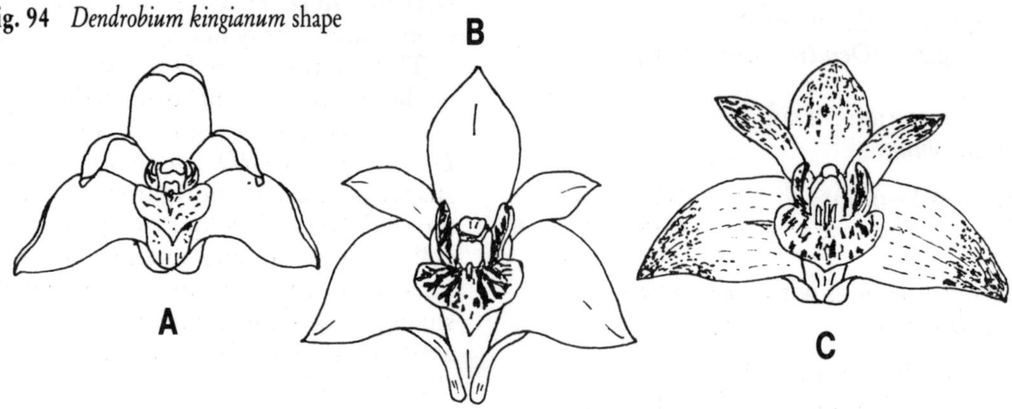

Fig. 95 *Dendrobium tetragonum* shape

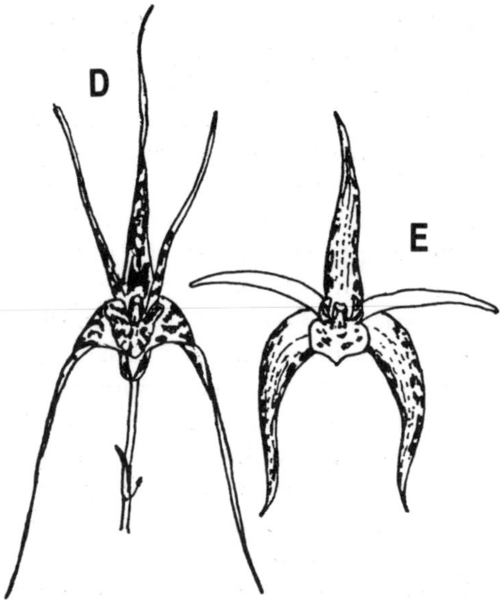

Fig. 96 *Dendrobium falcorostrum*/*D. speciosum* / *D. jonesii* shape

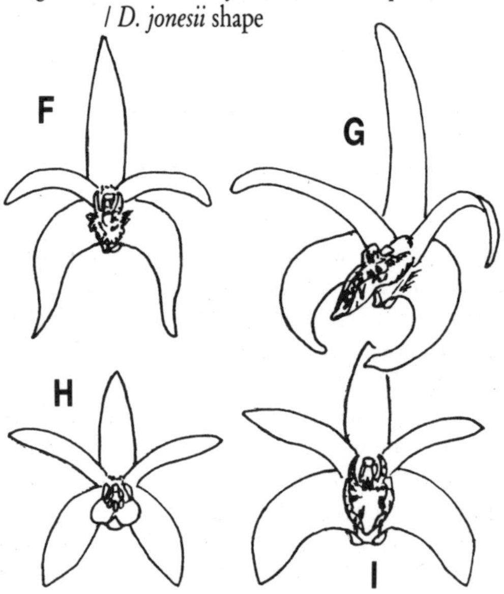

- **A** *Dendrobium macropus* subsp. *howeanum*
- **B** *Dendrobium kingianum*
- **C** *Dendrobium* Hastings
- **D** *Dendrobium tetragonum* var. *giganteum*
- **E** *Dendrobium* Jamie Upton
- **F** *Dendrobium* Eureka
- **G** *Dendrobium speciosum*
- **H** *Dendrobium falcorostrum*
- **I** *Dendrobium* Wonga

basically the *D. kingianum* shape, it does not appear to be shape dominant. Its major influence on its progeny appears to be its lip and colour. e.g. *D.* Hastings, *D.* Golden Fleck. (The first *D.* Golden Fleck made used *D. macropus* subsp. *howeanum*.)

When using the shape-dominant *D. tetragonum* with *D. kingianum*, the progeny are midway between the two shapes. If *D. tetragonum* var. *giganteum* is used, the progeny flowers more than once a year (*D.* Ellen).

D. Double Two (*D. x gracillimum* X *D. adae*) can be included in this group. *D. macropus* subsp. *gracilicaule* and *D. adae* have dominated the shape and lip. *D. speciosum* has helped in the good upright raceme and number of flowers.

D. John Upton, a cross between *D.* Kim and *D.* Ellen, has *D. kingianum* in both parents. The majority of the progeny are rich in *D. kingianum* colour and shape.

Other hybrids taking this shape are *D. x delicatum*, *D. x gracillimum*, *D.* Aussie Freckles, *D.* Aussie Hero, *D.* Aussie Love, *D.* Aussie Mist, *D.* Bardo Rose, *D.* Blushing Rose, *D.* Cawongla, *D.* Craig Hewitt, *D.* Delemma, *D.* Ella Rose, *D.* Ella Victoria Leaney, *D.* Gillian Leaney, *D.* Green Goddess, *D.* Jenni, *D.* Joy Wray, *D.* King Falcon, *D.* King Goblin, *D.* Kit Murdoch, *D.* Nathan Wall, *D.* Peach Glow, *D.* Penny Rose, *D.* Red Baron, *D.* Robyn Kaylene, *D.* Susan, *D.* Tawhai, *D.* Telekon, *D.* Tooloom, *D.* Yondi.

126 *D.* Elfin

127 *D.* Fiesta 'Lipstick'

128 *D.* Gracious Falcon

129 *D.* Gloucester Sands 'Jenni'

130 *D.* Goblin

131 *D.* Hilda Poxon 'Jill', using *D. tetragonum* 'No. One'

132 *D.* Hilda Poxon, using *D. tetragonum var. giganteum*

133 *D.* Golden Fleck

134 *D.* Golligosh

135 *D.* Graham Hewitt

136 *D.* Green Goddess

137 *D.* Green Star

138 *D.* Harold Hirsch 'Jill'

139 *D.* Hastings

140 *D.* Hot Shot

141 *D.* Ingleside 'Strawberry Fair'

142 *D.* John Rose 'Cherry Ripe'

143 *D.* Ira's Star 'Elanora'

144 *D.* Ira Butler

145 *D.* Jamie Upton 'Stardust'

146 *D.* John Upton

147 *D.* Judy Leroy 'Jill'

148 *D.* Karani

149 *D.* Kim 'Lollipop'

150 *D.* Kingrose

151 *D.* Lemon Glow

152 *D.* Jombock 'Jill'

D. tetragonum shape

The *D. tetragonum* shape is shown in Fig. 95. *D. tetragonum* is very dominant in flower shape and size, and mostly dominant in colour (except when crossed with *D. kingianum*). The raceme habit is arching to almost pendulous. The markings of the flowers are mostly passed on to progeny, some being quite heavily spotted or suffused. The lip characteristic is only partially dominant except in some clones.

In some hybrids the influence of *D. tetragonum* dominates the number of flowers produced.

When the northern Queensland form of *D. tetragonum* is used, it is not unusual for the hybrid to flower many times during the year.

In the following hybrids only the labellum favours the other parent and there is some loss in flower size: *D.* Star Imp (upright raceme from *D. adae*), *D.* Numkumbil, *D.* Allyn Star, *D.* Star Delight, *D.* Star of Gold, *D.* Hilda Poxon.

D. tetragonum has been totally dominant in shape and size in the following hybrids: *D.* Ku-Ring-Gai, *D.* Kungara (very large flowers) and the unregistered *D.* Andrew Persson X *D.* tetragonum (very large flower, broad mid-lobe to lip and upright raceme).

D. Blushing Star, a cross between *D. tetragonum* and *D. x suffusum* is marked with *D. kingianum* colouring and is not completely *D. tetragonum* shape.

D. Jamie Upton, a cross between *D. tetragonum* and *D.* Golden Fleck (*D. macropus* subsp. *howeanum* X *D. fleckeri*) has produced an arching raceme with *D. tetragonum* shaped flowers and lip but mostly with the *D.* Golden Fleck green to apricot colouring (from *D. fleckeri*). Of the 25 or so clones of this cross that I have flowered, only one was of a shape similar to *D. macropus* subsp. *howeanum*. The markings on the flowers varied but the wide flat mid-lobe of the lip was constant, a dominant characteristic of the parent clone of *D. tetragonum*.

Other hybrids taking this shape are *D.* Aussie Glow, *D.* Aussie Ira, *D.* Bright Spark, *D.* Golden Glory, *D.* Kim Heinze, *D.* Pauline Rankin, *D.* Peach Star, *D.* Rutherford Star, *D.* Sun Star.

D. falcorostrum/*D. speciosum*/*D. jonesii* shape

Fig. 96 shows the *D. falcorostrum*/*D. speciosum*/*D. jonesii* shape. There is no specific dominant species, but rather the accumulation of many characteristics both dominant and recessive. *D. jonesii* and *D.* sp. aff. *jonesii* are of equal importance in this shape, although no more dominant than the two other species. Furthermore *D.* sp. aff. *jonesii* generally has little influence on size.

The most noticeable characteristics passed on to their progeny are the habit of the raceme, number of flowers and shape of the lip. When crossed with *D. kingianum*, *D. macropus* subsp. *gracilicaule* and *D. tetragonum*, these species are dominant.

D. fleckeri, when used with this group, has influenced many hybrids with its colour and lip. It was a successful parent in *D.* Peter and *D.* Eureka although not dominant.

The strong and delightful perfume of *D. falcorostrum* is coming through in some of its progeny.

Generally the progeny with this shape are from the crossing of:

- two non-dominant species for shape (*D.* Eureka, *D.* Peter and *D.* Sunglow);
- two primary hybrids both having one dominant species in their parentage (*D.* Peter Hewitt);
- a hybrid with *D. kingianum* at least twice in its parentage crossed with *D. tetragonum* (*D.* Emma);
- a primary hybrid between two dominant species crossed with a non-dominant species (*D.* Colin, *D.* Harold Hirsch);
- a hybrid with the dominant *D. kingianum* crossed back to a non-dominant species for shape (*D.* Wonga).

Other hybrids taking this shape are *D.* Andrew Persson, *D.* Aussie Bonanza, *D.* Aussie Frontier, *D.* Aussie Majestic, *D.* Aussie Starlight, *D.* Delicate Falcon, *D.* Lemon Glow, *D.* Lynette Banks, *D.* Miara, *D.* Nerang, *D.* Nooro, *D.* Star of Riverdene, *D.* Tweed, *D.* Wirruna.

Section *Rhizobium* hybrids

The species mostly used are *D. cucumerinum*, *D. linguiforme*, *D. mortii*, *D. pugioniforme*, *D. schoeninum*, *D. striolatum* and *D. teretifolium*. Generally the resultant hybrids appear to be midway between the parents in vegetative growth and near enough between the flowers. No dominance is obvious at this stage, and most hybrids are florific.

D. Michael Jupp (*D. schoeninum* (as *D. beckleri*) X *D. linguiforme*) is a delightful hybrid producing upright racemes with about 10 upside-down flowers, about 30 mm across the lateral sepals, and slightly greenish in colour and purple striping at the base of the segments.

D. Virginia Jupp (*D. linguiforme* X *D. teretifolium*) is a very strong grower with long racemes carrying many white flowers. Both plant form and flowers are midway between parents.

Other hybrids in this category are *D.* Aussie Cascade, *D.* Burie, *D.* Dainty Cascades, *D.* Duffy, *D.* Fay Moore, *D. x foederatum*, *D.* Glenn Wall, *D.* Goose Bumps, *D. x grimesii*, *D.* Ida Mary, *D.* Jiggi, *D.* Julie Skillicorn, *D.* Limestone, *D.* Rosemary Jupp, *D.* Stephen, *D.* Super Ruffles and *D.* Wesley Pink.

Section *Monophyllaea* hybrids

This is a section of only three species. One hybrid has been made, *D.* Golligosh (*D. schneiderae* X *D. monophyllum*), which is a delightful miniature with long-lasting, heavy-substance flowers. The pseudobulbous stems are approximately 2 cm long and fairly close together. From 10 to 12 cream-yellow flowers with an orange-yellow labellum are carried on an arching raceme. Flowers are near to *D. schneiderae* in shape.

Hybrids between section *Rhizobium* and section *Dendrocoryne*

Plant form is about midway between the parents. The pseudobulbous stems favour the *Dendrocoryne* shape except that they are narrower and therefore have a tendency to become pendulous. Leaves are inclined to be tubular to terete and frequently the top leaf grows up the centre of the tube, often dying off when too much water lies in the tube for too long. Also, the apical flowering node is not effective, so the plants generally flower from the secondary nodes. The few primary hybrids made so far are not very florific, with some flowering only every second or third year. When a primary hybrid is taken back to a *Dendrocoryne* section species, there is, at this early stage, evidence that flowering is quite regular and reasonably florific (*D.* Wesley Rose).

The flowers are the reason for continuing on this kind of hybridizing. They are charming, quite large, and have a long and attractive labellum.

D. Pink Pencil (*D. kingianum* X *D. teretifolium*) grows pendulously, tied to a block. Stems are long and slender, often producing aerial growths from the side nodes so that they appear to branch like the rhizome of *D. teretifolium*. The flowers are mostly pale pink, about 50 to 60 mm, across the lateral sepals and have a long slender labellum with undulate edges.

D. Elfin (*D.* Dainty Cascades X *D. kingianum*) has a vegetative form approximately midway between the parents with some tubular leaves. Quite a small plant, it produces racemes of two or more pale-green flowers with purple tips to the segments and purple on the reverse side. The upside-down flowers are *D. kingianum* shaped and approximately 30 mm across the lateral sepals. Most clones do not flower regularly, although I have seen one that is quite florific and it is a delight, the flowers lasting about 3 weeks.

D. Wesley Rose (*D.* Pink Pencil X *D. speciosum*) has a few almost tubular leaves (but most are not) and the habit of the stems is reasonably upright and not as slender as *D.* Pink Pencil. The flowers are about 50 mm across the lateral sepals, the colour is creamish to white flushed with pink. About four to eight widely opening flowers are carried on the racemes.

D. Amaroo (*D. mortii* (as *D. tenuissimum*) X *D. falcorostrum*). Registered by C. Brandon in 1986. I have not seen it in flower.

Hybrids between section *Monophyllaea* and section *Dendrocoryne*

So far only 2 hybrids have been registered in this category, but a few more have been made and are growing well.

In the subgenus *Athecebium* we need to look to *D. monophyllum* for the all important characteristic of length of time the flowers last. This will in the early stages reduce flower size, due to the tiny flowers of *D. monophyllum*, but they are flowers of excellent shape and of a good yellow colour.

D. Orana (*D. adae* X *D. monophyllum*) has a delightful yellow flower that lasts for up to 6 weeks in good condition, also as *D. monophyllum* has a short pseudobulbous stem, the size of the plant is greatly reduced. The result is a good-looking smallish plant, with long racemes of small (30 mm) dark-yellow, long-lasting flowers.

In *D.* Karani (*D. kingianum* X *D. monophyllum*), the flowers are also long lasting and, by using the dominant *D. kingianum*, the colour and shape are like this species but all the parts, including vegeatative, are smaller.

Hybrids between the subgenera *Athecebium* and *Dendrobium*

In an attempt to improve both plant and flower form, numerous hybrids have been made between the subgenus *Athecebium* and the subgenus *Dendrobium*; the resultant progeny are some of the best Australian native hybrids yet produced. The species used so far are *D. affine*, *D. agrostophyllum*, *D. antennatum*, *D. bifalce*, *D. bigibbum*, *D. canaliculatum*, *D. discolor*, *D. falcorostrum*, *D. fleckeri*, *D. johannis*, *D. jonesii*, *D. kingianum*, *D. monophyllum*, *D. speciosum*, *D. teretifolium* and *D. tetragonum.*

D. tetragonum has been used the most, no doubt because of its size, however its dominance has resulted in a reduction in the number of flowers. *D. bigibbum* has shown its dominance in colour while *D. speciosum* and *D. canaliculatum* have generally improved the number of flowers. *D. canaliculatum* is also showing dominance in both shape and size of plant.

D. Peewee (*D. bigibbum* X *D. tetragonum*) is a very popular hybrid. The flowers are approximately midway between the parents for shape. The *D. bigibbum* is dominant for both colour and shape and the *D. tetragonum* is dominant for shape. The colours are from dark magenta to lilac-purple depending on the parents used. The use of *D. tetragonum* has reduced the number of flowers; approximately four to six flowers are an average. Plants are reasonably cold tolerant, but do better with some heat in winter, say a minimum of 10°C.

In a *D.* Peewee I made in 1980 I used the short pseudobulbous stem form *D. bigibbum* var. *compactum* with a large broad-lipped northern NSW form of *D. tetragonum.* The progeny are very dominant in *D. tetragonum* flower shape, except the petal width is greater, and they are dominant in 'compactum' plant size and flower colour. This plant first flowered in March 1983 and carried four flowers on a 75 mm pseudobulbous stem, only 3 years from pollination, or 1 year 7 months from the flask. Numerous plants have flowered since and the average flower size is 75 mm and they are long-lasting. The plants are cool growing, a minimum of 10°C in winter and a decided rest is advisable to keep the plant healthy and strong. The mature plant does not generally exceed 150 mm in height.

D. Andrew Upton (*D.* Judy Leroy X *D.* Star of Gold) comes from two warm-growing species (*D. affine* (as *D. dicuphum*) X *D. canaliculatum* crossed with a *Dendrocoryne* section hybrid). The plants are reasonably small, up to 150 mm. The raceme is upright and carries up to eight flowers. The flowers, 60 mm wide, are mostly pale green with mauve markings on the lip and are long-lasting and grow under cool conditions.

D. Jombock (*D.* Judy Leroy X *D.* Colin) is a cross between *D.* Judy Leroy as above and another *Dendrocoryne* section hybrid, which is proving to be an excellent parent in this kind of hybridizing. The hybrid flowered within 3 years of pollination and produced an upright raceme of seven pure-white, star-shaped, 50 mm flowers (1982). Since then numerous plants have flowered, they range from white through pale pink to quite a strong red-purple. The largest flower is 70 mm wide and the tallest plant is 260 mm. They are very florific and carry up to 30 flowers. Cool growing. Mostly winter flowering.

D. Duke (*D.* Judy Leroy X *D. tetragonum* 'No. 1') once again uses *D.* Judy Leroy, but this time with the dominant species for shape, *D. tetragonum.* The progeny are very star-shaped but have great charm. They are from pure white to a deep yellow with orange tips to the segments, up to 65 mm wide and long-lasting. Individual flowers last up to 7 weeks and one plant was in flower for at least 3 months. Plants up to 200 mm tall. Mostly winter flowering. Cool growing with a little warmth through the coldest months of winter, a minimum of 10°C.

D. Warringah (*D. bigibbum* var. *compactum* X *D. speciosum*) appears to be a fairly slow grower until it reaches flowering size. I made it in 1979 and so far have only seen three plants of this cross in flower. By using the small pseudobulbous stem form of *D. bigibbum* the plant is kept small. (The largest pseudobulbous stem so far is 150 mm high.) The flowers are a delicate shade of pink with some darker flushes, their substance is excellent and they are carried on an upright raceme. Flower shape is approximately midway between the parents, the petals being quite wide, approximately 50 mm across. Eight, 11 and 16 flowers were carried on the racemes. Plants are cool growing and will tolerate winter temperatures down to 5°C. Winter to spring flowering.

D. Elegant Heart (*D. Peewee* X *D. speciosum*), four plants flowered for the first time in May to July 1986. Three to six large (80 mm long) flowers of good substance on an arching to upright raceme, I anticipate approximately 15 flowers on a well-grown mature plant. Flowers are red to purplish on the apical two-third of the perianth and a white-cream on the basal one-third. I made this hybrid in December 1981, taking just over 4 years to flower, the total germination only amounted to about 30 plants. The flowers last approximately 3 weeks, no doubt they will last longer on a mature plant. An excellent hybrid, with a lot of potential, both on the show bench and for the florist.

D. Reg Burns (*D. affine* (as *D. dicuphum*) X *D. fleckeri*) was made and registered by Bob Zellar in 1984. The first I saw of this orchid was when it was displayed in The Orchid Society of NSW Spring Festival in September 1984 and won the Champion Australian Native Hybrid Orchid. Since then I have seen three more plants, two when my wife and I were judging the Manning River Orchid Society Spring Show when it won the best in its class (September 1985). The flowers, up to 14, are delicate and a warm creamy-green colour, with rose-red at the base of the segments forming a ring. Mid-lobe of the labellum is white. Flowers are carried on an arching raceme and are approximately 50 mm across the lateral sepals. Pseudobulbous stems are up to 30 cm. Winter to spring flowering.

D. Our Native (*D. johannis* X *D. speciosum*) was made and registered by McFarlane in 1977. It produces numerous racemes, up to 45 cm, of yellow-green flowers on sturdy pseudobulbous stems which are approximately 30 cm long. Flowers are about 55 mm across the petals and the racemes bear approximately 35 flowers. Winter to spring flowering. Cool growing but requires a little warmth in the coldest winter months, say a minimum of 10°C.

D. Memoria Lloyd Bradford (*D. canaliculatum* X *D. speciosum*), *D.* Justin Wall (*D.* Lowana Nioka X *D. speciosum*) and *D.* Natalie Wall (*D.* Lowana Nioka X *D.* Star of Gold) all have *D. canaliculatum* in their parentage and are producing smallish pseudobulbous stems with flowers very strong in *Spatulata* shape with upright racemes. Colours mostly green and yellow with purple tips. All reasonably cool growing.

D. Big Fleck (*D. bigibbum* X *D. fleckeri*). The first hybrid flowered in May 1986. The plant form is similar to *D. fleckeri.* The flowers are midway between the parents in shape, are a mauve colour and up to 60 mm wide, the mid lobe of the labellum is 12 mm across. It is cool growing and needs a minimum of 10°C in winter.

D. Lorikeet (*D.* Peewee X *D.* Ellen). The first hybrid flowered in July 1986 only

3 years from pollination. The flowers were a pale mauve; mostly *D. tetragonum* shape and were 75 mm long.

Other hybrids in this category are *D.* Aussie Angel, *D.* Aussie Ochre, *D.* Ayr Queensland, *D.* Berry, *D.* Collette Fleur Simpson, *D.* Elsie Cox, *D.* Esme Poulton, *D.* Gold Stripe, *D.* Justin Wall, *D.* Maron, *D.* Natalie Wall, *D.* Rosella, *D.* Star Twist and *D.* Willowie Gold.

Section *Spatulata* hybrids

Twenty-nine of the thirty hybrids within sections of the subgenus *Dendrobium* are within the section *Spatulata*, because this is the only section containing more than two species. There are eight species in the section *Spatulata* and all are horticultural subjects, so it is not surprising that of the 15 *Dendrobium* hybrids raised prior to 1960 all except two contained a species from this section. In six both parents were species from the section *Spatulata*. Only one of the hybrids was made and raised in Australia; this was *D.* Peter Petersen, made by James L. Bourner of Cairns, north Queensland in 1956. The rest were made in England, Singapore, Holland, Java, and USA (Hawaii 3).

Most of the very early hybrids in this section are seldom seen in Australia now; one notable exception is *D.* Peter Petersen (*D. discolor* X *D. nindii*). There have been many remakes of this hybrid, and some are very florific with colourful progeny. It is a very proud, tall plant (frequently around 1 m in southern Australian collections) with long racemes of 30 to 40 flowers. Sepals and petals are slightly twisted, widely opening, and are in varying shades of a warm lilac-purple. Minimum winter temperature of 14°C and a decided rest are necessary. A natural hybrid has also been reported.

A very popular hybrid in this section is *D.* Gloucester Sands (*D. canaliculatum* X *D. discolor*); the pseudobulbous stems are considerably reduced by *D. canaliculatum*. Numerous flowers are carried on tall upright racemes and are midway between the parents, extremely colourful in brown to purple.

D. Rachelle Simpson (*D. canaliculatum* X *D. johannis*) is similar to *D.* Gloucester Sands but generally smaller in all parts and flowers tend more to the browns.

D. Lowana Nioka (*D. canaliculatum* X *D. antennatum*) is a beautiful pale-green flowered hybrid with purple markings on the labellum. Flowers are generally similar to *D. antennatum* in size and colour but midway between the parents in shape.

D. Verninha (*D. antennatum* X *D. discolor*) is a large plant with an arching raceme. The only plants of this hybrid that I have seen were made using *D. discolor* var. *broomfieldii*, and they carried many green to golden-yellow flowers with red-brown suffusions.

D. Minnie (*D. carronii* X *D. johannis*) is a very pretty hybrid with a compact plant form and an upright raceme carrying approximately 20 pink and brown flowers with a golden labellum.

The other hybrids in this category are *D.* Allyn King, *D.* Allyn Prince, *D.* Allyn Sands, *D.* Allyn Supreme, *D.* Black Mickey Mouse, *D.* Camp Island, *D.* Champagne, *D.* Cherub, *D.* Dryander Mountain, *D.* Glenys King, *D.* Gloucester Dawn, *D.* John's Charm, *D.* Johulatum, *D.* Kaipu, *D.* Moses Mark, *D.* Red Wings, *D.* Sauterne, *D.* Son of Gloucester, *D.* Tusco, *D.* Valda, *D.* Wee Cheow Lay and *D.* Yellow Curls.

Section *Phalaenanthe* hybrids

The hybrid between the two species of the section *Phalaenanthe*, namely *D bigibbum* and *D. affine* (as *D. dicuphum*), was named *D.* Merritt Island in 1969. It has the appearance of a smallish *D. bigibbum* but is more delicate.

Hybrids between section *Phalaenanthe* and section *Spatulata*

Thirty of the thirty-one hybrids between sections of the subgenus *Dendrobium* are between the sections *Phalaenanthe* and *Spatulata*. No doubt the reason for this is that all species in these two sections are popular with orchid enthusiasts and of great horticultural value. Like their parents, the hybrids require good light, ample air movement, high humidity, a rest in the winter and a minimum temperature of approximately 12°C. Few of the very early hybrids are available now in Australian collections.

D. x superbiens is a natural hybrid between *D. bigibbum* and *D. discolor*. It is a popular plant and is regularly seen in collections. Although it is a very tall plant and requires lots of room, it is very florific. A large mature plant will carry many racemes of approximately 20 pale-pink to deep mauve-purple flowers. Some clones flower almost continuously.

D. Judy Leroy (*D. canaliculatum* X *D. affine* (as *D. dicuphum*)) is a small plant with a long, upright raceme of numerous (25 to 30) white flowers with delicate pink markings. An excellent parent as it passes on its very florific, long-lasting flowers and upright raceme characteristics to its progeny.

D. Mini Pearl (*D. canaliculatum* X *D. bigibbum*) is another very florific hybrid with a reasonable small and compact plant habit. The flowers are pink to dark mauve and the shape is about midway between its parents. It generally carried 12 to 20 flowers.

D. Eleebana (*D.* Judy Leroy X *D. bigibbum* var. *compactum*), when made using the small pseudobulbous stem form of *D. bigibbum*, produces a small and compact plant with longish racemes of delicate pink to mauve long-lasting flowers.

The other hybrids in this category are: *D.* Albanense, *D.* Albertine, *D.* Australia, *D.* Blue Gloucester, *D.* Blue Tiki, *D.* Bluey, *D.* Cherry Dance, *D.* Chieno, *D.* Dang Toi, *D.* D'Bush Chic, *D.* Donie, *D.* Dusky Girl, *D.* Frances Henton, *D.* Gloucester Gem, *D.* Gold Baby, *D.* Gresford, *D.* Lloyd Stainton, *D.* Lowana Tiki, *D.* Matilda, *D.* Moira Stewart, *D.* Penang Angel, *D.* Princess Stephanie, *D.* Rajah, *D.* Saskia and *D.* Three Star.

Hybrid between section *Phalaenanthe* and section *Trachyrhizum*

The only hybrid between section *Phalaenanthe* and section *Trachyrhizum* is *D.* Pink Ballerina, its parents are *D. bigibbum* and *D. agrostophyllum*. The plant is quite large like both its parents. The arching to pendulous raceme carries six to eight mauve flowers. It is cool growing.

Hybrid between section *Phalaenanthe* and section *Eleutheroglossum*

Between the section *Phalaenanthe* and the section *Eleutheroglossum* only one hybrid has been made (*D. fellowsii* X *D. bigibbum*), although at present it is unnamed and does not, therefore appear in the list of hybrids registered. The plant is like a small *D. bigibbum* and the flower colour also favours this species, however the flower shape and habit resembles *D. fellowsii*. So far it has only carried one or two flowers on a short raceme.

Future hybrids

The future of hybridizing in Australia using only indigenous species and their progeny holds a lot of promise. At present it is in the creative stage and will be another 10 years, at least, before any regular advances of a high standard can be seen. There have already been a number of breakthroughs, including new colours, larger flowers, better-shaped flowers, cool-growing hybrids using part-tropical species, and compact-growing plants with upright racemes, but these are isolated cases; only time will judge the real progress.

Australia is a large country with many different climatic conditions, and the hybridist needs to consider this. The native orchid growers in the north of Queensland find it is difficult to flower the delightful species *D. kingianum* because of their tropical weather. They need a warm-growing hybrid that looks like *D. kingianum*. The reverse applies with the enthusiast in the extreme south wanting to grow the warm-growing types in a shade-house. Nobody today wants to pay the excessive heating costs of a greenhouse.

One of the characteristics not yet achieved on a regular basis is a good habit of raceme, with flowers looking at the beholder instead of facing downwards. It is necessary, if *Dendrobium* orchids are to compete with all other orchids, to increase the duration and frequency of flowering. The future must contain hybrids that flower more than once a year and in all months of the year.

The length of time the flowers last is an important characteristic. Initially it will be necessary to look to *D. affine*, *D. agrostophyllum*, *D. bigibbum*, *D. canaliculatum*, *D. johannis*, *D. trilamellatum*, etc. in the subgenus *Dendrobium*. This will produce some warm growing hybrids, so it will be necessary to breed this factor out if cool growing long lasting hybrids are required.

The hybridist must consider the florist trade. At present, thousands of racemes of dendrobiums are imported into Australia from Asia. In the future, flowers of equal quality and durability have to be produced to compete, and these blooms exported from Australia. The advantage over the Asian dendrobes is that Australian plants can be grown in cool conditions.

Consideration should be given to producing small, compact-growing plants that are very florific and easily grown so that they can be used as a decorative indoor plant. Between flowerings they could be stood out in the garden to recuperate.

The potential of the Australian native *Dendrobium* hybrid is great, and we have the material to work with. It is necessary to bring all the good characteristics together so that we will achieve a desirable hybrid of outstanding excellence.

A number of the hybrids are pictured in plates 99–201.

6

Pollination and seed raising

The Universal English Dictionary defines pollination and fertilization as:

Pollination The fertilization of the ovules of plants with pollen.

Fertilization Act or process by which the reproductive cell is impregnated by the male cell.

The genus *Dendrobium* is hermaphroditic—the male pollinia and the female stigma are present in the one flower. Pollination takes place when the pollinia is placed into the stigmatic cavity. If the pollinia and the stigma are of the same flower, or another flower on the same plant, then self-pollination has taken place (a selfing). If they are from different clones of the same species, then cross-pollination has taken place (a sibling cross).

The ovary contains ovules, which develop into seeds, and is connected to the stigmatic cavity by a narrow channel.

The stigmatic cavity contains a sticky secretion, frequently referred to as the stigmatic fluid, and once the pollinia has been pressed into the fluid it will not fall out.

At the time of pollination the ovules are only in a rudimentary state, so the placing of the pollinia into the stigmatic cavity acts in the first instance as a stimulant to induce the ovules to mature. The pollen grains then germinate in the stigmatic fluid and produce long slender tubes, called pollen tubes. These grow through the narrow channel from the stigmatic cavity to the ovary. As the pollen tube elongates, two male gametes are produced, called the sperm. The sperm moves down the pollen tube and as the tip of the tube enters the ovule so its contents are discharged. One sperm nucleus unites with the egg nucleus to form the zygote, and the other sperm nucleus unites with the other nuclei in the ovule to form the endosperm—the food storage tissue for the developing embryo.

Australian dendrobiums growing in their native habitat are mainly pollinated by bees, wasps or birds. The pollinating process usually starts by the bird or insect visiting the flower for a specific purpose, such as in search of nectar or because of the flower's perfume. On the way into the flower a small amount of a sticky substance is smeared on to the insect's back or on the bird's beak. When leaving the flower the insect or bird dislodges the anther so that the pollinia falls on to the sticky substance. This rapidly dries, holding the pollinia firmly in position. In the next flower visited, as the pollinator backs out, the pollinia are deposited into the stigmatic fluid, generally being dislodged by catching on the rim of the stigmatic cavity or under the rostellum. This process is helped by the pollinator's struggles and by the labellum holding the pollinator close to the stigma.

In the genus *Dendrobium*, the process to achieve pollination is not as complex as in some genera, but nevertheless is most effective. Pollinating agents are only known for a few of the native dendrobiums. *D. smillieae* is pollinated by a bird and its flowers are well suited to this form of pollination, as they are tubular, odourless, brightly-coloured, of heavy substance and waxy in appearance, and have an abundant supply of nectar. The pollinia are a slate grey to

153 *D.* Kungara

154 *D.* Ku-Ring-Gai

155 *D.* Lowana Nioka

156 *D.* Lorikeet

157 *D.* Memoria Lloyd Bradford

158 *D.* Karen Nair

159 *D.* Minnie

160 *D.* Merritt Island

161 *D.* Michael Jupp

162 *D.* Mini Pearl

163 *D.* Narraweena

164 *D.* Nugget

165 *D.* Nunkumbil

166 *D.* Orana

167 *D.* Peach Glow

168 *D.* Our Native

169 *D.* Peter

170 *D.* Peter Hewitt 'Port Wine'

171 *D.* Peewee

172 *D.* Penny Rose

173 *D.* Peewee 'Harlequin'

174 *D.* Reg Burns

175 *D.* Penny Ann

176 *D.* Peter Petersen

177 *D.* Phil Deane

178 *D.* Pink Ballerina

179 *D.* Pink Pencil

180 *D.* Red Baron 'Formal'

181 *D.* Regal Rose 'Royalty'

182 *D.* Ring of Fire

brown colour, matching the beak. If the pollinia were yellow it would, no doubt, be noticeable to the bird and rubbed off.

A large number of the known orchid pollinators are bees and the main characteristics of flowers pollinated by bees and wasps are a fresh and generally sweet odour; a prominent lip; and colour from purple, blue or violet to yellow, green and white. They have nectar guides in the form of coloured lines mainly on the labellum. Although a large number of the Australian *Dendrobium* species fit this category, their ability to supply nectar is not very noticeable and in some possibly non-existent. Occasionally there is a shallow nectary at the base of the column foot. What really attracts bees needs further investigation, maybe it is edible tissue.

Pollination generally accelerates senescence; the petals and sepals start to wilt and wither within a day or two of pollination taking place. The actual fertilization of the ovules doesn't take place for some considerable time, weeks or even months, depending on the species. The ovary enlarges in a fairly short time. This, in the first instance, is caused by the ovules maturing to be ready to receive the pollen grains. The wilting flower ceases to attract pollinators and perfume emission ceases, ensuring that the pollinator visits other blooms and does not waste time visiting an already-pollinated flower. The column swells, changes its shape and seals the pollinia into position, protecting it against removal or damage. In the sections *Phalaenanthe* and *Spatulata*, the flowers do not wilt and wither as quickly after pollination as in other species.

D. mirbelianum and *D. stuartii* appear to be self-pollinating to a varying degree. I have seen clones of both species that are self-pollinating and not self-pollinating. *D. stuartii* is on occasions cleistogamous. Some species with known pollinating agents are given in Table 3.

Table 3 Dendrobium species and their known pollinators

Species	Pollinating agent	Reference
D. bigibbum	Bee of family *Colletidae* or *Halictidae*	Jones (1983)
D. canaliculatum	Bee of family *Colletidae* (*Hylaeus ruficeps* and *H. theodorei*)	Jones (1983)
D. jonesii	Bee (*Homalictus* sp.)	Jones and Gray (1977)
D. lichenastrum	Native bee (*Trigona carbonaria*)	R. Smythe (1970c)
D. linguiforme	Bees (*Euryglossina hypochroma* and *Euryglossina lynettae*)	Jones (1974)
D. linguiforme	Fly (*Syrphus viridiceps*) Wasps — Thynnid wasps	Gilbert (1958)
D. luteocilium	Native bee (*Ceratina* sp.)	R. Smythe (1970c)
D. pugioniforme	Bee	Jones (1983)
D. speciosum	Native bee (*Trigona carbonaria*)	R. Smythe (1970b)
D. smillieae	Bird — common yellow honeyeater (*Meliphaga flava*) of north Queensland	R. Smythe (1970a)
D. teretifolium	Bee (*Trigona carbonaria*)	Jones (1983)

Artificial pollination

Artificial pollination in the genus *Dendrobium* is a relatively simple process, merely entailing the removal of the pollinia from beneath the anther cap and the placement of that pollinia into the stigmatic cavity. This process is illustrated in Fig. 97. The steps involved in pollination are as follows:

- Select the parent to bear the capsule and remove its pollinia, using a toothpick or similar instrument. (This is commonly referred to as the pod parent.) In some of the smaller-flowered species it may be necessary to remove the labellum of the capsule-bearing parent to make the stigmatic cavity more accessible. I prefer not to do this, but to hold the labellum out of the way with my fingers while inserting the pollinia.
- To obtain the pollinia from the pollen parent, hold a clean piece of paper, or a plastic spoon, beneath the apex of the column (this is to catch the pollinia should it fall). With the other hand, take a clean toothpick or similar pointed instrument, gently touch the fluid in the stigmatic cavity, then lift the anther cap with the toothpick; at the same time collect the pollinia from the chambers

A Column, side-view
B Column, front-view
C Flower, front-view
D Pollinia
E Column with anther lifted, showing pollinia
F Column, showing pollinia in position in the stigmatic cavity
G Side-view of column cut in half, showing pollinia in position in the stigmatic cavity

Fig. 97 Artificial pollination *Dendrobium kingianum*

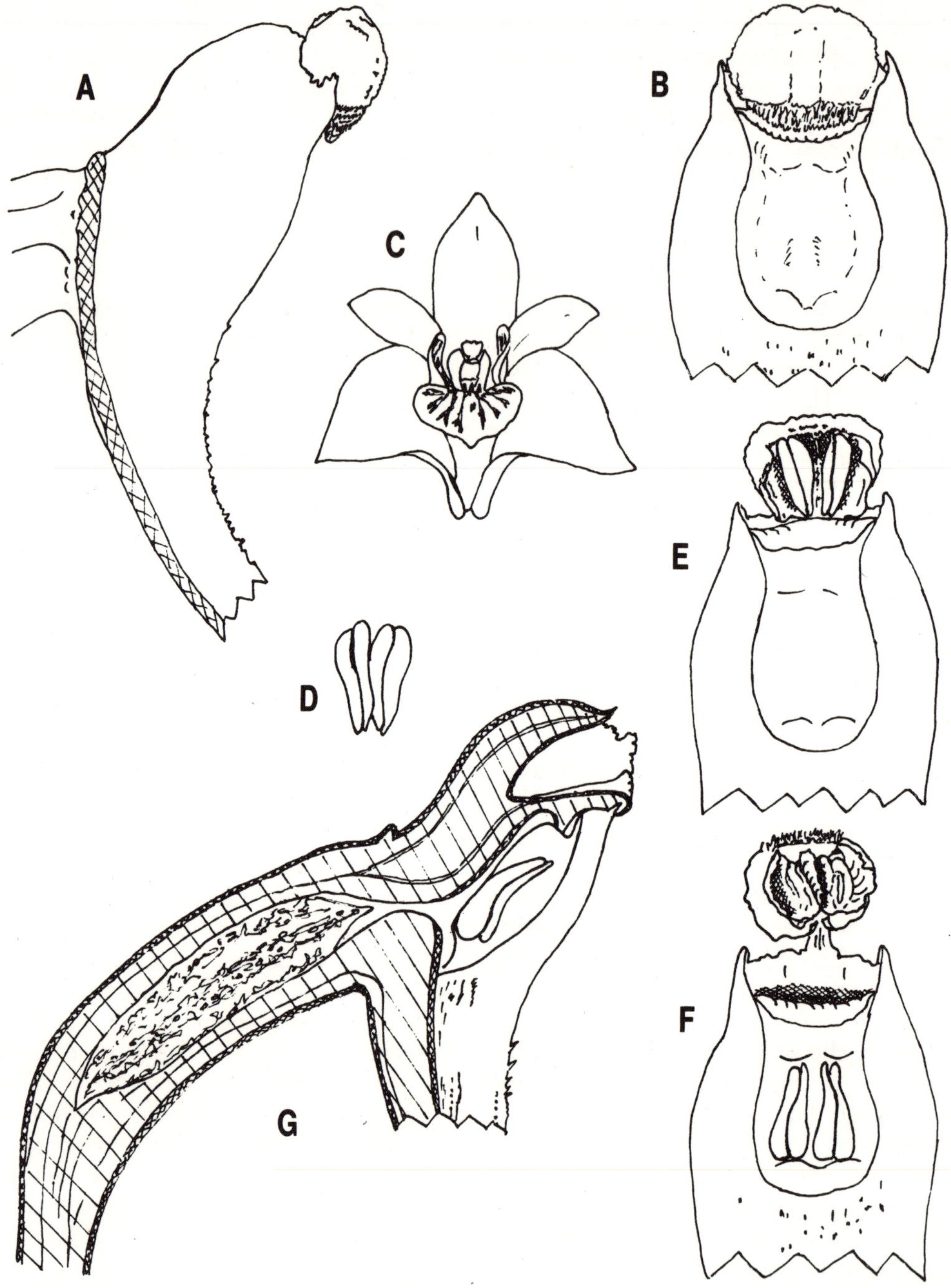

inside the anther by touching them with the sticky stigmatic fluid on the toothpick. The pollinia will normally stick onto the toothpick and stay there, but always have something beneath the pollinia to catch it should it drop. Frequently if you lift the anther too quickly the pollinia is suddenly ejected, as it catches on the rostellum and topples over and out. Do not touch the pollinia with your fingers.

- Place the pollinia collected from the pollen parent into the centre of the stigmatic cavity of the capsule-bearing parent, ensuring that you gently push the pollinia right into the stigmatic fluid.

- Label the flower pollinated, and make a record of the pollination, with all relevant details. It is a waste of time to hydridize without thought.

Once the hybrid programme has been thought out, fertilization chances are improved if flowers are selected from the centre of the raceme. The flowers at the basal end are too old by the time the terminal flowers open, and frequently the terminal flowers are small and poorer in shape and colour.

A flower must be mature before it is pollinated. Often the reproductive parts of the flower do not mature at the same time. The maturation of the pollen masses and the stigma should coincide for perfect pollination. Old pollen, a stigma beyond the receptive stage, sudden changes in temperature or even wrong temperature can sometimes hinder pollination.

Great care should be given to the choice of parent plants. In a set hybridizing programme, the plants to be used during the forthcoming flowering season are known, and these can be given special treatment to ensure they are strong and healthy and able to carry the capsules, and that the seed produced will be top quality. Preparation of the parent plants does not start when the flower opens, but many months before, in some cases after the previous season's flowers have died. Plants that are vigorous and in a good growing medium with plenty of nourishment are ideal parents and greatly assist capsule development.

A small plant of, say, three pseudobulbous stems, a seedling blooming for the first time or a weak, sick plant should not be used to bear a capsule. The energy it will require to bear a capsule and continue to grow satisfactorily may be too much. (Such plants may be used as pollen parents.) Once the plants for the hybridizing programme have been chosen, the stronger of the two should be used to bear the capsule, or (if they are both healthy and vigorous looking) both can bear capsules.

If one of the two parents has flowers much smaller than the other, it is advisable to take the pollen from the larger flower and place it into the stigma of the smaller flower. (If the pollinia is too large, then cut it to size with a clean razor blade.) The pollen tubes may be too short to reach the ovary if it is done the other way round.

The genus *Dendrobium* has four pollinia, but it is not necessary to use all four in the pollination of one flower, although I prefer to do so, or at least use two. One pollen mass can produce a capsule full of seed.

If it is desirable to pollinate two flowers that do not flower at the same time, it is quite easy to store the pollen by wrapping it in a piece of dry tissue paper, then placing it in a dry vial or glass tube and store it in a screw top glass jar on a shelf in the refrigerator. (Many hybridists store their pollen in a dessicator, but I have not found this necessary.) Pollen stored my way has remained viable for up to 1 year with a varying amount of success; in one instance using the 1-year-old pollen of *D.* sp. aff. *jonesii* and placing it into the stigmatic cavity of *D. tetragonum* var. *gigantum*, good germination was achieved. I frequently use pollen that is 6 months old with excellent results.

To check the viability of pollen, place a small piece into the stigmatic cavity of a flower you do not wish to use. After about 2 to 3 days, remove some of the stigmatic fluid from the flower, mix it with a single drop of water and place it between two glass microscope slides. The pollen grains with tubes just starting to germinate will be seen under approximately 250 times magnification if the pollen is viable. Testing the viability of pollen can ensure that a special flower is not wasted by attempting to pollinate it with dud pollen.

Capsule management

The perianth of a *Dendrobium* flower wilts within a few days of pollination. After the sepals, petals and labellum have wilted, they should be carefully cut off at their base. Great care should be taken in this operation or else the newly-forming capsule will come off too. Do not damage the column, which at this stage will be changing shape. The perianth segments are cut off to prevent possible fungal or other infection of the column and because they provide a great place for insect pests to hide. Capsules maturing in the hot humid months are particularly susceptible to fungal attack, so water should be kept off them.

In incompatible matings the capsule will frequently appear to be progressing well and then turn yellow and drop off, or the flower may shrivel and drop off after a few days. In the section *Dendrocoryne*, capsules will fall off after approximately 1 month if the parents are incompatible. In the sections *Phalaenanthe* and *Spatulata*, capsules may stay on for as long as 4 months before yellowing and falling off. Examination of these capsules usually shows that fertilization of the ovules did not take place.

The seed capsule needs constant attention while maturing. Insects, particularly grasshoppers and caterpillars, love them. They will eat out all the developing seed in the capsule, leaving just the shell. Capsules with little or no seed inside never seem to be attacked. Mice also eat out the contents of a capsule. Sudden falls or rises in temperature, or exposure to a draft, can cause cessation in the development of the pollen tubes. Put all plants with seed capsules into a glasshouse and keep the temperature as constant as possible, at a minimum of 16°C.

During the development of the capsule the plant should be kept clean and growing healthily. When necessary I use pressure-pack insecticides and fungicides over the capsules as they are easy to use and do not need mixing. Do not hold them close to the capsule or plant when spraying. Read the instructions on the pack.

The development of a capsule requires the production of much extra energy by the plant and it is necessary to help it keep up strength by giving extra, but weak, feeds. I use a little blood and bone sprinkled on the growing medium and watered in, as well as a weak fertilizer high in phosphorus, potassium, calcium, and magnesium.

Never allow a plant with a capsule on it to dry out completely, nor keep it too wet. It seems beneficial to keep the plant slightly on the dryish side.

The ripening or maturing time of a seed capsule varies greatly in the genus *Dendrobium*. In Australia it can vary from 55 days in some clones of a species of the subgenus *Athecebium* to approximately 280 days in section *Phalaenanthe* of the subgenus *Dendrobium*.

The list below gives the suggested optimum harvesting time for greenpod culture and the maturing time of a capsule—the first sign that the capsule has commenced to dehisce. These times are compiled from my hybridizing records covering the last 25 years and are taken from capsules that have produced good germination. Hybridists using this information must take into account, and compensate for, their own local conditions. Capsule opening can vary greatly between different clones of a species so the periods given should be used as a guide only.

The capsule bearing parent is usually responsible for determining the harvesting or maturing times.

It is important to keep records. I do, and they are very valuable indeed. Not only do they tell how long it takes a particular parent to mature its capsule, but they become vital documents when considering which parent to use. As the years go by you get to know the dominant and/or recessive characteristics of each plant.

As a seed capsule matures it will generally change from green to yellow, and as it dries it will turn brown. But no definite rule can be given on the colour of the capsule at the

Table 4 Harvesting and maturing times of seed capsules

	Suggested optimum harvesting time for greenpod culture (days)	Maturing time—first signs of dehiscence occurring (days)
D. adae	65– 75	85–100
D. aemulum	60– 65	65– 75
D. affine	130–140	150–160
D. agrostophyllum	90–100	110–120
D. antennatum	160–170	160–190
D. baileyi	60– 70	75– 85
D. bifalce	140–145	150–160
D. bigibbum	160–185	200–280
D. bowmanii	55– 60	65– 80
D. calamiforme	60– 65	70–115
D. canaliculatum	95–100	100–140
D. cancroides	65– 75	80–100

time of opening. The temperature, light, humidity, state of the plant, etc. can have a great bearing on this. Frequently capsules in the subgenus *Athecebium* open at the apical end while still green. Capsules matured in strong sunlight will turn to quite a dark purple early in their development. (A capsule on *D.* x *foederatum* was dark purple from its initial development and never did turn yellow, but then it dried to dark brown.) Unless conditions are fairly constant during development (even temperature, light, etc.) capsules will open erratically.

It will be noticed from Fig. 98 that the capsule opens at the tip first and then gradually splits away from the suture (a strip where the margins of the carpels join), along the whole length of the carpel. The sides of the carpel roll back on to its top surface tearing themselves away from the suture, the action is almost like a zip fastener.

Dispersal of the seed is, in most cases, a controlled action; it does not spill out all at once. The period of dispersal can be from a few days to weeks, depending on the climatic conditions. This ensures that the right conditions will occur at least on some of the days while the capsule is opening. If there was no breeze at all, the seed would spill out and not float further than the close vicinity of the parent, instead of being carried by the wind or by air currents to the far corners of its habitat.

The movement and regulation of the seed from the capsule is helped by the hygroscopic action of long fine hairs interspersed among the seed. When studying these hairs and seeds under the microscope, the hairs (after a while) begin to wriggle and then continuously move, no doubt due to the moisture from your breath.

Orchid seed is very tiny. It is understood to be the smallest in the plant kingdom, and is almost without any real food reserves. *Dendrobium* seed comprises an embryo and a thin seed-coat or testa. (Fig. 99) In all Australian *Dendrobium* seed that I have examined, the embryo is yellow and the seed coat is a papery thin translucent white to cream colour. Seed varies greatly in size; *D. insigne* has by far the largest seed.

Table 4 (con't)

	Suggested optimum harvesting time for greenpod culture (days)	Maturing time — first signs of dehiscence occurring (days)
D. carronii	90– 95	100–110
D. carrii	120–140	150–170
D. cucumerinum	55– 60	65– 80
D. discolor	150–160	180–205
D. fairfaxii	60– 65	70–115
D. falcorostrum	80– 90	90–105
D. fellowsii	120–130	140–180
D. fleckeri	65– 75	80– 90
D. insigne	65– 70	70– 80
D. johannis	100–120	130–150
D. johnsoniae	160–180	200–220
D. jonesii	60– 65	70– 80
D. sp. aff. *jonesii*	50– 55	60– 80
D. kingianum	55– 60	60– 90
D. lichenastrum	70– 75	90– 95
D. linguiforme	60– 65	70– 80
D. lobbii	50– 55	60– 65
D. luteocilium	50– 55	60– 65
D. macropus (Norfolk Island)	60– 70	80– 85
D. macropus subsp. *gracilicaule*	65– 75	80– 85
D. malbrownii	55– 60	65– 85
D. mirbelianum	100–110	120–160
D. monophyllum	140–150	160–170
D. moorei	65– 75	80– 85
D. mortii	50– 55	55– 65
D. nindii	150–155	155–200
D. prenticei	70– 75	90– 95
D. pugioniforme	50 60	55– 75
D. racemosum	60– 65	70– 80
D. rigidum	75– 85	90–105
D. schneiderae	65– 70	70– 80
D. schoeninum	55– 65	70– 80
D. smillieae	140–150	160–200
D. speciosum	65– 75	85– 90
D. speciosum var. *pedunculatum*	70– 80	90– 95
D. striolatum	60– 65	75– 85
D. stuartii	140–150	160–200
D. teretifolium	60– 65	70–115
D. tetragonum	70– 80	80– 90
D. tetragonum var. *giganteum*	75– 85	90–105
D. toressae	70– 75	90– 95
D. tozerensis	50– 55	60– 65
D. trilamellatum	85– 95	115–130
D. wassellii	65– 70	75– 80

Fig. 98 Seed capsules

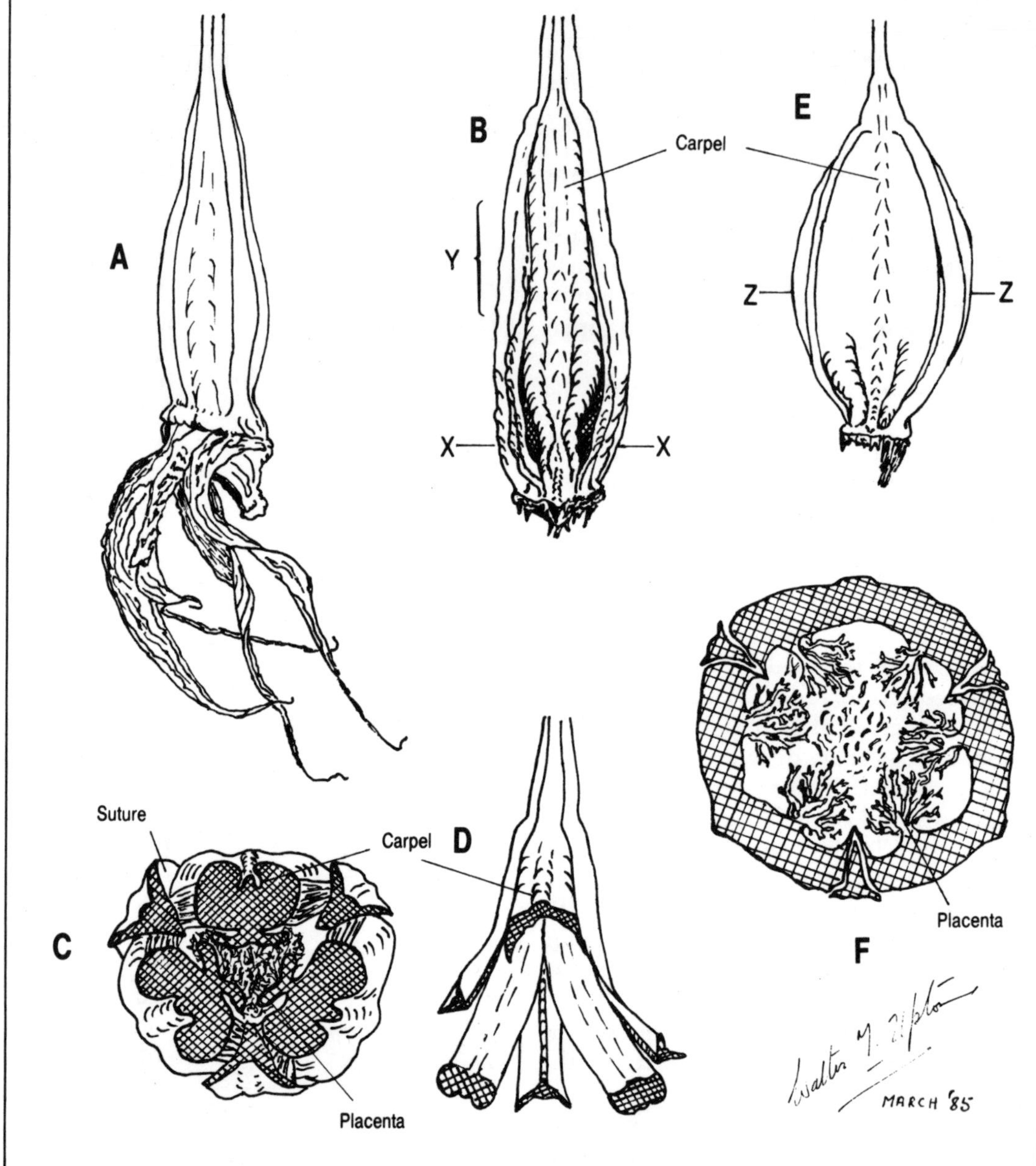

A *Dendrobium* Peewee X *D.* sp. aff. *jonesii* 18 days after pollination. Withered sepals and petals are ready to be cut off 3X

D. Peewee X *D.* Star of Gold 78 days after pollination

B Capsule opening at the apex 3X

C Cross-section at X-X, looking into capsule showing placenta and seed 6X

D Base of capsule, cut open at Y, seed removed, showing the sutures and carpels 3X

D. agrostrophyllum 110 days after pollination

E Seed capsule just commencing to open 3X

F Cross-section at Z-Z showing placenta and some of the thousands of seeds 6X

Seedling management

In their natural habitat, the seeds are borne by air currents to wherever they may settle, but their chances of survival are very small. As the seed is very tiny and has no real food reserves, it will not generally germinate unless a beneficial symbiotic fungus invades it and provides the food needed for that initial growth. For further information on this interesting phase of the development of orchid seed read Dressler (1981).

Dendrobium seedlings can be grown by

Fig. 99 *Dendrobium* seeds and protocorms

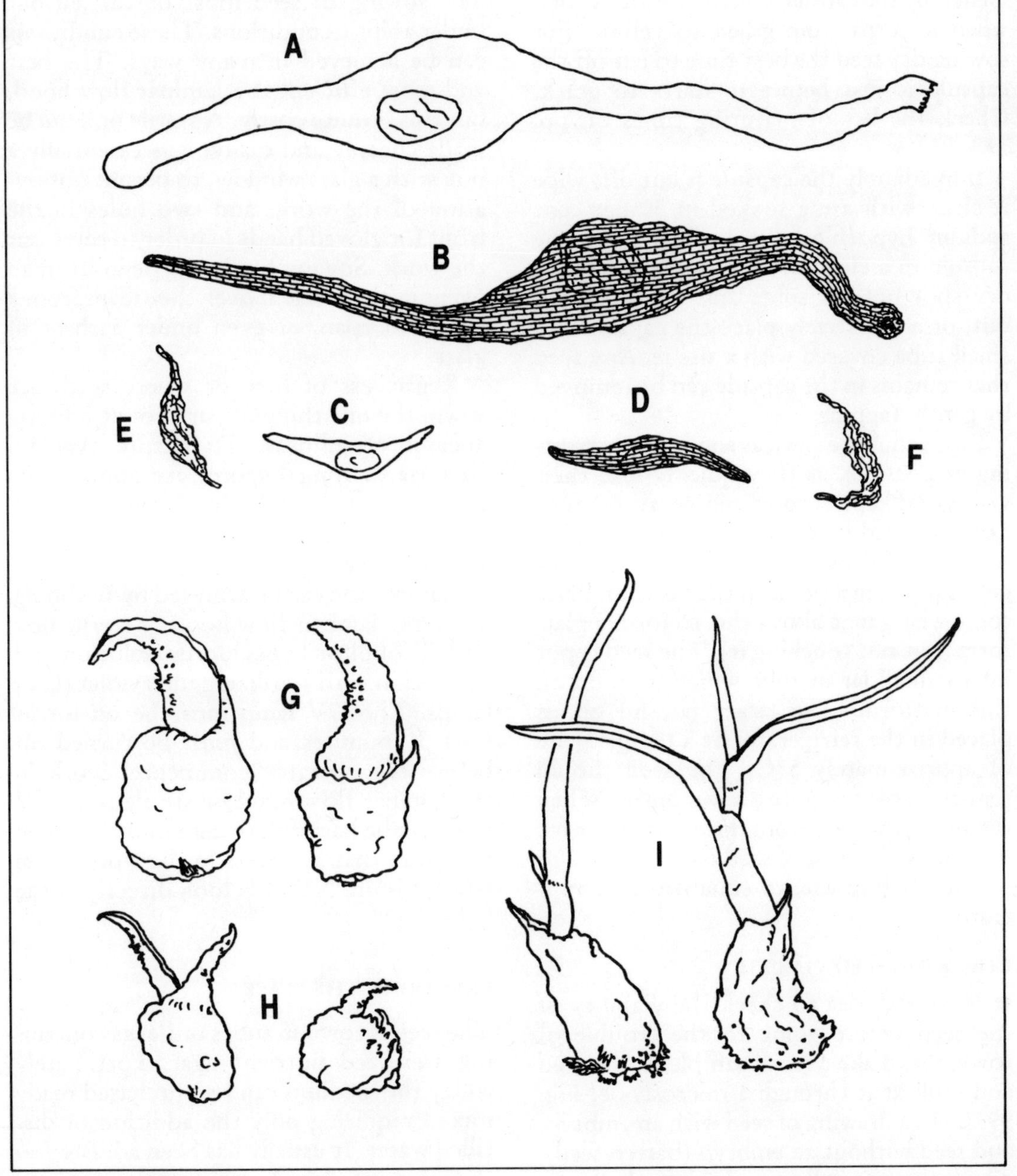

Seed of *D. insigne* 70X
A Showing embryo
B Showing net-like seed-coat
Seed of *D. agrostophyllum* 70X
C Showing embryo
D Showing net-like seed-coat
E & F Distorted barren seed
Protocorms of *D. agrostophyllum*
G 110 days after sowing seed 15X
H 130 days after sowing seed 10X
I 155 days after sowing seed 10X

this natural method by sowing seed on, or close to, the roots of healthily growing plants. The success rate is generally poor. In a large collection it is not unusual for tiny seedlings to appear at the base of plants if seed capsules are left on plants to mature.

It is, however, possible to raise *Dendrobium* seedlings asymbiotically (without using a fungal relationship), but it is necessary to provide them with all the sugars, minerals, salts, etc., in the form they need to make them grow. The procedure is very time consuming and requires a high degree of sterility, both of the growing media and of the seed, and a very clean sterile work area. But is really quite simple, if approached with a cool mind and keen determination.

In *Orchid Biology. Reviews and Perspectives II*, edited by J. Arditti (1982), there is a complete manual on 'Orchid Seed Germination and Seedling Culture'. The requirements are explained very briefly in the following paragraphs.

COLLECTION OF SEED

Generally the capsule is ready for harvesting when it turns from green to yellow. For sowing dry seed the best time to cut off the capsule is just before it starts to crack. Check the list of maturing times on pp. 178–9.

Immediately the capsule is cut off, wipe it clean with a rag soaked in 50 per cent sodium hypochlorite solution. Place the capsule in a clean dry envelope in a warm dry spot until it splits and the seed falls out, or alternatively place the capsule in a small tube covered with a tissue. Any seed that remains in the capsule can be removed by gently tapping.

Seed should be sown as soon after harvesting as possible, as they quickly lose their viability if kept at room temperature. Seed can be stored in a screw-top glass jar with some anhydrous calcium chloride or silica gel wrapped in a tissue in the bottom. Place some wire gauze above this to form a platform (but not touching it). The seed is put into a small jar or tube which is stood on this platform. The larger jar, lid on, is placed in the refrigerator at a temperature of approximately 5°C. The seed should remain viable for quite a few months. When removing the seed from the refrigerator for use, allow the tube of seed to stand for 2 to 3 hours before use, to equalize the temperature.

CHECKING SEED VIABILITY

It is a good idea to check the viability of the seed before going to the trouble of sowing it. Take a small sample of the seed and look at it through a microscope. Fig. 99 shows a drawing of seed with an embryo, and seed without an embryo (barren seed). The embryo is yellow and the seed-coat or testa is white to cream. The barren seed has no embryo and the seed-coat is generally distorted.

STERILE WORKING AREA

The sowing of seed must be carried out under aseptic conditions. These conditions can be achieved in many ways. The best and most efficient is a laminar flow hood, but this is quite costly. A sterile box can be made cheaply and easily; it is essentially a box with a glass window, to permit observation of the work, and two holes in the front for gloved hands in order to carry out the work. Sowing can be carried out in an oven, in the kitchen over the steam from a large saucepan, or even under a sheet of glass.

Regardless of how or where seeds are sown, the one thing of paramount importance is cleanliness. To ensure that no bacteria or fungal spores are going to be sown with the seed, all instruments, etc. used must be sterilized, as well as the working area.

Sterilization can be achieved by swabbing down the laminar flow hood or sterile box with alcohol or hypochlorite solution and irradiation with sterilizing ultraviolet (UV) lamps. The UV lamp must be on for at least 30 minutes and *must* be turned off before the operator commences work in the laminar flow hood or sterile box. UV light can be harmful, so care must be taken to ensure that no one is in the work-room when it is on. NEVER look directly at the light.

PREPARING THE FLASKS

The seed is sown in tubes or flasks containing sterilized nutrient agar. Fortunately today the medium can be purchased ready mixed requiring only the addition of distilled water. It usually has been adjusted to the correct pH. A satisfactory growing medium for Australian *Dendrobium* is Knudson's Formula C which contains:

Calcium nitrate	$Ca(NO_3)_2 4H_2O$	1.00 g
Potassium phosphate	KH_2PO_4	0.25 g
Magnesium sulphate	$MgSO_4 7H_2O$	0.25 g
Ammonium sulphate	$(NH_4)_2SO_4$	0.50 g
Ferrous sulphate	$FeSO_4 7H_2O$	0.025 g
Manganese sulphate	$MnSO_4 4H_2O$	0.0075 g
Sucrose (sugar)		20.00 g
Agar		18.00 g
Distilled water		1000 mL

183 *D.* Pixie 'Biscuit'

184 *D.* Star Imp

185 *D.* Tawhai

186 *D.* Telekon

187 *D.* Tooloom

188 *D.* Rosella 'Pillar Box'

189 *D.* Silvester

190 *D.* Star of Gold 'Jill'

191 *D.* Star of Riverdene 'Jill'

192 *D.* Twinkle

193 *D.* Tweed

194 *D.* Warrambool

195 *D.* Warringah 'Jill'

196 *D.* Wesley Pink

197 *D.* Wesley Rose 'Ruffles'

198 *D.* Wirruna

199 *D.* Wonga

200 *D.* Yondi 'Dip Falls', photo S. Batchelor

201 *D.* Bardo Rose 'Jill'

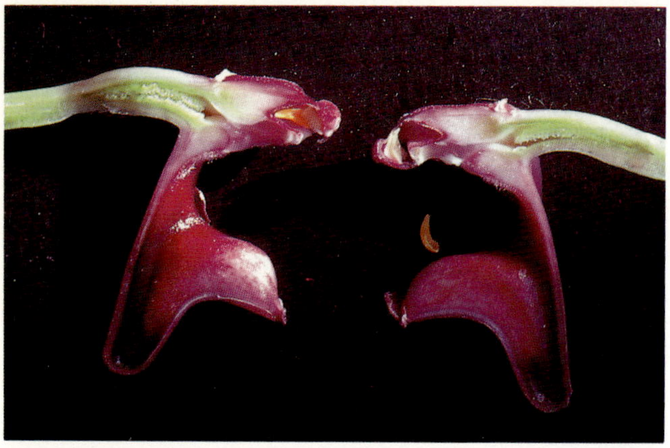

202 *D. bigibbum*, ovary, column and spur cut in half

203 *D. bigibbum*, capsule, showing reduction of column

204 *D.* x *superbiens* hybrid capsule; note ribs folding back 4½ months

205 *D. bigibbum* subsp. *bigibbum* var. *superbum* capsule cut in half

206 *D. prenticei*, one capsule just opening

207 *D. prenticei*, capsule fully open after 4 days

208 *D.* x *superbiens* hybrid, inside capsule showing carpel join

209 *D. bigibbum* capsule opened out

The medium is poured into flasks or bottles, about 1 to 2 cm in depth. The flasks are then stoppered and placed in an autoclave or pressure cooker and sterilized at 105 Kpa (15 pounds pressure per square inch) for 20 minutes. At this point the sterile area is prepared and all sterile instruments, flasks, etc. are placed inside so that from this point on everything is as sterile as possible.

STERILIZING THE SEEDS

Once a seed capsule has opened, it is assumed that the seed is contaminated with bacteria and fungus spores. Thus it must be sterilized to kill the live bacteria and fungus but, at the same time, not kill the seed. Calcium hypochlorite, 10 g, dissolved in 150 mL of water and filtered, is a good sterilizing agent, or a 10 per cent solution of household bleach.

Place the seed in a small test-tube with only a little of the sterilizing solution and leave for about 10 to 15 minutes. To ensure thorough surface sterilization, shake the tube vigorously every few minutes. This will get rid of any air bubbles that may attach themselves to the tiny seeds, and provide hiding places for bacteria or fungal spores. In persistent cases the addition of a drop of household detergent will act as a wetting agent and ensure that all the seeds are sterilized.

After 10 to 15 minutes the sterilizing solution is carefully extracted from the tube with an eye dropper, and sterile distilled water is added to rinse the seed. Sometimes I do not bother to rinse in sterile distilled water, but transfer the seed directly from the sterilizing solution to the flask; this seems to work just as well. If left in the sterilizing agent too long, much of the seed can be lost due to burning.

SOWING THE SEED

Ensuring that no bacteria or fungal spores contaminate the seed or media during sowing is a problem. With a laminar flow hood or a sterile box it is not too difficult to keep unwanted spores out, but in the kitchen (or similar situation) much more dexterity and skill is required. Having sterilized and rinsed the seed, suck it up into a sterile inoculating needle, eye dropper or similar, remove the stopper from the flask and quickly place a drop of the seed saturated distilled water onto the nutrient agar. Close the flask ensuring that the stopper is still sterile. Spread the seed over the nutrient agar by flicking the sterile distilled water across it until evenly distributed. Do not place too much seed in the flask, as a lot of it will die due to overcrowding and insufficient nutrient. This, in turn, makes it much more difficult to replate or reflask at a later date.

SOWING GREEN CAPSULE (GREENPOD) SEED

Greenpod seed is not completely ripe. The seed capsule is removed before it opens but after fertilization. The list in Table 4 (pp. 178–9) gives the approximate times to remove capsules. Hybrids made from parents with differing harvesting times usually favour the capsule-bearing parent.

After harvesting the seed capsule, thoroughly clean it with soap and water, scrubbing with a small brush. Rinse with water and then submerge it in 30 per cent household bleach for 30 minutes. Take the capsule out of the household bleach in the sterile area, and rinse in sterile distilled water. Hold the stem end of the capsule and, with a sterile scalpel, slice off the top. Ascertain if there is any free seed in the capsule, if so hold the capsule over the flask and tap the pod with the scalpel and so release some seed into the flask. Do not place too much seed in the flask and ensure that it is spread evenly over the surface. If there is no free or loose seed, slice the capsule with the sterile scalpel along the side of a suture and then again along the suture on the other side, taking out a slice (like a piece of melon). Scrape the seed from this slice into the flask. If there does not appear to be any seed in the capsule, and it is an important plant, scrape the whole placenta into the flask hoping that perhaps one seed is viable and still attached to it. The interior of a perfect green capsule is completely sterile so there is no need for sterilization.

The advantages of green capsule culture are numerous. Time is saved by harvesting the capsule earlier, there seems to be a much greater number of seeds and, of course, it is not necessary to sterilize the seed.

STORAGE OF FLASKS

After sowing, all flasks need to be moved to a place where they can be left undisturbed until the seeds germinate and the seedlings grow. A bright, warm position is quite satisfactory, but not in direct sunlight. A greenhouse is not satisfactory, as there is usually too great a fluctuation in temperature (unless well ventilated). The perfect conditions for growing adult plants do not suit flasked seedlings.

Flasks should be regularly inspected. If they are grown in a very dry atmosphere, the nutrient agar cracks and shrinks away from the sides of the flask, necessitating replating. Contaminated flasks should be immediately removed and the contents destroyed. Before opening a flask that has mould in it, sterilize it to kill the mould spores. Breathing these spores can be harmful.

I place my flasks in a rack with 40 watt fluorescent lights approximately 30 cm above them, either white light or Grolux tubes seem to be satisfactory. The tropical species require more light than the cool-growing dendrobiums. Too much light may bleach the seedlings. Immediately after they are sown I place three sheets of newspaper on the top of the flasks until the embryos start to swell (about 10 days in the *Dendrocoryne* section, sometimes longer in the *Phalaenanthe* section); at this point I remove one sheet of newspaper. When the protocorms appear, usually at approximately 3 to 6 weeks, I remove another sheet of newspaper. By the time the root hairs have appeared and the leaf is noticeable, I remove the last sheet. The flasks receive 12 hours of fluorescent light per day. Ideally the temperature for the tropical species should remain fairly constant and should not drop below 16°C or go above 20°C. A fluctuation of more than 7°C seems to retard growth. The temperate species are more tolerant of cooler temperatures and of greater fluctuations. The temperature can go down to 10°C (without greatly affecting growth.

REFLASKING OR REPLATING

From a few to several weeks after the seed has been sown, the embryos will turn green and growth will commence; a protocorm at first, then a leaf and so the seedling will develop. Fig. 99 shows the stages of growth and plates 210 and 211 show the actual protocorm and seedlings in the flasks.

The process of reflasking (commonly referred to as replating) is the transferring of tiny seedlings from the germinating or mother flask to a flask which contains different nutrients for more effective root and leaf development. Obviously this operation takes longer than sowing seed and it is therefore much more difficult to maintain a sterile situation without a sterile box or laminar flow hood.

I like to replate when the seedlings are about 5 to 8 mm high and the root hairs are showing; this is from about 10 weeks after sowing, but no general rule can be given on how long this may take. After replating I place one sheet of newspaper over the flask for about 2 weeks while the seedlings settle down.

An important point to remember in replating is that the seedlings will not remain small so too many should not be put in the flask. Overcrowding can soon occur and the smallest plants may even die. Seedlings should be graded according to size. A second replating may in some instances be beneficial.

A replating fork or a pair of long tweezers is used to transfer the seedlings. A replating fork can be made out of a 2 mm stainless steel rod flattened at one end with a hammer for approximately 1·5 cm and bent forward making a quarter circle. A V shape is then cut down the centre to give two pointed fork prongs.

REMOVAL OF SEEDLINGS FROM FLASKS

When the seedlings are large enough, usually at least 25 mm high, they are ready to be removed from the flask. The length of time taken to get to this stage varies greatly, depending on the conditions, temperature, light, season, and upon the species or hybrid. Some vigorous growers are ready 9 months from sowing, others are very slow and may require a second replate to boost them on a bit. Some very slow growers take up to 2 years.

Remove the tiny seedlings from the flask with great care, treat very gently and do not handle more than necessary. Thoroughly rinse in clean water to remove all traces of nutrient agar, any that remains could help fungus activity. If removing in winter, I add a little warm water to the rinsing bowls, just sufficient to take the chill off the water (this particularly applies to the tropical dendrobiums). Although it may not be necessary, I soak my tiny seedlings, after rinsing, in a bowl of approximately 4 litres water (with the chill off in winter) containing two drops of Previcur (a systemic fungicide), two drops of Formula 20 (a plant growth stimulant) and half a teaspoon of Aquasol (a soluble fertilizer with an N:P:K. of 23:4:18) for 10 minutes. The seedlings are then laid on clean newspaper for approximately 2 hours so that any surplus moisture can drain off.

The next step is to grade the seedlings from largest to smallest. The very largest can be potted straight into 50 mm pots, the rest go into community pots or trays, keeping the various sizes together so they can be managed more effectively. If the sizes are all mixed up, then later when the largest are ready to be potted, the small ones will have to be unnecessarily disturbed and then potted again into another community pot.

I prepare my community tray or pot by placing moist bark, medium grade, in the bottom third, then seedling growing medium (see next section) right to the top of the tray, or even slightly higher to ensure better air movement around the seedlings. When potting numerous flasks of seedlings at the same stage I use large plastic community trays 50 mm deep and 30 cm by 25 cm. Anything up to 150 tiny seedlings are placed in one tray. A trench is made across the width; the depth of the trench is determined by the root length. The seedlings are placed in a row along one side of the trench, then the medium is pushed up to them and around them so that they are firmly in position. A second trench is then made about 25 mm from the first one, and so on. If some of the seedlings are in a clump that are difficult to separate, I plant them in these clumps as they are easier to separate when they have grown and are ready for planting into individual pots.

On completing the potting I give the seedlings a good watering using a mister nozzle, which settles the growing medium around them. This is followed by a quick spray (using a hand spray) with a weak mix of a systemic fungicide for the control of damping off and for root rot caused by *Pythium* and *Phytophthora*. The pots and trays are then placed in the greenhouse and are not watered or touched again for about 7 to 10 days. At very humid times of the year this could be 14 days. If there is any sign of the seedlings shrivelling, a slight misting is beneficial. After this the seedlings are never allowed to completely dry out, but the frequency of watering will depend on the conditions under which they are grown. The medium must be kept just moist but not wet, as the seedlings should be in continuous growth and not given a resting period. In the very hot weather frequent mistings during the day are necessary, but never late in the day. Do not allow the water to remain on the seedlings at night.

SEEDLING GROWING MEDIUM

A standard seedling growing medium is:

- two parts fine bark, 3 mm to 6 mm (soaked in urea or Nitrosol for 2 days);
- one part charcoal size No. 3, about 3 mm to 6 mm;
- one part perlite (horticultural grade), about 3 mm;
- half part German peat moss;
- 1 kg blood and bone to 4½ buckets of mix;
- 100 g dolomite to 4½ buckets of mix;
- 14 mL of Previcur to a bucket of water;
- Bucket size 9 litre (2 gallon).

The ingredients are mixed thoroughly and moistened, while being mixed, with the Previcur and water. This mix is allowed to stand for a few days before use. The growing medium is always used in a 'just moist' condition.

GROWING CONDITIONS FOR YOUNG SEEDLINGS

The tropical species require warm greenhouse treatment with a minimum winter temperature of 16°C and if possible not above 29°C in the summer, plenty of air movement around the plants and high humidity. The community pots are kept in fairly heavy shade until they are obviously growing, then the light is increased. When the tropical species are transplanted to 50 mm pots and settled down, they require good light. The leaves become a light green if the right light is given. I achieve the right degree of light by having two tier benches. The community pots are placed on the lower bench away from the glass then, once plants are growing well, moved to the brighter light nearer the glass. The 50 mm pots are put on the top bench.

The temperate species are much more tolerant of temperature range. A minimum winter temperature of 6°C is quite acceptable although a small amount of warmth does help seedlings planted in the winter but the conditions must be airy and humid. Light conditions are the same as for tropical species while in community pots, but when in 50 mm pots the light should be less intense for temperate species.

FEEDING

Young *Dendrobium* seedlings benefit from a regular feeding programme. They must be kept in healthy vegetative growth and never allowed to rest until they are flowering size. As previously stated they must not ever be allowed to dry out, particularly when using a definite feeding programme. I use both organic and inorganic fertilizers, preferring proprietary brands that have a N:P:K rating of, as near as possible, 30:10:10. It is not considered advisable to use only one particular brand of fertilizer but to alternate between many different brands, all high in nitrogen, so that the tiny seedlings get a varied diet and have a greater likelihood of receiving the majority of the trace elements. Frequent feeding with a weak fertilizer is better than less-frequent feeding with a strong fertilizer. I feed twice a week at a third the dilution rate shown by the makers. NEVER feed a sick seedling and always water well before feeding.

PESTS

Aphids, thrips and red spider mite seem to be the worst offenders, and quickly drain the seedlings' strength. If the seedlings are kept in as near a perfect as possible growing atmosphere, and are regularly inspected, these pests should not be a problem. For control refer to pests in the chapter on culture. Snails, slugs and caterpillars love the new growth, but can be eliminated by a regular spraying programme.

7

Culture

The Australian and Norfolk Island species of the genus *Dendrobium*, grow naturally in a very wide range of habitats, within a climate varying from tropical to cold-temperate, and at an elevation from sea level to 1600 m. They can be found growing in full sun on rocks almost too hot to place your hand on or in darkish, steamy rainforests where they are continuously wet. But all situations have one common feature, excellent drainage (except that of the terrestrial *D. lobbii* which grows in swampy terrain).

Species groups

For cultural purposes the species can be divided, into three types. Reference to each species is essential to assess their individual requirements needed for your growing conditions. The statistics on temperature in various areas of Australia will assist (p. ix).

TROPICAL

North Queensland and Northern Territory

Minimum winter temperature 12°C, high light intensity and heavy watering while in active growth.

Species are *D. affine*, *D. antennatum*, *D. baileyi*, *D. bifalce*, *D. bigibbum*, *D. canaliculatum*, *D. cancroides*, *D. carronii*, *D. discolor*, *D. insigne*, *D. johannis*, *D. johnsoniae*, *D. lobbii*, *D. luteocilium*, *D. malbrownii*, *D. mirbelianum*, *D. nindii*, *D. smillieae*, *D. stuartii*, *D. x superbiens*, *D. trilamellatum*, *D. tozerensis*.

SUBTROPICAL

Highlands of North Queensland

Minimum winter temperature 6°C, reasonably high light intensity and heavy watering while in active growth.

D. adae, *D. agrostophyllum*, *D. carrii*, *D. fellowsii*, *D. fleckeri*, *D. jonesii*, *D.* sp. aff. *jonesii*, *D. lichenastrum*, *D. monophyllum*, *D. prenticei*, *D. racemosum*, *D. rigidum*, *D. toressae*, *D. wassellii*.

Although most of these grow quite well in shade-house conditions in the Sydney area of NSW, they do just that little bit better with some warmth in mid-winter.

TEMPERATE TO COOL TEMPERATE

South Queensland to Tasmania

Minimum winter temperature 2°C, bright light and heavy watering while in growth.

D. aemulum, *D. bowmanii*, *D. calamiforme*, *D. cucumerinum*, *D. x delicatum*, *D. fairfaxii*, *D.* sp. aff. *fairfaxii*, *D. falcorostrum*, *D. x foederatum*, *D. x gracillimum*, *D. x grimesii*, *D. kingianum*, *D. linguiforme*, *D. macropus* subsp. *macropus*, *D. macropus* subsp. *gracilicaule*, *D. macropus* subsp. *howeanum*, *D. monophyllum* (from southern end of distribution), *D. moorei*, *D. mortii*, *D. pugioniforme*, *D. schneiderae*, *D. schoeninum*, *D. speciosum*, *D. striolatum*, *D. x suffusum*, *D. teretifolium*, *D. tetragonum*.

The north Queensland forms of *D. speciosum* and *D. tetragonum* do better with a little bit of warmth in mid-winter.

Horticultural stages

For cultural purposes, the species can be divided into two groups according to flowering times; four growth stages are given for each group. All species cannot be locked into exact cultural boxes or horticultural stages, and the periods will overlap.

SPRING TO SUMMER FLOWERING

Stage 1 Growth period
September to April
(spring, summer, autumn)

Stage 2 Maturation
April to May
(autumn)

Stage 3 Rest
May to July
(late autumn, winter)

Stage 4 Flowering
August to November
(late winter through to spring)

AUTUMN TO WINTER FLOWERING

Stage 1 Growth period
August to March
(late winter to early autumn)

Stage 2 Maturation
March to April
(autumn)

Stage 3 Flowering
April to July
(autumn, winter)

Stage 4 Rest
June to August
(winter)

The months and seasons given are those in the average areas where the Australian dendrobes actually grow in nature. It is therefore necessary to make adjustments for your own area.

As geographical conditions differ greatly throughout the world, it is difficult to give actual months for the horticultural stages to suit all growers of Australian dendrobes. An obvious problem is that the months and seasons are totally opposite in the northern and southern hemispheres. Another problem is that seasons are not as clearly defined or as long in some areas as in others and therefore artificial means are necessary to compensate. In some areas in the tropics at a height of, say, 2500 m, it can almost be said to be spring all the year round. On the other hand, in the extreme northern countries the winter season is very long, with low temperatures, and spring, summer and autumn are very short.

Plant requirements during the stages given are as follows.

Stage 1
Develop plant to ensure good sturdy growth. Give plenty of light, moisture, fresh air and moderately high temperatures.

Stage 2
Mature plant by increasing light and reducing moisture.

Stage 3
Cool period for spring to summer flowering plants, so withhold water; slight shrivelling of pseudobulbous stems beneficial.
Flowering period for autumn to winter flowering plants, so water sparingly but not on flowers, give plenty of light.

Stage 4
For spring to summer flowering plants start feeding and watering gradually. Repot after flowering.
For autumn to winter flowering plants withhold water and give a decided rest. Hang up in the top of the glasshouse. Repot in the spring. If plants are kept moist and warm during autumn and winter, this will prevent the maturing of the pseudobulbous stems and new plants (aerial growths) will be encouraged instead of flowers. The autumn to winter flowering hardcane dendrobes respond well to a rise in temperature in early spring from say 16°C (60°F) to 26°C (80°F); this brings the plants into early growth and helps ensure that they mature by early autumn and are ready for flowering.

Housing

Like most ephiphytic orchids, the genus *Dendrobium* prefers shade-house conditions where it receives good light, free circulation of air and sufficient moisture. Of course the individual plant requirement of these factors will vary. Unfortunately in areas where the temperature requirement cannot be met, a greenhouse is necessary to maintain as near optimum conditions as possible.

POSITION

Whether you grow your dendrobiums in a shade-house or greenhouse, certain constant factors must be allowed for in selecting the position in which to locate the structure. It is desirable that:

- An open area is selected, not one that is surrounded by tall trees or buildings. Remember that young trees grow up, or the neighbours may extend their house or outbuildings.
- Early morning sunlight is unobstructed.
- The area is not in shade in the winter. In countries far from the tropics the position of the winter sun can be low in the sky most of the day, and this is when more light is necessary.
- It is not in a draughty position.
- There is plenty of room to extend at a later date.
- A good flow of fresh air is certain.
- The position will allow for a structure at least 3 m high.
- Adequate drainage is possible.
- A water supply is available and electricity if heat is required.

As many of the dendrobes grow into very large plants it is advisable to build a large and airy structure. Also, a small, snug shade-house or greenhouse is not conducive to a good and adequate circulation of air.

D. discolor has pseudobulbous stems that frequently exceed 2 m in height and a large well-grown specimen can be well over 2 m across. *D. speciosum*, *D. x delicatum*, *D. x gracillimum* (to name a few) also grow into large specimens, and plants about 1·5 m across are not an uncommon sight at orchid shows in Australia. *D. teretifolium* is a pendulous plant and a large specimen can hang down at least 2 m.

BENCHES

Sturdy wire mesh is most suitable for benches as it allows for plenty of air circulation and is cleaner—it does not have lots of places in which pests can breed.

My preference is for benches about 760 mm high and not more than 910 mm wide. Some of the shade-loving species can then be grown under the benches. On the side farthest away from the sun, a three-tier bench could be used as this enables the maximum amount of sunshine to reach the maximum number of plants.

As many Australian dendrobes grow so well on slabs, it is useful to cover the whole of the west wall from roof to ground, or at least from bench to roof, with a wire mesh frame. The angle of the frame should be such so as to avoid drips of water from one plant on to another—say about 10° to 15° from vertical.

FLOORS

The most satisfactory material for floors is ashes or a bark mulch, as both maintain a moist atmosphere when necessary. Cultivation of ferns and moss under the benches helps create the right atmosphere for orchids.

COVERINGS

It is difficult to continually change the shade cloth so as to install either a lighter or heavier grade at different times of the year. Thus a 50 per cent shade cloth is recommended as the most satisfactory. The ideal solution would be to have 60 per cent in mid-summer, 40 per cent in spring and autumn, and no covering at all during the winter months (though even then you would have to take into consideration the angle of the sun). If growing dendrobes in the tropics, at least 50 per cent shade cloth, even in winter, would in most cases be necessary. If roofing slats are used these should run due north and south to ensure that as the sun moves from east to west the plants receive alternate light and shade and do not get too much of one or the other.

In the northern hemisphere it is suggested that the west and north sides be totally enclosed and in the southern hemisphere the west and south sides. The material used would depend on the structure and its surroundings, it could be brick or any other solid wall.

LIGHT

The light requirement varies greatly. The tropical dendrobes require in the vicinity of 21 500 to 32 300 lux (2000 to 3000 foot candles), whereas the temperate dendrobes require in the vicinity of 16 200 to 21 500 lux (1500 to 2000 foot candles). It must be remembered that the light intensity that an orchid can tolerate is dependant on the relative humidity, air circulation and temperature. Morning light is more important than afternoon light.

ARRANGEMENT

Fig. 100 Hang sun-loving plants from, and close to, the roof, ensuring that the smaller ones are nearest to the morning sun or to the east and grading to the largest so as not to take away too much sunlight from the plants on the benches. Hang these plants so that excess water draining from them does not drip onto the pots below, otherwise it will disturb the growing medium or maybe a new growth or inflorescence will damp off through too much water. Place shade-loving plants on or below the benches, depending on the degree of light they require. In greenhouses watch for drips from the rafters, etc. and avoid placing plants in these spots. Hang those dendrobiums mounted on slabs on a frame furthest away from the morning sun.

Group the various cultural types together, so that watering can be carried out correctly for each. Micro-climates can be created by increasing shade or covering, or by installing a small fan, etc. Overcrowding must be avoided as this hampers free air circulation, makes it difficult to insure that

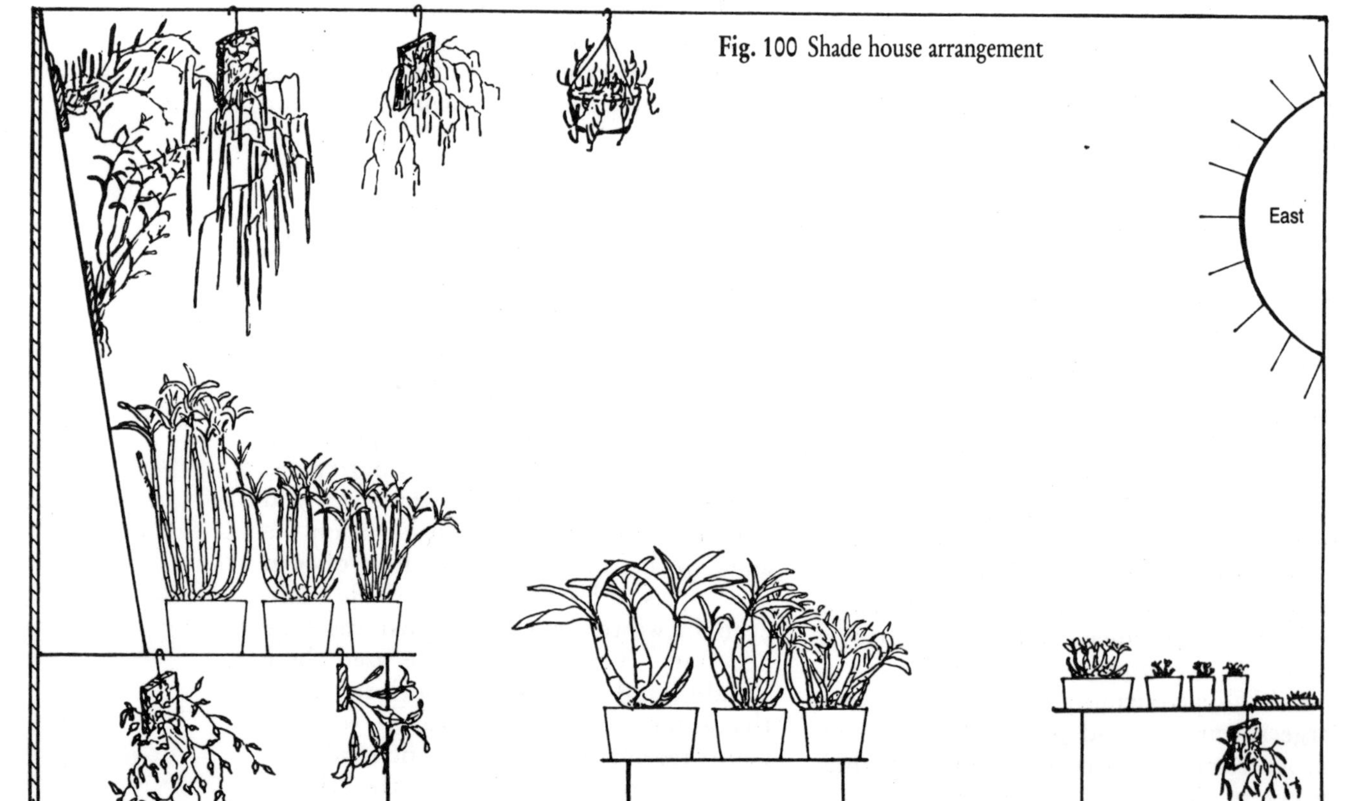

Fig. 100 Shade house arrangement

all plants get watered, and encourages pests such as snails and slugs.

GREENHOUSES

Essentially, if you live at near sea level in an area south of latitude 35°S or north of latitude 35°N, you need a greenhouse for all except the temperate Australian dendrobes, although the temperate Australian dendrobes need that extra bit of warmth when the temperature drops below 2°C. If you live at near sea level in an area south of about 28°S or north of about 28°N, you will need a greenhouse to supply the temperature requirements of tropical Australian dendrobes. If you live in the high ranges, even in the tropics, you also require a greenhouse to grow warmth-loving dendrobes.

During the hot summers it is necessary to maintain a high degree of humidity in a greenhouse, and if the correct amount of light is present it is almost impossible to put too much water into the greenhouse. Overhead watering does not do any harm if plants are not in direct sunlight. In areas where the atmosphere is extremely dry during the hot summer months, a drop in humidity can be prevented by placing a moisture-retaining substance, such as coke or ashes, in shallow trays below the plants. This can be kept wet. Numerous foliage plants can also help provide a good atmosphere in the greenhouse.

Build as large a greenhouse as possible to ensure there is a large volume of air, as this makes for evenness of temperature, which is most important in a very changeable climate. A lot of the northern Australian species are not accustomed to great changes in temperature. Ensure a constant flow of air in a greenhouse; this will help get rid of excessive moisture including condensation on the inside of the glass.

Greenhouses are commonly divided into three sections:

- a clear glass section for sun-loving plants like *D. bigibbum*;
- a section with light shading;
- a section with heavy shading for the shade-lovers and tiny seedlings.

Growing media

The essential requirements of a medium are that:

- It must retain a certain amount of moisture.
- Water should moisten the medium and then drain away as quickly as possible.
- It must not breakdown too quickly—it should last at least 2 years.
- It must have the correct pH, within the range 5 to 6.
- It must support the plant.
- The ingredients must be free of dust and fine particles.
- It should be moist, not dry, when it is being used.

The following are some suggested potting mixes for adult plants:

1.
   - three parts pine or sequoia bark, medium to large (1 cm to 2 cm) pieces soaked in urea for 2 days;
   - two parts tan bark (if unobtainable, may be omitted);
   - three parts charcoal size No. 5 (about 1 to 2 cm pieces);
   - one part wood chips (fresh material from, say, a chainsaw) not sawdust;
   - one part Hawkesbury sandstone broken into marble-sized pieces. (any sandstone that doesn't crumble too much is suitable);
   - two cups dolomite and three cups blood and bone to ten 9 litre (2 gallon) buckets of mix.
2.
   - three parts pine bark (medium-size pieces);
   - one part charcoal (medium-size pieces);
   - two parts tree-fern fibre;
   - one cup dolomite and two cups blood and bone to six 9 litre (2 gallon) buckets of mix.
3.
   - one part pine bark (medium-size pieces);
   - one part river gravel (medium-size pieces);
   - two cups blood and bone to six 9 litre (2 gallon) buckets of mix.
4.
   - one part cork bark (6 mm to 19 mm pieces);

- one part charcoal (6 mm to 19 mm pieces);
- one part sandstone (marble-size pieces);
- one cup dolomite and two cups blood and bone to six 9 litre (2 gallon) buckets of mix.

Any charcoal used should be hard, not soft and crumbly.

There are many different potting mixes; some growers use bark on its own, particularly the sequoia bark from the United States. But, regardless of the specific ingredients preferred, the essential factor necessary to good healthy growth is that it is an open mix, thus allowing air to the roots and good drainage.

The medium should be sterilized as a means of eliminating harmful organisms. Most damping-off diseases and root rots are controlled by sterilization. Sterilization can be by the use of heat (provided by steam) or by chemicals. Total sterilization is not required as orchids depend on some useful organisms in potting mixes. Disinfection by chemicals is possibly the easiest method if a large quantity of medium is to be sterilized. The medium should be well mixed, loosened, manured and watered before applying chemicals. Suggested chemicals are formalin or methylbromide. If unfamiliar with their use, guidance should be sought from an experienced grower.

If only a small quantity of medium is to be sterilized, a large pressure cooker is suitable. Sterilize at 105 Kpa (15 pounds per sq inch) for 10 minutes.

Potting

The best time to repot is in early spring, when new growth is just starting to show. Plants should not be watered for about a week prior to repotting, so that they will come out of the pot easily and the old potting mix, if dry, will fall away from the roots causing less damage.

Before potting or repotting, the plant should be thoroughly examined. All dead and dying pseudobulbous stems, leaves, flower racemes and roots should be removed and all cuts sealed with sulphur or similar. Strip all bracts from the pseudobulbous stems (these are great hiding places for pests). Check thoroughly for pests and kill them. If a plant has a bad infestation of scale, clean this off very thoroughly (use an old toothbrush) and dip the plant into an insecticide (refer to the section on pests). Check for mealy bugs on the roots and, if present, completely eliminate them. Remember, rot can easily get in through old dead roots, etc. left on a plant.

Before commencing any potting, make sure you have all the equipment necessary—various sizes of pots, stakes, ties, labels, pen or pencil, sulphur, clean tools, potting stick and a small Bunsen burner or jar of disinfectant (to sterilize tools after each cut). If you are using old pots, make sure they are completely clean; after cleaning the pot a good soak in household bleach (one part bleach to 20 parts water) is considered effective. If crocks are to be used in the bottom of the pot, make sure these are clean also. Terracotta pots and crocks should be soaked in clean water before use. All media to be used should be moist, not dry.

Examine the medium before use to ensure it is free of dust and dirt. The fine particles in bark find their way to the bottom of the pot where it becomes a solid muddy mess, clogging up the drainage holes and causing the medium to become soggy and sour. The result is no live roots and a very unhealthy plant.

After you have thoroughly examined the plant, decide if you are going to:

- pot it on into a larger pot, possibly to make a specimen;
- divide it into two or more plants;
- create a new growth arrangement in a new pot.

  If it is a fairly large plant and the centre or rear half has lots of old and dead pseudobulbous stems, cut off the old stems from the centre and take about five or six actively growing pieces, of about three pseudobulbous stems each, from the outside. Place them in all directions radiating from the centre of the pot, and plant them to ensure that it will make a good specimen in a few years with new growths all around the outside. The *Dendrobium* Bardo Rose in Plate 201 was treated this way.

After dividing the plant some pieces may be considered to be more suitable to slab culture, because of their habit of growth. Determine the size of pot according to the size of the piece, remembering it only needs sufficient room for 2 years growth. Use the smallest pot possible. DO NOT overpot.

When potting, the main points are:

- pot firmly;
- ensure perfect drainage;
- ensure that the plant is placed high in the pot. DO NOT cover the basal growth nodes (the eyes at the bottom of the stems) with growing medium. The crown of the plant sits on top of the medium. If covered, the basal growth nodes will most probably rot.

POTTING PROCEDURE

Having selected the piece to pot, proceed as follows.

1. Crock the bottom one-third of pot or place a smaller plastic pot upside-down over the drainage hole. *Dendrobiums* do not like a soggy medium and therefore frequently do not use the medium in the centre bottom half of the pot. There are seldom many roots in this area and the upturned pot keeps this area open. Use a 50 mm upside-down plastic pot in a 100 mm pot and a 65 mm plastic pot upside-down in a 125 mm pot, or use a similar ratio. The drainage hole in terracotta pots never seems to be large enough, so it should be enlarged. If using plastic pots ensure the holes are not covered over with a thin film of plastic (this frequently happens when the die used by the manufacturer is getting old) and also possibly increase the size of these holes.
2. Take the plant to be potted in one hand and hold it in the centre of the pot with the base of the pseudobulbous stem about 13 mm from the top of the pot. With the free hand place the potting medium around the roots until it reaches the base of the pseudobulbous stems. It is necessary to frequently shake the pot or tap it firmly on the side to ensure the medium is well packed into the pot. Take the potting stick and firmly push the medium down into the pot, from the outside inwards, adding more medium as necessary. You should be able to lift a well-potted *Dendrobium* by its pseudobulbous stems without the pot falling off!
3. If the plant is loose in the pot, it should be staked and tied until quite firm. If the plant wobbles around, the new roots have no chance of going down into the medium as the new root tips are rubbed and broken by the continuous movement of the plant.
4. Water the plant thoroughly to help settle the medium, then return it to the orchid house, preferably to a shady position for a week or two to reduce shrinkage of the pseudobulbous stems. Mist over the leaves every sunny day, until the plant is active again, then treat it normally. If you have cut all growing roots off during the repotting, do not commence watering the plant until new roots have grown.

Many of the dendrobes (mostly lithophytic growers) are most amenable to being potted in terracotta saucers; examples are *D. bigibbum* var. *compactum, D. carrii, D. jonesii, D. kingianum, D. lichenastrum, D. linguiforme*, and *D. schneiderae*. First, using a 6 to 10 mm masonry drill, make several holes in the bottom of the saucer, at least five to a 100 mm saucer. Place some larger pieces of the growing medium in the bottom of the saucer, then place the plant on top of these so that the crown is about 10 mm from the rim. Tie the plant firmly onto the medium through the holes in the bottom of the saucer. The positioning of the plant in the saucer will depend on the piece to be potted, generally plants like *D. linguiforme* would be potted near to the edge of the saucer to allow them to then go forward across the saucer. Next pack the medium in around the roots, firmly bringing its level to the crown of the plant.

POTTING ON

When a plant that is healthy and growing well has become too large for its pot, it can be transferred by placing the whole contents into a larger pot without touching the old medium and roots. This can be done at any time of the year without causing any setback at all. The larger pot is first prepared by placing medium over the crocks to bring the plant to within 13 mm of the top of the pot; new medium is then packed between the plant in its old medium and the rim. (It is advisable to take away any crocks from the bottom of the old medium.) Great care should be taken to make sure that the plant is really healthy and that the old medium is still quite good. Do not make a habit of potting on. In many cases it is only putting off the day when you have to break up a plant.

Slab culture

Most members of the genus *Dendrobium* are very amenable to slab culture, particularly those that ramble or climb and those that need excellent drainage and plenty of fresh air to the roots. A plant established on a slab is probably the easiest to look after. It generally doesn't take up much room and it can be hung almost anywhere without difficulty.

Most suitable slabs are made from:

- cork tree bark (*Quercus suber*)
- seasoned hardwood (old palings, floorboards, etc.)
- tree-fern fibre.

If using tree-fern fibre slabs, it is advisable to soak the slab in a weak solution of lime for a few days and thoroughly wash the lime out before attaching the plant to it. This reduces acidity in the slab.

FIXING PLANTS TO SLABS

Select the appropriate sized piece of slab, taking into account the future size of the plant. A plant tied onto a slab and growing well usually stays on its host for many years.

Prepare the plant in the same way as for potting. Examine it to ensure that it will be placed the correct way up on the slab. The leaves of all orchids are arranged so that only the upper surface is exposed to the direct light. If the underside is continuously exposed it could harm the plant. Many pendulous dendrobiums start their growth upwards and, as the stem elongates and becomes too heavy, turn downwards.

Tie the plant very securely on to the slab, so as not to permit any movement between the plant and the slab. The roots cannot take hold if there is any movement; the green growing tips are very easily damaged if continuously rubbed due to an insecurely mounted plant. Materials suitable for tying include nylon fishing line, wire (preferably plastic coated, galvanized wire must not be used) or old nylon stockings (take the lady out first). Holes are drilled through the slab in the appropriate position (one on either side of the plant or one of its pseudobulbous stems) and the tie pushed through the hole and securely fastened. When the plant has been securely tied, cover any new roots with sphagnum moss and carefully tie them down, just firmly enough to allow their continued development.

Establishing the plant is frequently difficult and it is recommended that the plant and block are left on the bench until the roots maintain a hold on the fibre, or at least until the root action starts. Fibre blocks can remain wet too long if overwatered, particularly in the winter. In the vertically suspended block, the lower half should be watched, as it is here that the water collects (especially in old blocks) causing root rot.

In very dry areas, Spanish moss (*Tillandsia usneoides*) hung around the plants on slabs in the hot summer months is beneficial. A heavy gauge copper wire is ideal for a wire hook to hang up the slab. Do not forget to label the plant!

For some of the smaller dendrobes, such as *D. lichenastrum* and *D. toressae*, a simple way to attach them is to turn a terracotta pot upside-down and, with a vegetable glue, affix the plant across the drainage hole of the pot, the pot should be kept moist until the new roots cling to it.

Dendrobiums in the open

Many Australian dendrobes are adaptable to outdoor garden culture, either attached to trees or rocks, hung from an open framework or planted in garden beds. Particular climatic conditions dictate the species or hybrids that can be grown. Reference to the list of tropical, subtropical and temperate growing orchids on p. 187, and the characteristic and habitat conditions stated for the individual species will help you decide which plants can be grown under your conditions.

IN GARDEN BEDS

Consideration must be given to the location of the garden bed in relation to the position of trees and buildings so that the orchids receive the right amount of sunlight. A garden bed facing approximately towards the early morning sun is ideal so that the plants receive ample morning sun. Good air movement around the plants is essential.

The garden bed should be at least 38 cm deep with approximately 15 cm of large crocks and charcoal in the bottom to ensure good drainage. The medium the plants are grown in should be very open—refer to the suggested mixes for adult plants. Plates 213 and 215 show temperate-growing dendrobes growing in the author's garden in such a garden bed.

ON ROCKS

Many orchids grow on rocks in their native habitat. When attaching orchids to rocks, it is important that the plant is fixed tightly onto the rock. This can be achieved by driving nails into the rock to serve as attachment points for wire which is used to strap the orchid onto the rock. Do not leave all the old roots on the plant as this impedes the attachment to the rock by the new roots.

Once the orchid has established itself on the rock it will not require much attention. Many dendrobes grow well on rock and make an outstanding display in the spring. *Dendrobium speciosum*, *D. kingianum*, *D. linguiforme* and numerous hybrids using these and many other Australian *Dendrobium* species (such as *D.* Bardo Rose, *D.* Ellen, *D.* Hilda Poxon, *D.* Penny Ann, *D.* Star of Gold, etc.) are ideal for the temperate climate situation. Plate 219 shows *D. speciosum* growing on rocks bordering a driveway. The ideal situation for a rock garden is about 50 per cent shade with considerable air movement.

ON TREES

An orchid in full bloom growing happily on a tree is a pleasure to behold, after all it is the natural habitat for so many orchids. Not all trees are suitable—those that shed their bark or have very dense foliage are not satisfactory. Deciduous trees are unsatisfactory unless sheltered by an evergreen tree, as they have no protection for plants during the extremes of winter.

Some of the best trees are lilly pilly (*Acmena smithii*), paperbark (*Melaleuca* spp.), blueberry ash (*Elaeocarpus reticulatus*), figs (*Ficus* spp.), river oak (*Casuarina cunninghamiana*) (useful in a wet area) and lemon (*Citrus limon*). Tree ferns most suitable to grow dendrobes on are the slender tree fern (*Cyathea cunninghamii*) and the soft tree fern (*Dicksonia antarctica*).

Some deciduous trees that are satisfactory if sheltered by an evergreen tree are the frangipani (*Plumeria aculifolia*), jacaranda (*Jacaranda mimosifolia*) and the apricot and almond (*Prunus* spp.).

Consider seriously the positioning of the plant in the tree, the shape of the orchid will generally give you a clue to its requirements. Is its growth upright or pendulous, does it need to be on the top of the branch or underneath it, or does it need to be in the fork of the tree or on its trunk where it will get the right amount of light, etc. Care should be taken to trim off all old roots so that the orchid can quickly attach itself to the tree.

Tie the orchid onto the branch or trunk of the tree very tightly so that movement by the wind will not damage the new roots and so hinder progress. To establish a small plant or seedling on a tree it is sometimes necessary to use some sphagnum moss around the roots or alternatively a covering of paperbark to encourage their growth by keeping them moist. As soon as the roots start to grow, frequent waterings are necessary until the plant is established.

It is no use watering an orchid if the roots are not healthy and the tips are not growing. Water is mainly taken into the plant by the root tips. When a root is actively growing its tips are a green or yellow-green, sometimes a reddish colour.

Orchids growing on trees require little attention except for an occasional watering or feeding. If the tree is in a dry corner of the garden, or the winds dry the plant out quickly, then more frequent waterings will be required. Alternately the plant can be placed in the fork or crotch of a tree where the roots can burrow into medium placed there to protect them from drying out too fast. Watch the plant and you will know its requirements. For instance, if it is too dry its leaves will appear dull and the pseudobulbous stems will start to shrivel.

ON OPEN TRELLIS

Orchids attached to natural cork slabs or in pots can be hung onto an open trellis or galvanized weldmesh attached to a wall. Do not fix lattice or weldmesh right up against the wall as it is necessary to have free air circulation around the plants. A distance of 15 cm from the wall is ideal, as walls become very hot in the summer.

Consider the amount of shade required by the orchids. Some dendrobes will grow happily in full sun but require acclimatizing to this condition gradually.

ON AN OPEN BALCONY OR VERANDAH

There are many dendrobes that will grow well on the balcony of a home unit or on an open verandah. An aspect facing the early morning sun is the best. Ensure that a good flow of fresh air is available and that the orchid is not placed in a draught (a current of cold air directed continuously on to it). In the winter the orchid should not be in heavy shade, so move it to the outside edge of the balcony where it will receive more light, but not the hot mid-day sun. Cold and drying winds should also be avoided.

Humidity can be achieved by placing the orchid pots in a tray full of gravel, stones or charcoal, with water in the bottom to keep it continuously moist. Do not let the pot stand *in* the water.

Some of the smaller dendrobes, *D. linguiforme*, *D. lichenastrum*, *D. toressae* etc., can be attached to the long leafless stems of plants such as *Dracaena* and *Dieffenbachia*. These can be taken indoors in the winter; they make a delightful display when in flower and are quite a conversation piece.

Specimen plants

To produce a large specimen plant requires time and patience, as it takes quite a few years to achieve a satisfactory result—a plant everyone stands back and admires for its sheer beauty. A specimen plant doesn't just happen, it is made!

Firstly you must select the right type of plant. It should, in my opinion, have most, if not all, of the following requirements or suggestions.

- It must be a plant that grows vigorously.
- It must flower easily.
- It must produce lots of racemes to each pseudobulbous stem and must produce lots of flowers to each raceme.
- If it produces single flowers they must be produced in great numbers.
- It is preferable that it has upright to arching racemes.
- It is preferable that the flowers are carried well clear of the foliage.
- The flowers should stand up and look at the viewer. Even if it has a pendulous habit, the flowers should still look up, not hang down.
- It must have an appealing colour and be vibrant. A dull colour detracts.
- It must grow in all directions so that it can be seen as a specimen all round.
- It must produce multiple new growths each season; a plant that produces two or more new growths from each of its pseudobulbous stems is good.
- It must have the habit of producing new growths from the eyes of the old pseudobulbous stems.
- A big plant with small leaves is much better than a small plant with big leaves.
- It is preferable that it is a plant that will withstand plenty of sun because, when it becomes really large, it will take up too much room in a shade-house and will need to be placed out in the open. If placed under a tree, it may not get enough sun to flower profusely.
- It is not subject to pest or fungal attack (some plants are more susceptible than

others). A specimen with badly chewed or black-marked leaves is unsightly.

Some good subjects for this treatment are *D. x delicatum*, *D. x gracillimum*, *D. speciosum*, *D.* Bardo Rose and a few clones of *D. kingianum*.

Having decided on a particular plant to develop as a specimen (it will, no doubt, be a medium-sized plant to have proved itself), the first step would be to repot it (as early as possible in the growing season). Take it out of its old pot and select five or six actively growing pieces of about three to four pseudobulbous stems each. Cut these off and plant them in a pot, allowing only enough room in the new pot for 2 years growth. Place the newest growth of each piece to the outside and plant them in all directions radiating from the centre of the pot; this will ensure that in future all the new growths will be all round the outside of the plant. Do not make the mistake of immediately planting it into a large pot, it will grow much faster and better if potted on every second year. Also the roots can be better contained this way and when potting it on to the next sized pot it will have fresh medium on which to continue to grow.

When repotting every 2 years, cut out any dead pseudobulbs and cut the rhizome about two pseudobulbs from the centre so that all parts of the plant will have new growths including the centre. A specimen is not a specimen if it has a dead centre.

By the time you have potted it on about three times (6 years later), it should be a reasonable-sized plant and possibly be in (depending on the species or hybrid) a 200 mm to 250 mm pot. Weight is now important. A 350 mm pot of *D. speciosum* is very heavy. I use large pieces of polystyrene foam (as used in packaging) for crocks in the bottom of the pot, and light materials such as perlite in the medium. Also, a large plastic pot is much lighter than a terracotta pot.

The culture of the plant should be as near perfect as possible to ensure that it will always look good. Inspect it on a regular basis for problems such as pests, diseases. Keep it clean. Grow it in a sunny spot with light all around so that the growth will be even.

Watering

Success or failure in growing dendrobes depends largely on correct watering. It is no use watering an orchid if the roots are not healthy and the tips are not growing. Water is mainly taken into the plant by the root tips. When a root is actively growing its tips are a green, yellow green or sometimes a reddish colour.

Many factors have to be considered when watering dendrobes as they are very varied in plant form, flowering time, habitat, temperature requirements, period of active growth, etc. The type of orchid house, ventilation, container it is grown in or on, growing medium, condition of medium, weather, time of year, etc. are also important. The ideal of watering each plant individually becomes impracticable in a large collection. In these circumstances it is useful to group dendrobes that have like watering requirements. In grouping plants remember that large pots dry out at a slower rate than smaller ones and that plastic pots do not dry out as quickly as terracotta pots.

Rain water is best for plants (except in heavily polluted areas), but the local water supply is much more convenient and usually suitable. Never use impure or stagnant water.

The method of watering is a matter of choice. However, the plant must be watered thoroughly when it is watered, and must be allowed to dry out between waterings. As a general rule, it should not be left to dry out completely for any length of time.

Before watering, have a look through your collection to see if any orchids are still wet. It is a good rule never to water a wet plant; if there is any doubt, do not water. You can afford to be overgenerous with water only in the plant's active growing period. The weight of a pot is a good indication if it is wet or not.

Damping down (watering on the ground beneath the plants) is very important with orchids — the vapour that rises is absorbed by the leaves and aerial roots. A well-watered, actively-growing plant should be continually receiving and passing out water through the pores of the leaves. Leaves should be sponged over occasionally to clean them both on top and underneath. This wipes away the dust and algae, and keeps the pores open.

In nature the action explained in the previous paragraph is continually carried out at the right time of the year. For instance, the dendrobes that grow in the tropical rainforests of north Queensland receive a nightly wash and drink. During part of this growing season, late summer to autumn, it is the wet season, but many of them make their new growth during spring and early summer. During this period there is little or no rain, the days are generally warm and, as soon as the sun goes down, there is a distinct coolness, resulting in dew formation on everything including the leaves of the orchids. The dew stays until the sun returns the following day. These conditions are conducive to strong healthy growth with a vigorous root system. This is a very difficult environment to reproduce in a greenhouse, and anything near to it usually produces dead roots and black spots on the leaves. To achieve a satisfactory environment in the greenhouse, avoid a cold and draughty situation with pockets of stagnant and still air, and keeping as near a constant temperature as possible with continuous air movement.

WINTER WATERING

Watering in winter requires great care and consideration. Ideally the chill should be taken off the water. In winter, water from the tap is very cold and risky to use on some of the tropical and subtropical dendrobes. With a small collection it is best to dip plants into a bucket of tepid water (temperature to about 16°C). If possible, avoid overhead watering—this particularly applies to the greenhouse.

Do not water on wet, dull or very cloudy days, particularly as they are often followed by more rain or by a cold snap. In winter always water on a rising temperature (this will usually be in the morning). Also water only on a good drying day as plants should be dry before nightfall.

Although there is no set rule, in general once a week watering in winter is the maximum. Frequently some of the hardcane dendrobes can go a month without water and still not shrivel at all.

One of the main faults in culture is too much water during the winter months.

SUMMER WATERING

Watering in the summer is not as critical as in the winter. In the hotter summer months once or sometimes twice a day should be adequate. If the plant is in active growth and healthy, it is practically impossible to overwater it in mid-summer, but even so use common sense when watering.

Leaves can be damaged if water is allowed to lie on them. The sun's rays heat the water causing brown burn marks. Mist plants and damp their surroundings at least once a day during the hot summer days, particularly in the evenings.

Resting

Most dendrobes enjoy a good resting period, particularly the hardcane types (*D. bigibbum*, *D. discolor*, etc.). Hybrids, generally, do not seem to need such a long period of dormancy, but even so a dryer period during the winter is beneficial to their flowering.

Plants should be hardened to prepare them for resting. This can be achieved by reducing the amount of water, starting with the night watering in early autumn, and giving the plants more direct sunlight so that by winter they are receiving nearly full or, in some areas, full sunlight. The winter sunlight should be equivalent to the amount of sun received through the shading during the summer months. In the cooler areas where the sun is not hot enough to burn the leaves in the winter, all shading should be taken off the greenhouse or shade-house, and the tropical hardcane dendrobes hung up high. The hardening process in autumn and the resting in winter will develop hard plump pseudobulbous stems able to carry good racemes of flowers, and the resting will ensure the development of the flowers. Do not fertilize while the plants are resting. The minimum temperature for the tropical dendrobes is about 12°C while resting, and temperatures above 18°C for any length of time will start them into growth. If the temperature is kept too high, the plants will continue growing and produce premature growths, leading to the possibility of plant exhaustion.

210 Mother flask six weeks after sowing

211 *Dendrocoryne* section hybrid just replated at 12 weeks

212 *D. antennatum* attached to natural cork slab

213 *D.* x *delicatum* in garden bed

214 *D.* x *delicatum* specimen plant in hollow log

215 Australian dendrobiums in garden setting

216 Dendrobium Beetle and leaf damage

217 Larvae of beetle and stem damage

218 Pupation stage of beetle at stem base

219 *D. speciosum* on rocky bank bordering driveway

220 *Dendrobium* species and hybrids in a show display

221 *Dendrobium* species and hybrids in a show display

222 *D. wassellii* attached to tree fern slab

Watering during the resting period should be according to plant requirements. If the pseudobulbous stems shows signs of shrivelling or the leaves become soft and flabby, the plant requires water.

Plants are brought back into active growth in early spring by increasing the temperature in the greenhouse to about 16°C to 27°C and gradually increase watering, this will ensure that the new growths are then matured before autumn and ready for flowering.

Flowering

A healthy plant usually produces the best quality and greatest number of flowers.

Preparation for flowering starts when the previous flowers are cut. In potted plants consider whether repotting is necessary. If so, do this as early as possible to enable the plant to make an early start for the next year's growth and ultimate flowering. With some of the warmer growing, autumn to winter flowering dendrobes, it is advisable to leave repotting until early in the spring, as they rest after flowering.

Perhaps the most important single factor to aid flowering is the amount of light available. Species that grow in open situations require bright light to flower while those growing in heavy shade do not.

Water with great care, keeping plants just moist not wet all the time, as frequently new flower buds become soft and drop off because of excess water. The first sign of flowering is the swelling of the terminal or lateral nodes. These nodes are situated in the leaf axils, generally in the upper part of the pseudobulbous stem. With the terete and succulent leaved types (*D. linguiforme*, *D. toressae*, etc.), the flowering nodes appear at the base of the leaf. If excess water lies in the leaf axils for any length of time, the flowering nodes may wither. With a small collection it is safest to dip pots in water instead of watering overhead at this time. Generally, as buds begin to swell, slightly more water can be given.

Pests must be kept away to prevent marked and chewed flowers. Do not use liquid sprays on the buds or spotting may occur. Avoid damage to flowers from leaves or stems hitting them as they are blown in the wind.

Dendrobes flowering in a shade-house, should, if practical, be placed under cover should rain threaten or develop. Ensure that the flowers still face in the same direction when moved. Any change in orientation of the flowers will cause the raceme to twist towards the light.

Bud drop can be attributed to many factors, most frequently to plants having 'wet feet'. Buds should, if possible, be kept dry. The bud drop problem is not very prominent in shade-house grown plants, as these are generally more hardy. It is in the greenhouse, where heating is required, that bud drop most frequently occurs, and can be attributed to one or more of the following reasons:

- High night temperatures without sufficient air movement or ventilation.
- Hot dry conditions.
- The use of icy-cold water.
- Plant being kept too wet, the roots are cold while the rest of the plant is not.
- Small plant potted in a large pot.
- Plant grown in a pot in a saucer and water left in the saucer.
- Cold draughts.
- Excessive condensation inside the greenhouse causing drips from the roof (these drips always seem to land on the flower bud).
- Feeding plants at the wrong time, particularly feeding a plant when the medium is dry.

To eliminate another important reason for bud drop, you should ensure that air movement is just sufficient to continuously move the flowers. Clean foliage, enabling the leaf to function properly and as near a constant temperature as possible, are also of great importance.

In many plants, bud drop is an inbred weakness or an expression of the plant's dislike of artificial conditions. But in other cases, climatic conditions do have an influence. If the pitfalls outlined above are avoided, bud drop should largely disappear.

As the inflorescence progresses, consideration should be given to staking it into the required position. Where are you going to display the orchid? If in a corner of a room, direct all the racemes to face one way; if for the middle of the room, direct the racemes to face all around. If you intend to exhibit the orchid at a show, stake it in such a way that the orchid flower will look at the judge. If two racemes are too near to each other, train them away.

Frequently, the weight of the flowers causes the pseudobulbous stem of some dendrobes to bend right over, so tie the stem securely to a stake. Turn a plant so that the inflorescence faces the strongest light and leave it there until all the flowers have opened. All the flowers will be directed the same way. If the plant is turned around, the inflorescence will turn towards the strongest light again, causing a twisted inflorescence, which is unsightly.

Generally, as with other orchids, the colour of the blooms can be influenced by giving the buds the right amount of light while developing. The whites should be flowered in plenty of shade to retain their purity of colour, while the reds, browns and yellows need up to full sunlight to obtain the greatest intensity of colour. Once the flowers have opened, do not leave them in the sun or the colour will be bleached out.

Dendrobes lose much of their strength during flowering and when the last flower falls the plant will have reached a weak point in its growth cycle. Lots of thought must be given before taking any action with the plant at this stage; in some cases premature repotting may be detrimental. The plant may have produced a new growth while flowering and this sight may inspire you to immediately start watering heavily. Don't, the plant has probably been kept comparatively dry while flowering and a sudden change from dry to wet conditions will probably cause the new growth to damp off. There is enough nutriment left in old pseudobulbous stems to maintain new growth for many weeks, so wait till new roots have formed before watering to any great degree.

Feeding

The general opinion is that at least some form of fertilization is necessary to ensure good healthy growth. The one important consideration is that care must be taken in application, whether the fertilizer is organic or inorganic, whether it is mixed with the growing medium at the time of potting, or whether it is applied by liquid or foliar feeding.

Some important points to remember are never feed a sick plant, never exceed the strength recommended for a fertilizer (if anything, apply at below strength), never feed unless the plant has been thoroughly watered first and never feed when the plant is dormant.

If using a manure in the growing medium (sheep or poultry manure or blood and bone are most common), be sure the medium is moist when potting.

The growing season as a general rule is the safe period for feeding. This extends from about September to April, and most dendrobes can be fed once a fortnight over this period. It is a good idea to water well for a day or so after feeding to get rid of excess salts.

It is recommended that a definite fertilizing programme be arranged and strictly kept to, to ensure that the type of dendrobes you grow receive a well-balanced diet. A plant could become deficient in a certain element, or elements, if only one proprietary brand or only one type (organic or inorganic) of fertilizer is used.

A good liquid organic fertilizer can be made by placing about three handfuls of fresh poultry manure in a large (9 litre) bucket. Fill with water and allow to stand for 2 to 4 weeks, occasionally giving it a stir. It will be ready for use when, by bacterial action, it has lost the cloudy appearance. Strain through a muslin cloth and add water until it is a very light brown in colour and free from any large solid particles.

Inorganic fertilizers can be purchased from any nursery—the outside of the packet will give details of the ingredients and the directions for use—but do not exceed the strength recommended by the makers. The major nutrient elements are nitrogen (N), phosphorus (P) and potassium (K). The labels on fertilizer packages usually give the total percentage of these three major elements plus, of course, the percentages of the minor elements. Do not be confused because they do not add up to 100 per cent. Only the nutrient elements are listed; omitted are oxygen, chlorine and water associated with these elements.

During the growing season, spring to summer, it is recommended that a high nitrogen fertilizer be used, say 30:10:10 (N:P:K). During autumn a fertilizer high in phosphorus and potassium, say 10:30:20 (N:P:K), is best. A pinch of blood and bone on top of the medium in very early spring is beneficial to the spring flowering and to the plant generally.

Pests

Anticipating pests can save many plants and flowers. There is nothing worse than to see flowers and plants that have been disfigured by some little insect, snail or slug.

Cleanliness in the orchid house is the biggest deterrent to pests. Do not leave old medium or leaves lying around. Keep benches clean and, if wooden, paint frequently. Do not have cracks and crevices where pests can hide and breed. Ensure good air movement around pots and keep ants away. Ants can carry and protect scale insects, aphids and mealy bugs.

Orchids grown under glass are more prone to attack from pests such as thrip, red spider mite, mealy bugs, aphids, etc.

SNAILS AND SLUGS

One of the best forms of control is to attack these at night by torch and dispose of them under foot, this method is also quite effective on wet, dull days.

To control snails and slugs, spray the whole plant with a mixture made up of 3 g arsenate of lead powder in 1 litre of water, (1/2 oz to 1 gallon). Dry bait made up of 10 g of Metaldehyde (fine powdered) to 500 g of bran (1/3 oz to 1 pound) should be set up in small heaps around the orchid house. Commercial snail bait pellets are quite effective.

The tiny bush snail is more difficult to eradicate. It inhabits the growing medium and often the first knowledge that it is present is when you notice that the new root tips are being eaten off. Mesurol at the strength mentioned on the packet washed down into the pot is usually effective.

APHIDS

Aphids are soft-bodied, sucking insects which vary in colour from light green to brown or black. They vary in size from 1 mm to 6 mm long. Aphid populations build up very quickly and attack buds, with resultant markings on the flowers. Spray with Malathion 50® or Metasystox® as directed on the package.

RED SPIDER MITE, TWO-SPOTTED MITE

Mites are minute insects, smaller than the head of a pin. They are slow moving and from yellow-green to pinky-red in colour. An infestation can be detected by silvery frosting on the underside of the leaves and some small white blotches on the surface of the leaves. In a bad attack, the leaves can turn black on the surface. They are a serious menace and difficult to eradicate once they take hold. Red spider mite prefers a dry atmosphere, so should not be a problem if you have a good humid growing atmosphere in your orchid house. Spray with Kelthane®, Pentac®, etc. as directed on the package. Two applications are given 7 to 10 days apart.

THRIPS

Thrips are very small insects that pierce the plant tissue and suck the juices of the cells. They are less than 1 mm long and the adults have two pairs of wings which are fringed with fine hairs. They are grey-to-black in colour and only just visible to the naked eye, but if you blow on the surface of the potting medium and they are there, you will see them running around. They like dry conditions. Evidence of their existence is seen by rusty patches on the leaves, disfigured and discoloured flowers and growths, and contorted flower buds. The first signs are frequently small, semi-transparent spots on the new growth of an orchid. Spray with Malathion 50®, Diazinon® or Metasystox® as directed on the package.

MEALY BUGS

These are small, sucking insects with a soft body covered with a white powdery substance. Although cousins of the scale insects, they do not form a protective scale covering. They hide in bracts and at the base of the leaf where it is closely appressed to the stem. Spray with Labaycid®, Malathion 50® as directed on the package.

SCALE INSECTS

Well over 350 species of scale insects have been classified in Australia. The males are minute, winged insects, while the larger, fleshy females are wingless and generally lose their antennae and legs and become immobile. Most of the scale-forming species settle down and feed (after a short period of active life) within their protective covering. Some are white and waxy, others are brown. Some can be easily wiped away with a wet cloth, others are quite tenacious. Sponging with a soft soap and water moves most scales, otherwise spray with a mixture of malathion (as directed on the package) and clear white oil. A number of applications may be needed depending on the particular scale involved.

CATERPILLARS

These do considerable damage; they eat holes in the leaves, chew right through a pseudobulbous stem or a raceme, and make a meal of the flowers. The most effective way to kill them is a quick squeeze with the fingers. Spray with Dipel®, Dipterex® or Carbaryl® as directed on the package.

ANTS

Although not injurious to orchids, they, nevertheless, bring other insects like scale onto the plants. They are difficult to destroy. The best place to destroy them is in the nest. Spray with Chlordane® as directed on the package.

GRASSHOPPERS

These pests create much damage, no stem being too big for them to chew through. During the spring and summer, nightly excursions to the orchid houses with a large pair of tweezers or forceps is the answer.

DENDROBIUM BEETLE (*Stethopachys formosa*)

This is not to be confused with the pumpkin beetle which is harmless to orchids and is much smaller than the Dendrobium beetle. The adult beetles are orange and black to yellow and black and are very hard, Plate 216. They will eat the surface of the leaves, the new growth and the flowers, causing great disfiguration.

The adult female lays numerous eggs down the centre portion of the new growth, they are very tiny, elongated and a greenish cream in colour. Incubation takes approximately 10 days, then the small larvae immediately eat their way down to the heart of the new growth and the damage is not noticed until it is too late. The larvae are whitish maggot-like individuals about 8 to 10 mm long, Plate 217.

Pupation usually takes place at the base of the plant among the surface roots. The pupae resemble thin, pure-white, spaghetti-like bundles mostly oval in shape about 10 to 20 mm in size, Plate 218. When the beetle emerges from the cocoon it immediately attacks the orchid. Flowers seem to be preferred.

The beetles are very active in late spring to early summer and are usually in pairs. When catching the beetle, hold a hand underneath, as beetles drop when your hand approaches them and they fall on to their back showing their underside which is difficult to see.

Chemicals which kill dendrobium beetles could damage the orchid causing burning, so it is necessary to hunt for and physically squash the beetles.

Diseases

FUNGUS

Damping off, spotting of flowers and other fungus-related diseases (like those causing a brown watery patch on the leaves) can be prevented to a great degree by plenty of free air circulation, good light, correct watering and (of course) cleanliness in the orchid house. Insect damage or bruising allows fungal spores to enter the plant, particularly in the hot humid months. Fungal attack can be controlled by Zineb®, Captan®, Benlate® or Natriphene®. Use as directed on the package.

BLACK ROT

Caused by two closely related genera of fungi namely *Phytophthora* and *Pythium*. These are most destructive fungi. They cause both stem or root rot and the first visible signs are small lesions on the stem or leaf, or a new lead becomes infected and the infection is seen as a brownish area with an advancing margin of slightly yellow colour. If the infection first occurs in the rhizome or roots, it spreads to other roots and upwards into the stem, at which stage the leaves usually fall off, or the leaf tip turns black followed fairly quickly by the rest of the leaf.

Control it by spraying with Fongarid®, Ridomil® or Previcur® (as directed on the package or bottle) when the humidity is high, as this is especially favourable to these fungi. Plants that have been infected should receive a drench of Fongarid® or Ridomil®.

VIRUS

The first sign of viral attack is usually a light green or yellow striping or mottling of the leaves of the new growth. Viruses can be transferred from one plant to another by insects or by failure to sterilize tools between each plant when potting. Doubtful plants should be moved out of the orchid house and expert opinion sought on the orchid's condition. If the orchid has a virus, the general opinion is to destroy the plant.

Appendix 1
Main months of flowering in native habitat

	Summer			Autumn			Winter			Spring		
	Dec.	**Jan.**	**Feb.**	**Mar.**	**Apr.**	**May**	**Jun.**	**Jul.**	**Aug.**	**Sept.**	**Oct.**	**Nov.**
D. adae								X	X	X	X	
D. aemulum								X	X	X		
D. affine				X	X	X	X	X	X			
D. agrostophyllum									X	X	X	
D. antennatum						X	X	X	X	X	X	X
D. baileyi		X	X									
D. bifalce						X	X					
D. bigibbum				X	X	X	X	X				
D. bowmanii		X	X	X	X	X	X					
D. calamiforme									X	X	X	X
D. canaliculatum								X	X	X	X	
D. cancroides	X	X										X
D. carrii									X	X	X	
D. carronii									X	X	X	
D. cucumerinum	X	X	X	X	X	X						X
D. x *delicatum*									X	X	X	
D. discolor								X	X	X	X	
D. fairfaxii									X	X	X	
D. sp. aff. *fairfaxii*								X	X	X		
D. falcorostrum									X	X	X	
D. fellowsii	X	X										X
D. fleckeri	X	X							X	X	X	X
D. x *foederatum*	X	X									X	X
D. x *gracillimum*									X	X	X	
D. x *grimesii*						X	X	X	X	X	X	
D. insigne	throughout the year											
D. johannis				X	X	X	X					
D. johnsoniae					X	X	X	X				
D. jonesii								X	X	X	X	X
D. sp. aff. *jonesii*											X	X
D. kingianum									X	X	X	X
D. lichenastrum	throughout the year											
D. linguiforme									X	X	X	X
D. lobbii								X	X	X		
D. luteocilium	X	X	X								X	X

	Summer			Autumn			Winter			Spring		
	Dec.	Jan.	Feb.	Mar.	Apr.	May	Jun.	Jul.	Aug.	Sept.	Oct.	Nov.
D. macropus subsp. *gracilicaule*									X	X	X	
D. malbrownii		X	X	X								
D. mirbelianum									X	X	X	
D. monophyllum									X	X	X	
D. moorei				X	X	X	X	X				
D. mortii										X	X	
D. nindii								X	X	X		
D. prenticei	throughout the year											
D. pugioniforme										X	X	X
D. racemosum										X	X	
D. rigidum	throughout the year											
D. x ruppiosum										X	X	X
D. schneiderae			X	X	X	X						
D. schoeninum									X	X	X	
D. smillieae									X	X	X	X
D. speciosum							X	X	X	X	X	
D. striolatum	X									X	X	X
D. stuartii		X	X									
D. x suffusum									X	X	X	
D. x superbiens			X	X	X	X	X	X				
D. teretifolium								X	X	X	X	
D. tetragonum southern forms										X	X	X
D. tetragonum northern forms					X	X	X	X	X	X	X	
D. toressae	throughout the year											
D. tozerensis	X	X									X	X
D. trilamellatum									X	X	X	
D. wassellii						X	X					
Norfolk Island												
D. macropus									X	X		
Lord Howe Island												
D. macropus subsp. *howeanum*										X	X	

Appendix 2
Dendrobium hybrids using only Australian indigenous parents registered to February 1988

Hybrid name	Parentage	Registrant and year of registration (originator in brackets)
D. Alan Printer	*D.* Andrew Persson X *D. kingianum*	Jupp 1969
D. Albanense	*D. x superbiens* X *D. discolor*[1]	Sanders 1948
D. Albertine	*D. bigibbum* X *D. antennatum*[2]	Nagrok, about 1940
D. Allyn King	*D.* Allyn Sands X *D. discolor*[1]	Jupp 1984
D. Allyn Prince	*D.* Rachelle Simpson X *D. discolor*[1]	Jupp 1984
D. Allyn Sands	*D.* Lowana Nioka X *D. johannis*	Jupp 1977
D. Allyn Star	*D. x gracillimum* X *D. tetragonum*	Jupp 1977
D. Allyn Supreme	*D.* Allyn Sands X *D.* Rachelle Simpson	Jupp 1984
D. Amaroo	*D. mortii* (as *D. tenuissimum*) X *D. falcorostrum*	Brandon 1986
D. Amber	*D.* Peter X *D. tetragonum*	Crawford 1987
D. Andrew Persson	*D. speciosum* X *D. falcorostrum*	Slade 1960
D. Andrew Upton	*D.* Judy Leroy X *D.* Star of Gold	Upton 1982
D. Aquarius	*D.* Bardo Rose X *D.* Star of Gold	Harris 1987
D. Arcadia	*D.* Eureka X *D.* Hastings	Banks 1986
D. Aussie Angel	*D.* Aussie Ira X *D. bigibbum*	Spence 1982
D. Aussie Aurora	*D.* Peter X *D.* Star of Gold	Spence 1985
D. Aussie Bonanza	*D.* Aussie Ira X *D. falcorostrum*	Spence 1983
D. Aussie Bounty	*D.* Bardo Rose X *D. falcorostrum*	Spence 1983
D. Aussie Cascade	*D. pugioniforme* X *D. schoeninum* (as *D. beckleri*)	Spence 1983
D. Aussie Charm	*D.* Ellen X *D.* Gillian Leaney	Spence 1987
D. Aussie Child	*D.* Hilda Poxon X *D.* Aussie Spectrum	Spence 1986
D. Aussie Choice	*D. monophyllum* X *D. speciosum*	Spence 1987
D. Aussie Delight	*D.* Eureka X *D.* Gracious Falcon	Spence 1983
D. Aussie Dream	*D.* Ellen X *D. fleckeri*	Spence 1987
D. Aussie Ecru	*D. jonesii* (as *D. ruppianum*) X *D.* Eureka	Spence 1985
D. Aussie Ember	*D.* Aussie Ira X *D.* Peter	Spence 1987
D. Aussie Flame	*D.* Aussie Bonanza X *D.* Aussie Freckles	Spence 1987
D. Aussie Freckles	*D. x suffusum* X *D. falcorostrum*	Spence 1982
D. Aussie Frontier	*D.* Aussie Gypsy X *D. speciosum*	Spence 1983
D. Aussie Gift	*D.* Aussie Ira X *D.* Hastings	Spence 1987
D. Aussie Glow	*D.* Aussie Ira X *D. fleckeri*	Spence 1983
D. Aussie Gypsy	*D.* Hastings X *D. jonesii* (as *D. ruppianum*)	Spence 1983
D. Aussie Hero	*D.* Eureka X *D.* Kingrose	Spence 1985
D. Aussie Ira	*D.* Ellen X *D. tetragonum*	Spence 1982

Hybrid name	Parentage	Registrant and year of registration (originator in brackets)
D. Aussie Jewel	*D.* Aussie Bonanza X *D.* Aussie Mist	Spence 1987
D. Aussie Joy	*D.* Peter X *D. falcorostrum*	Spence 1985
D. Aussie Love	*D.* Eureka X *D. x delicatum*	Spence 1983
D. Aussie Luck	*D.* Star of Gold X *D.* Aussie Mist	Spence 1987
D. Aussie Lust	*D.* Aussie Ira X *D.* Kingrose	Spence 1987
D. Aussie Majestic	*D.* Aussie Freckles X *D. falcorostrum*	Spence 1983
D. Aussie Magic	*D.* Eureka X *D.* Hilda Poxon	Spence 1987
D. Aussie Merit	*D.* Ella Victoria Leaney X *D. jonesii* (as *D. ruppianum*)	Spence 1987
D. Aussie Mist	*D.* Peter X *D.* Bardo Rose	Spence 1982
D. Aussie Nave	*D.* Aussie Ira X *D. jonesii* (as *D. ruppianum*)	Spence 1987
D. Aussie Ochre	*D. agrostophyllum* X *D. speciosum*	Spence 1982
D. Aussie Parade	*D.* Aussie Ira X *D.* Star of Gold	Spence 1987
D. Aussie Pride	*D.* Eureka X *D.* Bardo Rose	Spence 1983
D. Aussie Reward	*D. x delicatum* X *D.* Aussie Starlight	Spence 1987
D. Aussie Quest	*D.* Graham Hewitt X *D.* Gillian Leaney	Spence 1985
D. Aussie Sandy	*D. x delicatum* X *D.* Aussie Mist	Spence 1987
D. Aussie Sequins	*D.* Aussie Bonanza X *D. kingianum*	Spence 1988
D. Aussie Spectrum	*D.* Peter X *D. speciosum*	Spence 1983
D. Aussie Spice	*D. falcorostrum* X *D.* Hastings	Spence 1983
D. Aussie Springtime	*D.* Aussie Starlight X *D.* Gillian Leaney	Spence 1987
D. Aussie Starlight	*D.* Aussie Ira X *D. speciosum*	Spence 1982
D. Aussie Sunshine	*D.* Peter X *D.* Eureka	Spence 1982
D. Aussie Treasure	*D.* Ella Victoria Leaney X *D.* Eureka	Spence 1988
D. Aussie Treat	*D.* Ellen X *D.* Peter	Spence 1987
D. Aussie Utmost	*D.* Aussie Starlight X *D.* Kingrose	Spence 1986
D. Aussie Wonder	*D.* Ella Victoria Leaney X *D.* Ellen	Spence 1987
D. Aussie Zest	*D.* Eureka X *D.* Aussie Starlight	Spence 1986
D. Australia	*D. bigibbum* X *D. nindii*[3]	Nagrok, about 1940
D. Ayr Queensland	*D. canaliculatum* X *D. teretifolium*	Simpson 1983
D. Bardo Rose	*D. falcorostrum* X *D. kingianum*	Overall 1961
D. Big Fleck	*D. bigibbum* X *D. fleckeri*	Upton 1986
D. Barnard	*D.* Wyuna X *D.* Eureka	Brandon 1983
D. Berry	*D. kingianum* X *D.* Mini Pearl	Masahiro Oda 1983
D. Berserker	*D.* Hilda Poxon X *D.* Peter	Banks 1986
D. Black Mickey Mouse	*D.* Rachelle Simpson X *D. canaliculatum*	Suphachadiwongs 1986
D. Blue Gloucester	*D. x superbiens* X *D. nindii*[3]	MacPherson 1966
D. Blue Tiki	*D. johannis* X *D.* Judy Leroy	Limberlost 1970
D. Bluey	*D. x superbiens* X *D.* Peter Petersen	D'Bush Orc. Nurs. 1981
D. Blushing Rose	*D.* Bardo Rose X *D. x suffusum*	Cannon 1982[4]
D. Blushing Star	*D. x suffusum* X *D. tetragonum*	Cannon 1978
D. Blushing Sun	*D.* Sunglow X *D.* Blushing Star	Cannon 1986
D. Bright Spark	*D.* John Upton X *D. tetragonum*	Upton 1986
D. Brinawa	*D.* Hilda Poxon X *D. bigibbum*	Jarvis 1988
D. Burie	*D. cucumerinum* X *D. schoeninum* (as *D. beckleri*)	Brandon 1984
D. Camp Island	*D.* Peter Petersen X *D. discolor*[1]	MacPherson 1983
D. Carawah	*D.* Sunglow X *D.* Star of Gold	Brandon 1983
D. Cawongla	*D. kingianum* X *D. x suffusum*	Brandon 1983

Hybrid name	Parentage	Registrant and year of registration (originator in brackets)
D. Champagne	*D. discolor*[1] X *D. mirbelianum*	S'pore Bot. Gdns 1951
D. Cherry Dance	*D. bigibbum* X *D.* Mini Pearl	Nagasaka 1983
D. Cherub	*D.* Peter Petersen X *D. carronii*	D'Bush Orc. Nurs. 1984
D. Chieno	*D. discolor*[1] X *D. affine* (as *D. dicuphum*)	Aoki 1964 (Hayashi)
D. Colin	*D.* Ellen X *D. jonesii* (as *D. ruppianum*)	Upton 1979
D. Coljohn	*D.* Colin X *D.* John Upton	Upton 1986
D. Collette Fleur Simpson	*D. canaliculatum* X *D. monophyllum*	Simpson 1987
D. Colsboys	*D.* Jamie Upton X *D.* John Upton	Upton 1986
D. Colreka	*D.* Colin X *D.* Eureka	Upton 1988
D. Coral Sun	*D.* Gulginni X *D. tetragonum*	Harris 1987
D. Craig Hewitt	*D.* Emmy X *D. speciosum*	Upton 1981
D. Dainty Cascades	*D. pugioniforme* X *D. mortii* (as *D. tenuissimum*)	Upton 1984
D. Dang Toi	*D. x superbiens* (as *D. goldiei*) X *D. discolor*[1]	S'pore Bot. Gdns 1941
D. D'Bush Chic	*D. x superbiens* X *D. johannis*	D'Bush Orc. Nurs. 1981
D. Debra Wray	*D. x delicatum* X *D.* Kingrose	Wray 1986
D. Delemma	*D.* Emma X *D. x delicatum*	Upton 1986
D. Delicate Ellen	*D.* Ellen X *D. x delicatum*	Cannon 1986
D. Delicate Falcon	*D. falcorostrum* X *D. x delicatum*	Cannon 1979
D. x delicatum	*D. speciosum* X *D. kingianum*	Natural hybrid
D. Dilgry	*D. speciosum* X *D.* Bardo Rose	Merellen's 1987
D. Donie	*D.* Rachelle Simpson X *D. bigibbum*	Bearden 1984 (o/u)
D. Dot Sheen	*D.* Ellen X *D.* Aussie Bonanza	Batchelor 1987
D. Double Two	*D. x gracillimum* X *D. adae*	Jupp 1979
D. Dryander Mountain	*D.* Gloucester Sands X *D. discolor*[1]	MacPherson 1983
D. Duffy	*D. striolatum* X *D. pugioniforme*	Skillicorn 1984
D. Duke	*D.* Judy Leroy X *D. tetragonum*	Upton 1983
D. Dusky Girl	*D.* Judy Leroy X *D. canaliculatum*	Noble 1981
D. Eleebana	*D.* Judy Leroy X *D. bigibbum*	Upton 1982
D. Elegant Heart	*D.* Peewee X *D. speciosum*	Upton 1986
D. Elfin	*D.* Dainty Cascades X *D. kingianum*	Upton 1984
D. Elinjaa	*D.* Graham Hewitt X *D. tetragonum*	Roper 1987
D. Ella Rose	*D.* Ella Victoria Leaney X *D.* Blushing Rose	Upton 1986
D. Ella Victoria Leaney	*D. kingianum* X *D. jonesii* (as *D. ruppianum*)	Cannon 1980[4]
D. Ellen	*D. kingianum* X *D. tetragonum*	Schmidt 1928
D. Ellen Grace	*D. macropus* (as *D. gracilicaule*) X *D.* Ellen	Sheen 1987
D. Elsie Cox	*D. x superbiens* X *D. x delicatum*	Osment 1961
D. Emma	*D.* Gillian Leaney X *D. tetragonum*	Upton 1980
D. Emmy	*D. aemulum* X *D. kingianum*	Butler 1966
D. Esme Poulton	*D. bigibbum* X *D. kingianum*	Poulton 1978
D. Essie Banks	*D.* Hilda Poxon X *D. speciosum*	Banks 1983
D. Eureka	*D. fleckeri* X *D. jonesii* (as *D. ruppianum*)	Cannon 1980[4]
D. Fairy Floss	*D.* Penny Ann X *D. tetragonum*	Upton 1984
D. Fay Moore	*D.* Rosemary Jupp X *D. teretifolium*	Jupp 1984
D. Ferg Martin	*D.* Ellen X *D.* Penny Ann	Banks 1986
D. Fiesta	*D.* Colin X *D.* Kim	Upton 1984
D. Fiona Banks	*D.* Aussie Freckles X *D. kingianum*	Banks 1983

Hybrid name	Parentage	Registrant and year of registration (originator in brackets)
D. x foederatum	*D. calamiforme* (as *D. teretifolium*) X *D. rigidum*	Natural hybrid
D. Frances Henton	*D.* Dang Toi (as *D.* Albanense) X *D. nindii*[3]	Lee Kim Hong 1958
D. Gillian Leaney	*D. x delicatum* X *D. kingianum*	Leaney 1964
D. Gilpet	*D.* Peter X *D.* Gillian Leaney	Arnott 1984
D. Glenn Wall	*D. bowmanii* (as *D. mortii*) X *D. linguiforme*	Jupp 1984
D. Glenys King	*D.* Lowana Nioka X *D. canaliculatum*	King 1966 (MacPherson)
D. Gloucester Dawn	*D. mirbelianum* X *D. canaliculatum*	MacPherson 1966
D. Gloucester Gem	*D.* Gloucester Sands x *D. bigibbum*	MacPherson 1966
D. Gloucester Sands	*D. discolor*[1] X *D. canaliculatum*	MacPherson 1963
D. Goblin	*D.* Emmy X *D.* Ellen	Upton 1984
D. Gold Baby	*D.* Judy Leroy X *D. x superbiens*	Noble 1983
D. Gold Stripe	*D. bifalce* X *D. discolor*[1]	Limberlost 1973
D. Golden Cascades	*D. macropus* subsp. *graciilicaule* (as *D. gracilicaule*) X *D. jonesii* (as *D. ruppianum*)	Arnott 1984
D. Golden Fleck	*D. fleckeri* X *D. macropus* subsp. *gracilicaule* (as *D. gracilicaule*)	Cannon 1980[4]
D. Golden Glory	*D. fleckeri* X *D. tetragonum*	Purvis 1985
D. Golligosh	*D. schneiderae* X *D. monophyllum*	Upton 1985
D. Goose Bumps	*D. teretifolium* X *D. cucumerinum*	Jupp 1977
D. Gracia	*D. macropus* subsp. *gracilicaule* (as *D. gracilicaule*) X *D. adae*	Hurst 1986
D. x gracillimum	*D. speciosum* X *D. macropus* subsp. *gracilicaule* (as *D. gracilicaule*)	Natural hybrid
D. Gracious Cascades	*D. speciosum* X *D. x gracillimum*	Cannon 1982[4]
D. Gracious Falcon	*D. x gracillimum* X *D. falcorostrum*	Cannon 1982
D. Gracious Star	*D. tetragonum* X *D.* Gracious Falcon	Upton 1988
D. Graeme Banks	*D.* Sunglow X *D.* Hilda Poxon	Banks 1987
D. Graham Hewitt	*D.* Ellen X *D. speciosum*	Upton 1981
D. Green Goddess	*D. adae* X *D.* Susan	Upton 1986
D. Green Star	*D.* Jamie Upton X *D.* Golden Fleck	Upton 1986
D. Gresford	*D. affine* (as *D. dicuphum*) X *D. nindii*[3]	Jupp 1977
D. x grimesii	*D. calamiforme* (as *D. teretifolium*) X *D. linguiforme*	Natural hybrid
D. Gulginni	*D.* Sunglow X *D. kingianum*	Brandon 1983
D. Harold Hirsch	*D. x suffusum* X *D. speciosum*	Upton 1983
D. Hastings	*D. kingianum* X *D. fleckeri*	Cannons 1970
D. Hilda Poxon	*D. speciosum* X *D. tetragonum*	Grundon 1977 (Butler)
D. Hot Shot	*D.* Hilda Poxon X *D.* Golden Fleck	Upton 1988
D. Ingleside	*D.* John Upton X *D. speciosum*	Upton 1988
D. Ira Butler	*D.* Hilda Poxon X *D. jonesii* (as *D. ruppianum*)	Upton 1983
D. Ida Mary	*D. schoeninum* (as *D. beckleri*) X *D. mortii* (as *D. tenuissimum*)	Skillicorn 1985
D. Jamie Upton	*D. tetragonum* X *D.* Golden Fleck	Upton 1981
D. Jane Leaney	*D. x delicatum* X *D. speciosum*	Leaney 1964
D. Jenni	*D.* Yondi X *D.* Kingrose	Batchelor 1986
D. Jesmond Gem	*D.* Yondi X *D. speciosum*	Purvis 1987

Hybrid name	Parentage	Registrant and year of registration (originator in brackets)
D. Jiggi	*D. schoeninum* (as *D. beckleri*) X *D. teretifolium*	Brandon 1983
D. John's Charm	*D. discolor*[1] X *D. trilamellatum* (as *D. semifuscum*)	D'Bush Orc. Nurs. 1984
D. John Rose	*D.* John Upton X *D.* Bardo Rose	Upton 1986
D. John Upton	*D.* Kim X *D.* Ellen	Upton 1981
D. Johulatum	*D. discolor*[1] X *D. johannis*	Korsman 1983
D. Jombock	*D.* Judy Leroy X *D.* Colin	Upton 1983
D. Joy Wray	*D. kingianum* X *D.* Kingrose	Wray 1985
D. Judy Leroy	*D. canaliculatum* X *D. affine* (as *D. dicuphum*)	Limberlost 1961
D. Julatten	*D. kingianum* X *D.* Graham Hewitt	Roper 1987
D. Julie Skillicorn	*D. mortii* (as *D. tenuissimum*) X *D. striolatum*	Skillicorn 1982
D. Justin Wall	*D.* Lowana Nioka X *D. speciosum*	Jupp 1984
D. Kaipu	*D. mirbelianum* X *D. johannis*	Kirsch 1958
D. Kalang	*D.* Sunglow X *D.* Gracious Falcon	Brandon 1983
D. Karani	*D. monophyllum* X *D. kingianum*	Brandon 1983
D. Karen Nair	*D.* Jamie Upton X *D.* Wonga	Upton 1988
D. Karloo	*D. speciosum* X *D.* Gracious Falcon	Brandon 1983
D. Kathryn Banks	*D.* Ellen X *D. kingianum*	Banks 1983
D. Kevin Banks	*D. x delicatum* X *D. adae*	Banks 1983
D. Kim	*D.* Peter X *D. kingianum*	Bedford 1981
D. Kim Heinze	*D.* Blushing Star X *D.* Hilda Poxon	Cannon 1982
D. King Falcon	*D. falcorostrum* X *D.* Kingrose	Cannon 1978
D. King Goblin	*D.* Goblin X *D. kingianum*	Upton 1986
D. King Leaney	*D. kingianum* X *D.* Ella Victoria Leaney	Brandon 1984 (Ezzy)
D. Kingrose	*D.* Bardo Rose X *D. kingianum*	Butler 1966
D. King-Wong	*D. kingianum* X *D.* Wonga	Graham 1987
D. Kit Murdoch	*D. kingianum* X *D.* Susan	Cannon 1980
D. Kookaburra	*D.* Star of Gold X *D.* Ellen	Brandon 1986
D. Kungara	*D. tetragonum* X *D. jonesii* (as *D. ruppianum*)	Upton 1982
D. Ku-Ring-Gai	*D. tetragonum* X *D. x delicatum*	Friend 1971
D. Lemon Glow	*D.* Jamie Upton X *D. speciosum*	Upton 1986
D. Leonie Jane	*D. adae* X *D. falcorostrum*	Fetherston 1973
D. Liela Gilbert	*D. macropus* subsp. *gracilicaule* (as *D. gracilicaule*) X *D. aemulum*	Donovan 1987
D. Limestone	*D. bowmanii* (as *D. mortii*) X *D. cucumerinum*	Skillicorn 1985
D. Lloyd Stainton	*D. x superbiens* X *D. canaliculatum*	Lonne's (I. Lonne) 1986
D. Lorikeet	*D.* Peewee X *D.* Ellen	Upton 1986
D. Lowana Nioka	*D. canaliculatum* X *D. antennatum*[2]	Limberlost 1962 (Muir)
D. Lowana Tiki	*D. antennatum*[2] X *D.* Judy Leroy	Limberlost 1966
D. Lynette Banks	*D.* Eureka X *D. speciosum*	Banks 1983
D. Manning	*D. speciosum* X *D.* Willowie Gold	Merrellen's 1987
D. Margery Angeleri	*D.* Aussie Bonanza X *D.* Golden Fleck	Banks 1986
D. Maron	*D. affine* (as *D. dicuphum*) X *D. tetragonum*	Haylock 1983

Hybrid name	Parentage	Registrant and year of registration (originator in brackets)
D. Matilda	*D.* Dang Toi X *D. canaliculatum*	Korsman 1982
D. Memoria Eric Gordon	*D. x delicatum* X *D.* Peter	Spence 1985
D. Memoria Lloyd Bradford	*D. canaliculatum* X *D. speciosum*	Spence 1985
D. Merritt Island	*D. bigibbum* (as *D. compactum*) X *D. affine* (as *D. dicuphum*)	Dewey 1969
D. Miara	*D.* Hilda Poxon X *D. falcorostrum*	Roper 1985
D. Michael Jupp	*D. schoeninum* (as *D. beckleri*) X *D. linguiforme*	Jupp 1980
D. Mighty Mite	*D. canaliculatum* X *D. carronii*	Richella's 1987
D. Mini Pearl	*D. canaliculatum* X *D. bigibbum* (as *D. compactum*)	Yagi 1974
D. Minnie	*D. carronii* X *D. johannis*	D'Bush Orc. Nurs. 1984
D. Minuet	*D.* Penny Ann X *D.* Goblin	Upton 1986
D. Moira Stewart	*D. x superbiens* (as *D. goldiei*) X *D. mirbelianum*	Stewart 1960
D. Moses Mark	*D. antennatum*[2] X *D. mirbelianum*	Shee Shim 1961 (Voon Orchids)
D. Narraweena	*D.* Colin X *D.* Gillian Leaney	Upton 1988
D. Natalie Wall	*D.* Lowana Nioka X *D.* Star of Gold	Jupp 1984
D. Nathan Wall	*D. kingianum* X *D.* Double Two	Jupp 1986
D. Nerang	*D.* Wyuna X *D. speciosum*	Brandon 1984
D. Nooro	*D.* Eureka X *D. falcorostrum*	Brandon 1983
D. Northmead	*D. speciosum* X *D.* Ku-Ring-Gai	Banks 1986
D. Nugget	*D.* Eureka X *D. macropus* (as *D. gracilicaule*)	Upton 1988
D. Numkumbil	*D. tetragonum* X *D.* Eureka	Upton 1983
D. Ocker's Dream	*D.* Eureka X *D.* Ellen	Sobey 1987
D. Omega	*D.* Tully X *D.* Carawah	Brandon 1986
D. Orana	*D. adae* X *D. monophyllum*	Upton 1983
D. Our Native	*D. speciosum* X *D. johannis*	McFarlane 1977
D. Pauline Rankin	*D.* Hilda Poxon X *D. tetragonum*	Rankin 1986
D. Peach Glow	*D.* Golden Fleck X *D.* Hastings	Cannon 1981[4]
D. Peach Star	*D. tetragonum* X *D.* Hastings	Cannon 1982
D. Peewee	*D. bigibbum* X *D. tetragonum*	Cannons 1979
D. Penang Angel	*D.* Merritt Island X *D. canaliculatum*	Ooi Leng Sun Or. 1985
D. Penny Ann	*D. x gracillimum* X *D. kingianum*	Upton 1974
D. Penny Rose	*D.* Penny Ann X *D.* Bardo Rose	Upton 1986
D. Peter	*D. fleckeri* X *D. falcorostrum*	Bedford 1972
D. Peter Hewitt	*D.* Hilda Poxon X *D.* Hastings	Upton 1982
D. Peter Petersen	*D. nindii*[3] X *D. discolor*[1]	Bourner 1956
D. Phil Deane	*D.* Ella Victoria Leaney X *D. tetragonum*	Batchelor 1982 (Deane)
D. Pink Ballerina	*D. agrostophyllum* X *D. bigibbum*	Upton 1985
D. Pink Pencil	*D. kingianum* X *D. teretifolium*	Upton 1984 (Butler)
D. Pinterry	*D.* Ellen X *D.* Graham Hewitt	Roper 1987
D. Pixie	*D.* Ellen X *D. x suffusum*	Upton 1984
D. Princess Stephanie	*D. mirbelianum* X *D. bigibbum*	Limberlost 1962

Hybrid name	Parentage	Registrant and year of registration (originator in brackets)
D. Rachelle Simpson	*D. canaliculatum* X *D. johannis*	Simpson 1967
D. Rajah	*D.* Dang Toi X *D.* Champagne	Khoo Kay Ann 1961
D. Red Baron	*D.* Blushing Rose X *D. kingianum*	Upton 1984
D. Red Wings	*D. antennatum*[2] X *D. johannis*	Deane 1954
D. Regal Rose	*D.* Blushing Rose X *D. speciosum*	Upton 1988
D. Reg Burns	*D. affine* (as *D. dicuphum*) X *D. fleckeri*	Zellar 1984
D. Ring of Fire	*D.* Ku-Ring-Gai X *D.* John Upton	Upton 1986
D. Robyn Kaylene	*D.* Ella Victoria Leaney X *D. macropus* subsp. *gracilicaule* (as *D. gracilicaule*)	Haylock 1985 (Elmy)
D. Ronald Banks	*D.* Double Two X *D. speciosum*	Banks 1984
D. Rosella	*D. tetragonum* X *D.* Peewee	Cannons 1983
D. Rosemary Jupp	*D. striolatum* X *D. teretifolium*	Jupp 1975
D. x ruppiosum	*D. speciosum* X *D. jonesii* (as *D. ruppianum*)	Natural hybrid
D. Rutherford Star	*D.* Star Imp X *D. tetragonum*	Smith 1985
D. Rutherford Sunstar	*D. tetragonum* X *D.* Gracious Cascades	Smith 1987
D. Rutherford Supreme	*D.* Wonga X *D.* Hastings	Smith 1986
D. Rutherford Surprise	*D.* Hastings X *D. kingianum*	Smith 1984
D. Saskia	*D. x superbiens* X *D. bigibbum*	Barendsen 1951
D. Sauterne	*D.* Champagne X *D. johannis*	D'Bush Orc. Nurs. 1981
D. Shan Leaney	*D. x gracillimum* X *D. macropus* subsp. *gracilicaule* (as *D. gracilicaule*)	Leaney 1964
D. Silvester	*D. tetragonum* X *D.* Bardo Rose	Upton 1983
D. Son of Gloucester	*D.* Peter Petersen X *D. canaliculatum*	MacPherson 1982
D. Specio-kingianum	*D. kingianum* X *D. speciosum*	Lawrence 1892
D. Star Child	*D.* Ku-Ring-Gai X *D.* Star of Gold	Jupp 1984
D. Star Delight	*D.* Star of Gold X *D. jonesii* (as *D. ruppianum*)	Jupp 1980
D. Star Falcon	*D.* Star of Gold X *D. falcorostrum*	Harris 1987
D. Star Imp	*D. tetragonum* X *D. adae*	Jupp 1977
D. Star of Dawn	*D.* Star of Gold X *D. kingianum*	Jupp 1980
D. Star of Gold	*D. falcorostrum* X *D. tetragonum*	Jupp 1973
D. Star of Riverdene	*D.* Star of Gold X *D. speciosum*	Jupp 1981
D. Star One	*D. x delicatum* X *D.* Star of Gold	Jupp 1979
D. Star Twist	*D. tetragonum* X *D. johannis*	Jupp 1981
D. Stephen	*D. schoeninum* (as *D. beckleri*) X *D. striolatum*	Skillicorn 1983
D. x suffusum	*D. kingianum* X *D. macropus* subsp. *gracilicaule* (as *D. gracilicaule*)	Natural hybrid
D. Sunglow	*D. speciosum* X *D. fleckeri*	Cannon 1980 (Bedford)
D. Sun Sprite	*D. x gracillimum* X *D. x delicatum*	Jupp 1975
D. Sun Star	*D.* Sunglow X *D. tetragonum*	Cannon 1986
D. x superbiens	*D. bigibbum* X *D. discolor*	Natural hybrid
D. Super Ruffles	*D.* Rosemary Jupp X *D. mortii* (as *D. tenuissimum*)	Harris 1987
D. Susan	*D. falcorostrum* X *D. macropus* subsp. *gracilicaule* (as *D. gracilicaule*)	Bedford 1970
D. Sweet Ellen	*D.* Ellen X *D.* Susan	Cannon 1986
D. Tambar	*D.* Golden Fleck X *D.* Aussie Glow	R. Arnott 1986
D. Tarean	*D.* Star of Gold X *D.* Hilda Poxon	Brandon 1983

Hybrid name	Parentage	Registrant and year of registration (originator in brackets)
D. Tawhai	*D.* Kingrose X *D.* Bardo Rose	Wray 1985
D. Tarantula	*D.* Allyn Star X *D. tetragonum*	Skillicorn 1987
D. Telekon	*D.* Hilda Poxon X *D. kingianum*	Banks 1984
D. Three Star	*D. johannis* X. *D. x superbiens* (as *D. goldiei*)	Imura 1954 (Miranda)
D. Tom Jupp	*D. x delicatum* X *D.* Andrew Persson	Jupp 1974
D. Tooloom	*D. adae* X *D. kingianum*	Brandon 1983
D. Tully	*D.* Wyuna X *D.* Sunglow	Brandon 1983
D. Tusco	*D. nindii*[3] X *D. johannis*	Komatsubara 1957 (Inouye)
D. Tweed	*D.* Peter X *D.* Sunglow	Brandon 1983
D. Twinkle	*D.* Penny Ann X *D. falcorostrum*	Upton 1984
D. Ultravox	*D.* Bardo Rose X *D.* Andrew Persson	Banks 1984
D. Valda	*D. discolor*[1] X *D. carronnii*	Hatcher 1985
D. Verninha	*D. discolor*[1] X *D. antennatum*[2]	McAdam 1961
D. Vida Banks	*D.* Hilda Poxon X *D.* Wonga	Banks 1984
D. Virginia Jupp	*D. linguiforme* X *D. teretifolium*	Jupp 1969
D. Wakehurst	*D.* John Upton X *D.* Kungara	Upton 1988
D. Warooa	*D. adae* X *D.* Eureka	Brandon 1984
D. Warrambool	*D.* Ellen X *D. falcorostrum*	Upton 1983
D. Warringah	*D. bigibbum* X *D. speciosum*	Upton 1984
D. Warrior	*D.* Hilda Poxon X *D.* Ku-Ring-Gai	Banks 1984
D. Wee Cheow Lay	*D.* Verninha X *D.* Yellow Curls	Wee Sian Hee 1966 (S'pore Bot. Gdns)
D. Wesley Pink	*D.* Virgina Jupp X *D. teretifolium*	Upton 1985
D. Wesley Rose	*D.* Pink Pencil X *D. speciosum*	Upton 1984
D. Whispering Jack	*D.* Golden Fleck X *D.* Peter	Banks 1987
D. Whitehart	*D.* Nunkumbil X *D. adae*	Upton 1986
D. Willowie Gold	*D. discolor*[1] X *D. tetragonum*	Limberlost 1970
D. Winston Hills	*D.* Kevin Banks X *D. tetragonum*	Banks 1987
D. Wirruna	*D.* Hilda Poxon X *D. fleckeri*	Upton 1983
D. Woko	*D. speciosum* X *D.* Wonga	Merrellen's 1987
D. Wonga	*D. speciosum* X *D.* Hastings	Cannons 1979
D. Wyuna	*D. fleckeri* X *D.* Ella Victoria Leaney	Brandon 1983
D. Yellow Curls	*D.* Champagne X *D. discolor*[1]	S'pore Bot. Gdns 1956
D. Yellow Venture	*D. speciosum* X *D. adae*	Jupp 1979
D. Yondi	*D. kingianum* X *D.* Eureka	Batchelor 1982
D. Zip	*D.* Ku-Ring-Gai X *D. kingianum*	Skillicorn 1987

[1] as *D. undulatum* [2] as *D. d'albertisii* [3] as *D. tofftii* [4] Originator W. G. and J. Cannons

Appendix 3
Dendrobium hybrids including *D. phalaenopsis* and *D. schroederianum* in their parentage to March 1987

Hybrid name	Parentage	Registrant and year of registration
D. Agnes Cheok	*D.* Anouk X *D.* Lady Constance	Woodlawn 1958
D. Anouk	*D.* Pompadour X *D. phalaenopsis*	Vacherot-Lecoufle 1950
D. Arctic Gem	*D.* White Gem X *D. affine* (as *D. dicuphum*)	Richella Orchids 1985
D. Aussie Tracey	*D. johannis* X *D. phalaenopsis*	Spence 1982
D. Bailys Blue	*D. nindii* (as *D. tofftii*) X *D.* Pauline	Limberlost 1961
D. Beauty Queen	*D.* Lee Siew Sim X *D.* Pompadour	S'pore Orchids 1967
D. Betty Matsuda	*D. bigibbum* X *D.* Pompadour	Takiguchi 1957
D. Bill Yeakel	*D.* Liliha X *D. bigibbum*	Gallion 1968
D. Blossom Ross	*D.* E. V. Dulin X *D.* Louis Bleriot	Pickell 1960
D. Boey Chun Heng	*D. schroederianum* X *D.* Roger Sander	Kuah 1955
D. Boisseyense	*D. bigibbum* X *D. schroederianum*	Vacherot-Lecoufle 1926
D. Brigdjen	*D.* Saskia X *D.* Orchidwood	Moestopo 1959
D. Carol Hausermann	*D.* Tim Tam X *D.* Anouk	Hausermann's 1969
D. Colin Potter	*D. phalaenopsis* X *D. nindii* (as *D. tofftii*)	Lyon 1944
D. Coneac	*D.* Endeavour X *D. phalaenopsis*	Limberlost 1965
D. Cupie Doll	*D. phalaenopsis* X *D.* Lowana Nioka	Limberlost 1969
D. Daintree	*D.* Bailys Blue X *D. phalaenopsis*	Limberlost 1967
D. Dark Victory	*D.* Pompadour X *D. discolor* (as *D. undulatum*)	Conklin 1951
D. David Baver	*D. johannis* X *D. phalaenopsis* var. *compactum*)	Baver 1956
D. David Collier	*D.* Helen Fukumura X *D.* Garden	Richella Orchids 1985
D. David Sander	*D. schroederianum* X *D. nindii* (as *D. tofftii*)	de Saram 1944
D. Digger	*D.* Colin Potter X *D. discolor* (as *D. undulatum*) var. *broomfieldii*	Miller 1959
D. Dorothy Deane	*D.* Red Wings X *D. phalaenopsis*	Deane 1954
D. Endeavour	*D.* Liliha X *D.* Lady Constance	Limberlost 1961
D. Eugene	*D.* Liliha X *D. schroederianum*	Mujata 1954
D. E. V. Dulin	*D.* Garnet X *D. schroederianum*	E. T. Iwanaga 1954
D. Florence Foo	*D.* Albanense X *D. phalaenopsis*	Lee Kim Hong 1955
D. Full Moon	*D.* Louis Bleriot X *D.* Helen Fukumura	Miwa 1956
D. Garden	*D.* Lady Constance X *D.* Louis Bleriot	Miyamoto 1953
D. Garnet	*D. bigibbum* X *D.* Roger Sander	Muriel Bridges 1954
D. Garnouk	*D.* Garnet X *D.* Anouk	H. Iwanaga 1959

Hybrid name	Parentage	Registrant and year of registration
D. Gemwood	*D.* Lady Gem X *D.* Orchidwood	D'Bush Orc. Nurs. 1982
D. Gigi	*D.* Lady Constance X *D.* Garnet	Inouye 1959
D. Gladys Oumae	*D.* Liliha X *D. discolor* (as *D. undulatum*)	Oumae 1954
D. Gregor Duruty	*D. phalaenopsis* X *D. superbiens* (as *D. goldiei*)	Duruty 1957
D. Grove Gem	*D.* Lady Gem X *D. bigibbum*	D'Bush Orc. Nurs. 1982
D. Guynemer	*D.* Boisseyense X *D.* Pompadour	Vacherot-Lecoufle 1930
D. Hale Kulani	*D.* Helen Fukumura X *D.* Top Hat	Kimball 1958
D. Harry Wheeler	*D. nindii* (as *D. tofftii*) X *D.* Anouk	Toogood 1962
D. Heavenly	*D.* Garden X *D. bigibbum*	Richella 1980
D. Helen Banks	*D.* Helen Fukumura X *D. phalaenopsis*	R. Palmer 1963
D. Helen Fukumura	*D.* Lady Constance X *D. schroederianum*	Fukumura 1952
D. Hula Maid	*D.* Pauline X *D. discolor* (as *D. undulatum*)	Kushima 1981
D. Jean	*D.* Lady Constance X *D. x superbiens*	Mrs E. Yamasaki 1954
D. Joan Waddell	*D.* Pompadour X *D.* Louis Bleriot	Waddell 1961
D. Joy	*D.* Helen Fukumura X *D.* Dark Victory	Garth 1970
D. Joyful	*D. bigibbum* X *D.* Pauline	Stuart Low 1950
D. Judy Rutz	*D. phalaenopsis* X *D. antennatum*	Thoms 1985
D. Kailua	*D.* Garnet X *D. bigibbum* var. *compactum*	Rev. Yamada 1965
D. Lady Constance	*D.* Orchidwood X *D. phalaenopsis*	Weber 1947
D. Lady Gem	*D.* Lady Constance X *D.* White Gem	D'Bush Orc. Nurs. 1981
D. Lady Marie	*D.* Tim Tam X *D.* Lady Constance	George Black 1966
D. Leeanum	*D. schroederianum* X *D. x superbiens*	Natural hybrid
D. Lee Siew Sim	*D.* Pauline X *D.* Frances Henton	S'pore Orchids 1963
D. Lehua	*D.* Lady Constance X *D.* Helen Fukumura	Fukumura 1957
D. Liliha	*D.* Pauline X *D. phalaenopsis*	Kazamura 1948
D. Lim Kit Yee	*D.* Pompadour X *D. discolor* (as *D. undulatum*) var. *broomfieldii*	Lim Chooi Seng 1959
D. Lorraine Hanai	*D.* Dark Victory X *D.* Lady Constance	Joseph 1958
D. Louis Bleriot	*D. schroederianum* X *D. x superbiens*	Vacherot-Lecoufle 1929
D. Lovely Hawaii	*D.* Orchidwood X *D.* Lady Constance	McCoy 1959
D. Madame Butterfly	*D.* Mary Katherine X *D.* Helen Fukumura	Inouye 1958
D. Maui Queen	*D.* Anouk X *D.* Helen Fukumura	Fukumura 1959
D. Makassar	*D.* Louis Bleriot X *D.* Orchidwood	Perhimpunan 1959
D. Margaret Paxton	*D.* Pauline X *D. johannis*	S'pore Bot. Gdns 1950
D. Marianne de Jong	*D.* Pompadour X *D.* Roger Sander	de Jong 1951
D. Marisa	*D.* Sarie Marijs X *D. phalaenopsis* var. *alba*	H.S.H. Prince Chalerm Phol. 1956
D. Marten Hausermann	*D.* Tim Tam X *D.* Pompadour	Hausermann's 1968
D. Mary Katherine	*D.* Pompadour X *D.* Lady Constance	Inouye 1955
D. Mem Tu Pu	*D.* Pauline X *D.* Louis Bleriot	Peh Eng Kee 1970
D. Meng Aun	*D. phalaenopsis* X *D.* Yellow Curls	Lim Swee Aun 1964
D. Mini Stripe	*D. bairdianum* X *D. phalaenopsis*	Thompson 1981
D. Momi Fujimoto	*D. discolor* X *D.* Louis Bleriot	Miyamoto 1951
D. Myola	*D.* Lady Constance X *D. bigibbum*	D'Bush Orc. Nurs. 1982

Hybrid name	Parentage	Registrant and year of registration
D. Nah Hiok Hwee	*D.* Pompadour X *D. antennatum* (as *D. d'albertisii*)	Tay Boon Kim 1965
D. Naxos	*D.* Guynemer X *D.* Pompadour	Vacherot-Lecoufle 1952
D. Nora Potter	*D.* Colin Potter X *D. schroederianum*	Potter 1947
D. Omigosh	*D.* Heavenly X *D. phalaenopsis*	Richella's 1986
D. Ora Tatsumara	*D.* Colin Potter X *D.* Pompadour	Kodama 1955
D. Orchidwood	*D. bigibbum* X *D. phalaenopsis*	Orchidwood 1934
D. Oriel Mai	*D.* Anouk X *D.* Pauline	Cook 1963
D. Painted Doll	*D.* Judy Leroy X *D. phalaenopsis*	Limberlost 1966
D. Pauline	*D. phalaenopsis* X *D. discolor* (as *D. undulatum*)	Stuart Low 1932
D. Phalee	*D.* Leeanum X *D. phalaenopsis*	D'Bush Orc. Nurs. 1981
D. Polynesian	*D.* Pompadour X *D.* Orchidwood	McCoy 1951
D. Pompadour	*D.* Louis Bleriot X *D. schroederianum*	Vacherot-Lecoufle 1934
D. Puppchen	*D. schroederianum* X *D. discolor* (as *D. undulatum*)	—
D. Raimonde de Jong	*D.* Louis Bleriot X *D.* Roger Sander	de Jong 1951
D. Robyn Jones	*D.* Peter Petersen X *D.* Joan Waddell	Limberlost 1962
D. Rockhampton Prince	*D.* Anouk X *D. bigibbum*	Williams 1981
D. Roger Sander	*D. mirbelianum* X *D. schroederianum*	Sanders 1936
D. Rosy Tips	*D. canaliculatum* X *D. phalaenopsis*	Muir 1960
D. Rubicon	*D.* Guynemer X *D. phalaenopsis*	Vacherot-Lecoufle 1953
D. Ruth Macdonald	*D. phalaenopsis* X *D.* Saskia	Macdonald 1962
D. Sarie Marijs	*D. antennatum* (as *D. d'albertisii*) X *D. schroederianum*	Tjipeganti —
D. Startling	*D.* Painted Doll X *D. phalaenopsis*	Limberlost 1971
D. Suzanne	*D. phalaenopsis* X *D. tetragonum*	Willersdorf 1965
D. Taprobane	*D. x superbiens* X *D.* Pompadour	Ceylon Bot. Gdns 1961
D. Tau	*D.* Dark Victory X *D. discolor* (as *D. undulatum*) var. *broomfieldii*	Chang 1957
D. Tim Tam	*D.* Anouk X *D.* Pompadour	McCoy 1958
D. Toff	*D.* Peter Petersen X *D.* Colin Potter	D'Bush Orc. Nurs. 1981
D. Top Hat	*D.* Anouk X *D. schroederianum*	McCoy 1956
D. Top Notch	*D.* Top Hat X *D.* Anouk	Kushima 1958
D. Topsy de Jong	*D. schroederianum* X *D.* Pompadour	de Jong 1951
D. Tunku Josephine Jawahir	*D.* Louis Bleriot X *D.* Yellow Curls	Trevor 1970
D. White Gem	*D. affine* (as *D. dicuphum*) X *D. schroederianum*	Sanders 1946

Appendix 4
Reclassifying the genus *Dendrobium*

Dr F. G. Brieger's revision of the Subtribe *Dendrobiinae* as it applies to the genus *Dendrobium* in Australia (Schlechter's *Die Orchideen* ed. 3 (11/12) 636–752, 1981. This is the latest attempt at a reclassification of the genus *Dendrobium*. It is not currently accepted by botanists at Canberra or Kew.

Dendrobiinae

Series Genera

Dimorphophyllae

Genera

1. *Dockrillia*

Section *Dockrillia*
Sp. *D. cucumerina*
D. linguiformis
D. wassellii
Section *Teretifolia*
D. beckleri
D. mortii
D. pugioniformis
D. racemosa
D. rigida
D. striolata
D. tenuissima
D. teretifolia
Section *Lichenastra*
Sp. *D. lichenastra*
D. toressae

2. *Australorchis*

Sp. *A monophylla*
A. schneiderae

3. *Katherinea*

Sp. *K. carrii*

4. *Dendrocoryne*

Sp. *D. adae*
D. aemula
D. falcorostris
D. fleckeri
D. gracilicaulis
D. kingiana
D. moorei
D. ruppiana
D. speciosa
D. tetragona

5. *Latourorchis*

Sp. *L. bairdiana*
L. bifalce
L. johnsoniae

Dendrobia

Genera

6. *Dendrobium*

Subgenus *Dendrobium*
Section *Dendrobium*
Sp. *D. stuartii*

Subgenus *Stachyobium*
Section *Phalaenanthe*
Sp. *D. bigibbum*
D. dicuphum
Section *Ceratobium*
Sp. *D. antennatum*
D. carronii
D. discolor
D. johannis
D. mirbelianum
D. nindii
D. semifuscum
Section *Eleutheroglossum*
Sp. *D. canaliculatum*

7. *Trachyrhizum*

Sp. *T. agrostophyllum*

8. *Pedilonum*

Section *Pedilonum*
Sp. *P. smillieae*

Caulescentes

Genera

9. *Grastidium*

Section *Grastidium*
Sp. *G. baileyi*
G. cancroides
G. luteocilium
G. tozerensis

10. *Dichopus*

Sp. *D. insignis*

11. *Monanthus*

Sp. *M. malbrownii*

12. *Conostalix*

Sp. *C. lobbii*
(as *C. calcarata*)

Dr Stephan Rauschert carried out the new combinations omitted by Brieger in *Die Orchideen*, these being necessary for taxonomic reasons, in *Feddes Repertorium* 94: 433–471 Berlin 1983. Brieger had overlooked the older synonymous generic names of *Tropilis* and *Sayeria*, therefore Rauschert transferred the species involved from *Dendrocoryne* and *Latourorchis* respectively, to these genera. *D. canaliculatum* was transferred from the section *Eleutheroglossum* to the section *Spatulata* (Syn. *Ceratobium*) by Lavarack and Cribb in *Austrobaileya* 1982.

Appendix 5
Authors of plant names and publication abbreviations

Authors

Bail., F. M.	Frederick Manson Bailey
Bateman	James Bateman
Benth.	George Bentham
Bidw.	John Carne Bidwill
Binnend.	Simon Binnendijk
Blake, S. T.	Stanley T. Blake
Blume	Carl Ludwig Blume
Brieger	Dr F. G. Brieger
Br., R.	Robert Brown
Cady	Leo Isaacs Cady
Clements	Mark A. Clements
Clemesha	Stephen C. Clemesha
Colenso	Col. William Colenso
Cribb	Dr Phillip J. Cribb
Cunn., A.	Allan Cunningham
Dcne.	Joseph Decaisne
Dockr.	Alick William Dockrill
Endl.	Stephan Ladislaus Endlicher
Finet	Achille Eugène Finet
Fitzg.	Robert Desmond Fitzgerald
Gaud.	Charles Gaudichaud-Beaupre
Gilbert	Percival A. Gilbert
Green, P. S.	Peter S. Green
Hawkes, A. D.	Alex Drum Hawkes
Hill, W.	Walter Hill
Hook.	Sir Joseph Dalton Hooker
Hort.	Hortorum ('of the gardens'), frequently indicating of garden or unknown origin
Hunt, T. E.	Trevor E. Hunt
Krzl.	Fritz Kraenzlin
Kuntze	Otto Kuntze
Lavarack	Dr Peter S. Lavarack
Linden	Lucien Linden
Lindley	John Lindley
Lodd.	Conrad Loddiges
Lour.	João de Loureiro
Macleay	William Sharp Macleay
Maiden	Joseph Henry Maiden
Masters	Maxwell T. Masters
Miq.	Friedrich Anton Wilhelm Miquel
Moore, S.	Spencer le Marchant Moore
Muell., F.	Baron Sir Ferdinand Heinrich Jacob von Mueller
Nicholls	William Henry Nicholls
Rauschert	Dr Stephan Rauschert
Reichb. H. G.	Heinrich Gustav Reichenbach
Rendle	Alfred Barton Rendle
Rolfe	Robert Allen Rolfe
Rupp	Herman Montague Rucker Rupp
Schltr.	Friedrich Richard Rudolf Schlechter
Smith	Sir James Edward Smith
Sm. J. J.	Johannes Jacobus Smith
St. Cloud	Stanley F. Goesling-St. Cloud
Steud.	Ernest Gottlieb Steudel
Summerh.	Victor S. Summerhayes
Sw.	Olaf Swartz
Teijsmann	Johannes Elias Teijsmann
Veitch	John Gould Veitch
White, C. T.	Cyril Tennyson White
Williams	Benjamin Samuel Williams

Publications

Aust. Indg. Orchids	Australian Indigenous Orchids
Aust. Orchid Rev.	Australian Orchid Review
Aust. Orch.	Australian Orchids
Aust. Pl.	Australian Plants
Austrobaileya	Austrobaileya
Beitr. Syst. Pfl.	Beitrage zur systematischen Pflanzenkunde
Bot. Bull. Dept. Agric. Q/ld.	Botany Bulletin Department of Agriculture Queensland
Bot. Jahrb. Syst.	Botanische Jahrbucher fur Systematik Pflanzengeschichte und Pflanzengeographie
Bot. Mag.	Curtis's Botanical Magazine, London
Bull. Dep. Agr. Ind. N'eerl.	Bulletin du départment de l'agriculture aux Indes Néerlandaises, Buitenzorg
Bull. Misc. Inform. Kew	Bulletin of Miscellaneous Information Royal Botanic Gardens, Kew
Bull. Soc. Bot. France	Bulletin Société Botanique France, Paris
Die Orchideen	Die Orchideen, Rudolf Schlechter
Edwards's Bot. Reg.	Edwards's Botanical Register, London
Engl. Pflanzenreich	Engler's Das Pflanzenreich—*Orchidaceae-Monandrae-Dendrobiinae*
Exotic Bot.	Exotic Botany London
Feddes Repert.	Feddes Repertorium, Berlin
Fl. Aust.	Flora Australiensis
Fl. Ind. Bat.	Flora Indiae Batavae, Amsterdam
Fragm. Phyt. Aust.	Fragmenta Phytographiae Australiae
Freycinet Voy. Uranie	Louis de Freycinet—Voyage autour du monde sur les corvettes 'l'Uranie' et 'la Physicienne'
Gard. Chron.	Gardeners' Chronicle and Gardening Illustrated

Hamburger Garten-Blumenzeitung	Hamburger Garten-und Blumenzeitung. Zeitschrift fur Garten- und Blumenfreunde, Kunst-und Handelsqartner
J. Arnold Arbor.	Journal of the Arnold Arboretum, U.S.A.
J. Bot. Lond.	Hooker's London Journal of Botany
J. Proc. Linn. Soc. Bot.	Journal and Proceedings of the Linnean Society of London, Botany
Kongl. Vetensk. Acad. Nya Handl. Stockh.	Kongl. Vetenskaps Academiens Nya Handlingar, Stockholm
Lindenia	Lindenia. Iconographie des Orchidees Ghent and Brussels
Linnaea	Linnaea. Ein Journal fur die Botanik in ihrem ganzen Umfange
Man. Orchid. Pl.	A Manual of Orchidaceous Plants, Veitch & Sons
Mus. Bot. Lug.-Bat.	Museum Botanicum Lugdano-Batavum
Natuurk. Tijdschr. Ned.-Indie	Natuurkundig Tijdschrift voor Nederlandsch Indie
Nomencl.	Nomenclator Botanicus Ed. 2. Stuttgart
Nouv. Ann. Mus.	Nouvelles Annales du Muséum d'Histoire Naturelle, Paris
Nth Q/ld. Nat.	The North Queensland Naturalist
Occas. Pap. Q/ld. Fl.	Occasional Papers on the Queensland Flora
Orchadian	The Orchadian
Orchids New South Wales	The Orchids of New South Wales
Öst. Bot.	Öesterreichische Botanische Zeitschrift
Pax. Flower Gdn.	Paxton's Flower Garden, London
Proc. Linn. Soc. N.S.W.	Proceedings of the Linnean Society of New South Wales
Proc. Roy. Soc. Q/ld.	Proceedings of the Royal Society of Queensland
Prod. Florae Norfolkicae	Prodromus Florae Norfolkicae

Prodr. Fl. Novae Holl.	Prodromus Florae Novae Hollandiae et Insulae Van Diemen
Q/ld. Agric. J.	Queensland Agricultural Journal
Q/ld. Fl.	The Queensland Flora
Q/ld. Nat.	The Queensland Naturalist
Repert. Spec. Nov. Regn. Veg. Beih.	Repertorium Specierum Novarum Regni Vegetabilis. Beihefte
Rev. Gen. Pl.	Revisio Generum Plantarum. Wurzburg
Roy. Hort. Soc. Dict. Gdng.	The Royal Horticultural Society Dictionary of Gardening
Sth. Sci. Rec.	The Southern Science Record
Sydney Mail	The Sydney Mail
Sydney Morning Herald	The Sydney Morning Herald
Syn. Q/ld. Fl.	A Synopsis of the Queensland Flora
Victorian Nat.	The Victorian Naturalist
Votes Proc. Legis. Ass. Q/ld.	Votes and Proceedings of the Legislative Assembly, Queensland
Walp. Ann. Bot. Syst.	W. G. Walpers, Annales Botanices Systematicae — Leipzig
Wing's South. Science Rec.	Wing's Southern Science Record
Xenia Orchid.	Xenia Orchidaceae, Beitrage zur Kenntnis der Orchideen Leipzig

Appendix 6
Name changes of *Dendrobium* orchids as proposed November 1988

Since the completion of the manuscript for this work, *Native Orchids of Australia* by David L. Jones has been published (September 1988). The supplement at the back of Jones's book gives the very latest name changes and new taxa; these have been taken from the *Catalogue of Australian Orchidaceae* being prepared by Mark Clements for publication.

It remains to be seen if all or any of these name changes will be accepted — only time will tell.

Dendrobium speciosum is a very variable species and there are numerous intermediates between the six varieties in both plant and flower form as well as a considerable range in flower size for each variety. I find it difficult to accept the upgrading to specific rank of the varieties and will continue to regard them as varieties. Very little is known of their pollinators and I doubt if each species is pollinator specific. In my collection in Sydney all varieties become naturally pollinated quite commonly.

Dendrobium tetragonum is another variable species; I believe its varieties should be left at varietal rank, particularly as intermediates do occur and differences are minor. A form of this species very similar to the lower north coast of New South Wales form is found near Wyong, NSW.

Name in *Dendrobium Orchids of Australia*	Name in *Native Orchids of Australia*, supplement
D. bigibbum var. *venosum* F. M. Bail.	*D. x lavarackianum* M. Clements
D. sp. aff. *fairfaxii*	*D. dolichophyllum* D. Jones & M. Clements
D. sp. aff. *jonesii*	*D. jonesii* subsp. *bancroftianum* (H. G. Reichb.) M. Clements & D. Jones
D. jonesii Syn. *D. ruppianum* var. *blackburnii* (Nicholls) Dockr.	*D. jonesii* subsp. *blackburnii* (Nicholls) M. Clements & D. Jones
D. linguiforme var. *nugentii* F. M. Bail.	*D. nugentii* (F. M. Bail.) D. Jones & M. Clements
D. macropus subsp. *howeanum* P. S. Green	*D. comptonii* Rendle
D. speciosum var. *curvicaule* F. M. Bail.	*D. curvicaule* (F. M. Bail.) M. Clements & D. Jones
D. speciosum var. *grandiflorum* F. M. Bail.	*D. rex* M. Clements D. Jones
D. speciosum var. *hillii* Masters	*D. tarberi* M. Clements & D. Jones
D. speciosum var. *pedunculatum* Clemesha	*D. pedunculatum* (Clemesha) D. Jones & M. Clements
D. tetragonum var. *giganteum* Gilbert	*D. capitisyork* M. Clements & D. Jones
D. tetragonum var. *giganteum* Gilbert pale green colour form (Fig. 100)	*D. cacatua* M. Clements & D. Jones
D. tetragonum var. *tetragonum* A. Cunn. a form from the lower north coast area of NSW (Fig. 101)	*D. melaleucaphilum* M. Clements & D. Jones

Fig. 101 *Dendrobium tetragonum* A. Cunn.
var. *giganteum* Gilbert
A pale green colour form with white labellum,
from Mt Lewis, north Queensland

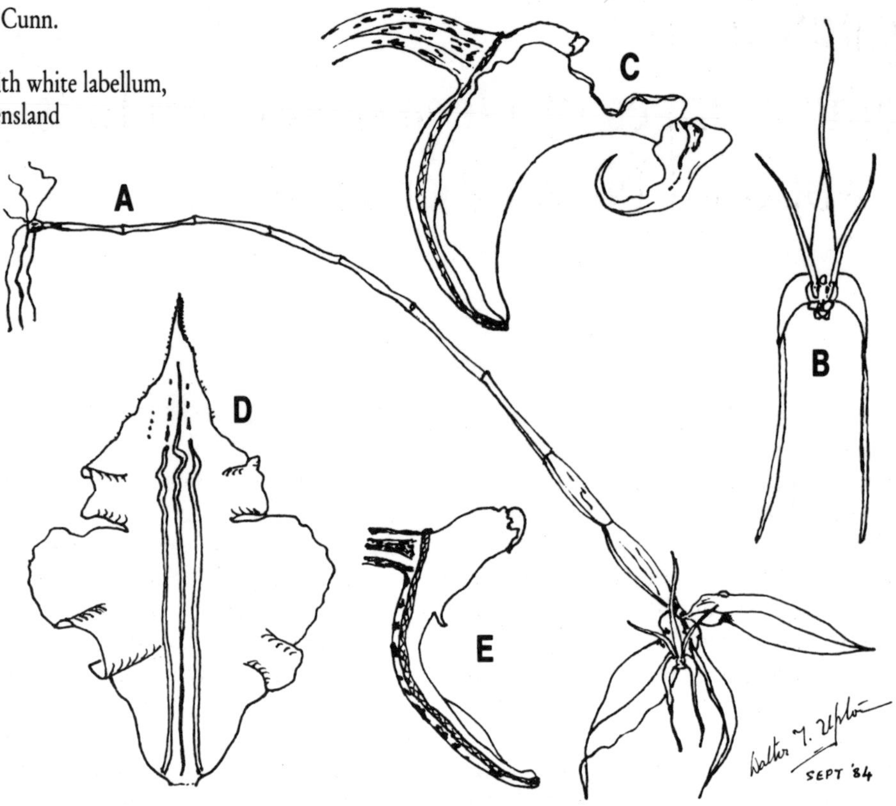

A Plant 1/5X
B Flower front-view 2/3X
C Column and labellum side-view 4X
D Labellum pressed almost flat 4X
E Column side; note appendage and red/purple markings on ovary 4X

Fig. 102 *Dendrobium tetragonum* A. Cunn.
var. *tetragonum*
A form from the lower north coast area of NSW

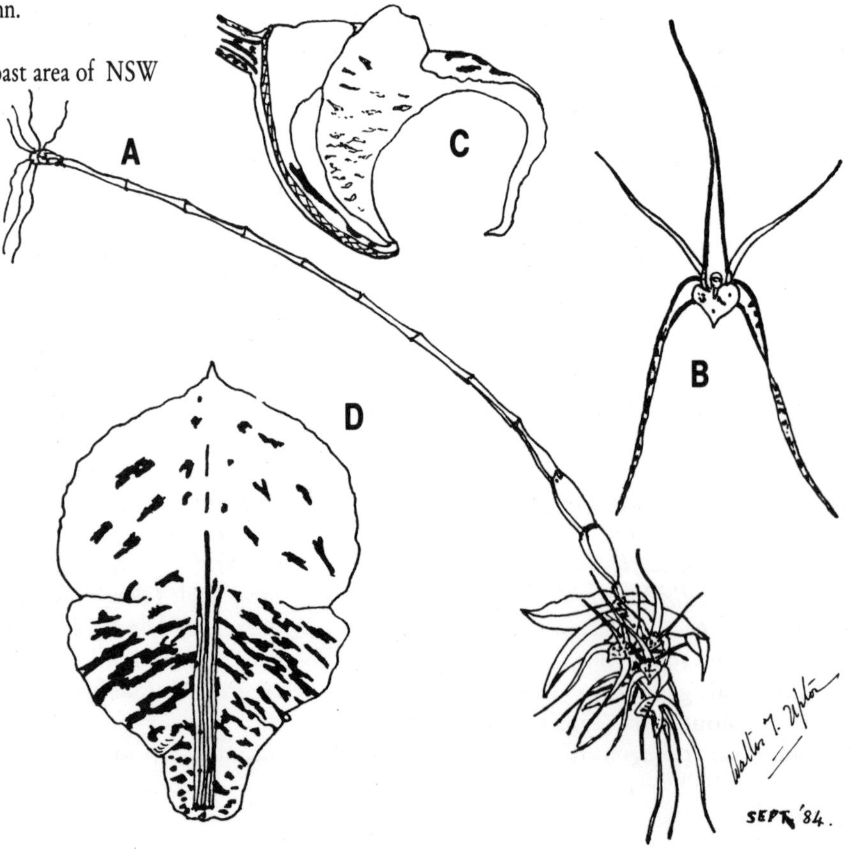

A Plant 1/5X
B Flower front-view 2/3X
C Column and labellum side-view 4X
D Labellum pressed flat 4X

Glossary

Aberrant	Differing from the normal.
Androecium	Male part of the flower.
Anther	Pollen-bearing part of the column.
Anthesis	Flowering—the period when the flowers are open.
Asymbiotic	Without symbiosis—referring to a method of germinating orchid seed without the presence of fungi.
Basionym	The original name. The first usage of the specific epithet.
Bract	The leaf-like or scale-like structure at the base of the flower or leaf, or below the leaves of the stem.
Callus	A crest or fleshy outgrowth of the lip.
Capsule	A dry seed vessel formed from a multi-carpelled ovary.
Carpel	A leaf-like structure which bears the ovules and seeds.
Caudicle	A slender extension of the pollinium. Part of the pollen mass, produced within the anther.
Chromosomes	Rod-like or bead-like bodies of definite number, which become visible during cell division and are made up of material known as chromatin which carries the genetic information.
Ciliate	Having hairs on the margin.
Cleistogamous	Refers to flowers which automatically self-pollinate without opening.
Clinandrium	That portion of the column directly beneath the anther. The anther bed.
Clone	The whole group of individuals derived from an original plant, whether produced vegetatively or by asexual reproduction by a single organism. All members of a clone are genetically identical.
Column	The central structure in an orchid flower, formed by the fusion of the sex organs.
Column foot	A ventral extension of the base of the column which has the lip attached at its tip.
Coriaceous	Of a leathery texture.
Dehisce	To burst open. The opening of the capsule, by a process of splitting along definite lines.
Diploid	Normal chromosome condition of a mitotically divided nucleus. The symbol 2n is used to denote diploid condition, n the haploid.
Disc	A fleshy structure on the labellum.
Embryo	The young plant while still surrounded by maternal tissues, or within the seed.
Endosperm	The nutritive tissue of certain seeds.
Ephemeral	Of very short duration.
Epiphyte	Any plant which grows upon another plant.

Filament	The slender, sterile portion of the stamen which bears the anther. Part of the column.
Filiform	Thread-like.
Funicle	The cord or thread which attaches an ovule or a seed to the placenta.
Fusiforme	Spindle-shaped, broadest at centre and tapered at both ends.
Gametes	Sex cells.
Genus (Pl. genera)	Group of related species. The taxonomic category above the species.
Haploid	With a single set of unpaired chromosomes in the nucleus.
Holotype	The specimen used or designated, by the author of a name, as the type of species.
Hygroscopic	Capable of absorbing moisture from the atmosphere; movement caused by such absorption.
Indeterminate	Of indefinite or protracted growth. A raceme in which the lower flowers open first.
Inflorescence	The flower or flower cluster of a plant (including the peduncle, bracts, rhachis, pedicels and flowers).
Keels	Ridges on lamina of labellum (callus).
Labellum	One of the three petals which is usually larger and differently shaped than the other two.
Lamina	The broad flattened portion of the labellum, petal or leaf.
Lithophyte	Any plant growing on rocks.
Locule	A chamber in a closed structure, as in fruit or anther.
Mentum	A chin-like extension at the base of the flower made up of the column foot and the lateral sepals.
Nectary	A nectar-producing structure or gland.
Ovary	The part of the flower which develops into the fruit.
Ovule	The structure which develops into the seed.
Pedicel	The stem which supports an individual flower
Peduncle	The stem of an inflorescence or solitary flower.
Peloric	An abnormality in which the labellum is like the other petals in form.
Perianth	A collective term for sepals and petals, together.
pH	Hydrogen ion concentration—the degree of acidity or alkalinity in plant systems.
Placenta	That portion of the ovary that bears the ovules.
Pollen tube	A tubular process developed from pollen grains after attachment to the stigma and growing towards the ovule.
Pollinium (Pl. pollinia)	Compact masses of pollen. The contents of the anther cell.
Polyploidy	Multiplication of number of chromosome sets normal to the haploid.
Proliferation	Bearing of offsets.
Protocorm	A germinating orchid seed, the body from which a shoot and roots are formed.
Pseudobulbous stem	A thickened stem.
Pubescent	Hairy, the short hairs being soft or downy.
Raceme	An unbranched indeterminate inflorescence with pedicellate flowers.
Resupination	Twisting or bending so that the labellum which is uppermost in the bud, is on the lower side when the flower matures.
Rhachis	The axis of an inflorescence, above the peduncle to which the pedicels are attached.
Rhizome	A horizontal stem which has roots and erect branches. In *Dendrobium* mostly made up of the bases of successive shoots.
Rostellum	A portion of the stigma which is specialized to assist in the transfer of pollen; the tissue which separates the anther from the fertile stigma. the fertile stigma.
Rugose	Covered with wrinkles.

Senescence	Physiological ageing.
Spur	A slender, tubular or sack-like projection from a flower part (usually a nectary) commonly formed by the base of the labellum.
Stigma	The sticky, receptive part of the column, which produces a sugary substance on which to place the pollinia.
Stipe	A strap-like extension of columnar tissue by which the pollinium are attached to the vissidium either directly or by the caudicle.
Striae	Fine longitudinal lines.
Suture	Line of junction of two united parts.
Symbiosis	An ecological relationship between members of two different species in which both obtain some benefit from the relationship.
Sympodial	A habit of growth in which each shoot has limited growth, new shoots usually arising from the base of the older ones.
Synonym	A scientific name which refers to the same taxon as another name. A name rejected in favour of an accepted correct name.
Taxon (Pl. taxa)	A taxonomic group of any rank.
Terete	Cylindrical, circular in cross-section.
Terrestrial	Growing on the ground.
Tetraploid	A polyploid with four times the basic haploid chromosome number.
Tomentose	Densely pubescent with matted hairs.
Triploid	A polyploid with three times the basic haploid chromosome number.
Tuberculate	Warty—covered with wart-like excrescences.
Type	The original representative of a species, genus or other taxon, upon which the distinguishing characters of the taxon are based.
Vestigial	Rudimentary, imperfectly developed or under developed.

Conversion chart

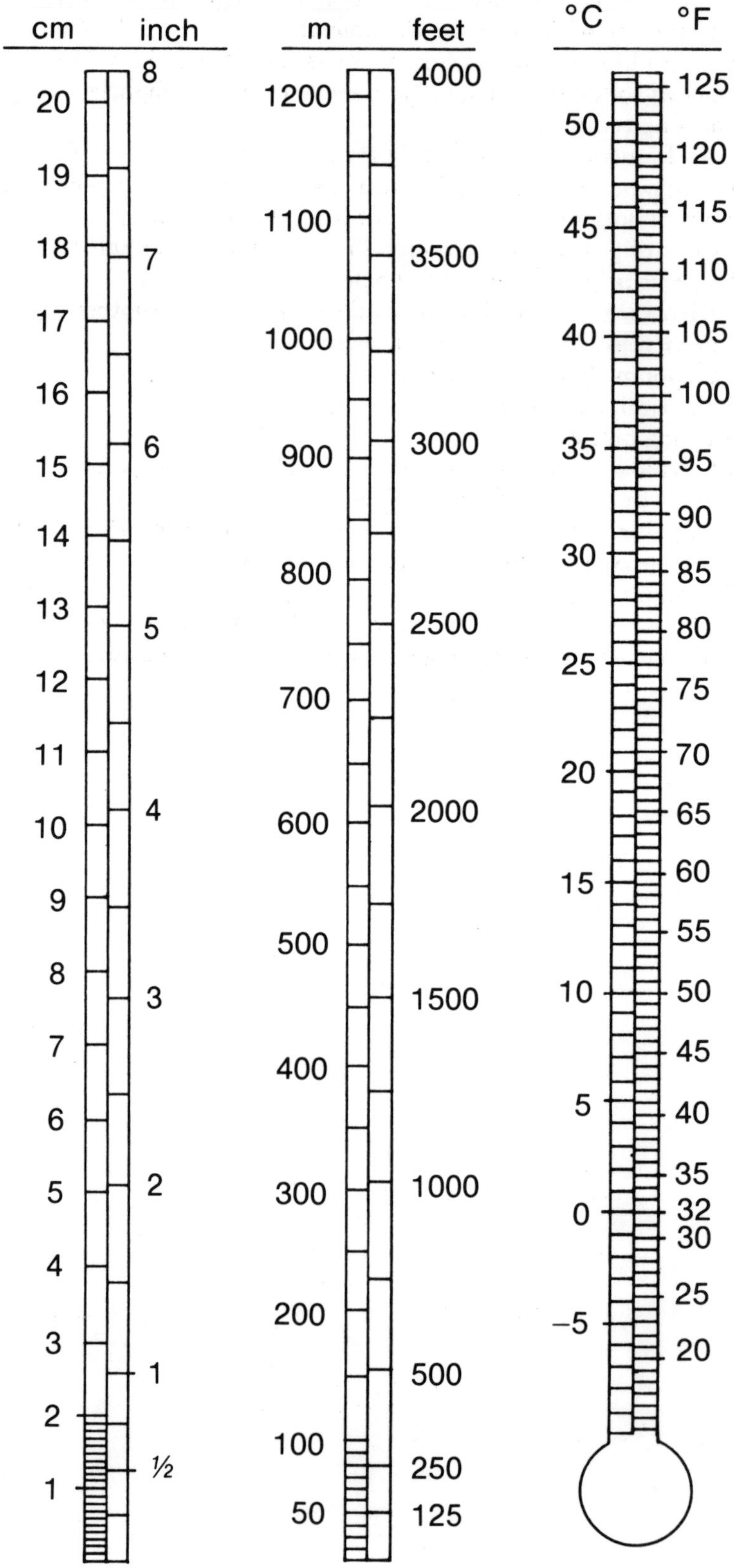

References and selected bibliography

Arditti, J. (1982). 'Orchid Seed Germination and Seedling Culture — A Manual' *Orchid Biology, Reviews and Perspectives II*, Cornell University Press, Ithaca and London.

Australian Almanac, The (1984). Angus and Robertson Ltd, Sydney.

Australian Encyclopaedia, The. The Grolier Society of Australia Pty Ltd, Lane Cove, Sydney.

Australian Orchid Review, Printcraft Press Pty Ltd, Sydney.

Bechtel, H., Cribb, P. and Launert, E. (1981). The Manual of Cultivated Orchid Species, The MIT Press, Cambridge, Massachusetts, USA.

Bedford, R. (1969). *A Guide to Native Australian Orchids*, Angus & Robertson Ltd, Sydney.

Blake, S. T. (1966). 'The Cooktown Orchid', *Australian Orchid Review*, Vol. 31, No. 1, pp. 29–32.

Cady, Leo and Rotherham, E. R. (1970). *Australian Native Orchids in Colour*, A. H. & A. W. Reed, Sydney.

Cambourne, C. (1936). 'The Orchids of Thursday Island', *Australian Orchid Review*, Vol. 1, No. 2, p. 27.

Clements, M. A. (1982). *Preliminary Checklist of Australian Orchidaceae*, National Botanic Gardens, Canberra.

Clemesha, S. C. (1969). '*Dendrobium schneiderae* F. M. Bail. and its variety *major* Rupp', *Orchadian*, Vol. 3, No. 4, p. 42.

Clemesha, S. C. (1978). 'A Review of *Dendrobium bigibbum* Lindl.', *Orchadian*, Vol. 6, No. 2, pp. 27–33.

Clemesha, S. C. (1981–1982). 'A Review of *Dendrobium speciosum* Sm.', *Orchadian*, Vol. 6, No. 11, pp. 247–262 and 'A New Variety of *Dendrobium speciosum* Sm. from Central Queensland', *Orchadian*, Vol. 7, No. 5, pp. 103–106.

Collins, R. D. (1979). 'Some Historical Notes on *Dendrobium nindii* W. Hill', *Orchadian* Vol. 6, no. 5, p. 116.

Cooper, D. (1981). *New Zealand Native Orchids*, Price Milburn, Wellington, New Zealand for the Wellington Orchid Society Inc.

Cox, J. Murray (1946). *A Cultural Table of Orchidaceous Plants*, The Shepherd Press, Sydney.

Cribb, P. J. (1983). 'A Revision of *Dendrobium* Sect. *Latouria* (Orchidaceae)', *Kew Bulletin*, Vol. 38(2), London.

Curtis, W. M. (1979). *The Students Flora of Tasmania*, Part 4A, Orchidaceae, Government Printer, Hobart, Tasmania.

Dockrill, A. W. (1964). *The Orchadian*, Vol. 1, p. 79.

Dockrill, A. W. (1969). *Australian Indigenous Orchids*, Vol. 1, The Epiphytes: The Tropical Terrestrial Species, The Society for Growing Australian Plants, Sydney.

Dressler, R. L. (1981). *The Orchids, Natural History and Classification*, Harvard University Press, Cambridge, Massachusetts, USA.

Firth, M. J. (1965). *Native Orchids of Tasmania*, published by the author, Devonport, Tasmania.

Fitzgerald, R. D. (1882). '*Austraian Orchids*', Government Printer, Sydney.

Gilbert, P. A. (1939). 'Birds and Orchids', *Australian Orchid Review*, Vol. 4, No. 2, p. 51.

Gilbert, P. A. (1958). '*Dendrobium linguiforme* Sw. Australian Dendrobiums'. *American Orchid Soc. Bulletin*, Vol. 27, p. 47.

Hallé, N. (1977). *Flore de la Nouvelle Calédonié et Dépendances*, Museum National d'Naturelle, Paris.

Handbook on Orchid Nomenclature and Registration (1985). International Orchid Commission, London.

Hawkes, A. D. (1965). *Encyclopaedia of Cultivated Orchids*, Faber & Faber Ltd, London.

Holttum, R. E. (1953). *Flora of Malaya*, Vol. 1, *Orchids*, Government Printing Office, Singapore.

Holttum, R. E., Brieger, F. G. and Cribb, P. J. (1979). 'A Proposal for the retypification of *Dendrobium* Sw. Nom. Cons.', *Taxon*, Vol. 28(4), 109, p. 418.

Jaede, C. A. (1936). 'The Orchids of Lord Howe Island', *Australian Orchid Review*, Vol. 1, No. 4, p. 13.

Johns, J. and Molloy, B. (1983). *Native Orchids of New Zealand*, A. H. & A. W. Reed Ltd, Wellington, NZ.

Jones, D. L. (1974). 'The Pollination of *Dendrobium linguiforme*', *Australian Orchid Review*, Vol. 39, No. 2, p. 70.

Jones, D. L. (1983). 'The pollination and occurrence of natural hybrids in Australian species of the genus *Dendrobium*'. *Proceedings of the 8th Australian Orchid Conference*, The Townsville Orchid Society.

Jones, D. L. and Gray, B. (1977). 'The Pollination of *Dendrobium ruppianum* A. D. Hawkes', *American Orchid Soc. Bulletin*, Vol. 46, p. 54.

Kitchen, D. R. (1970). 'A New Natural Hybrid', *Orchadian*, Vol. 3, No. 11, p. 151.

Lavarack, P. S. (1976). '*Orchids of the Iron Range Area*', Queensland National Parks and Wildlife Service.

Lavarack, P. S. (1978). 'Recent Discoveries in North Queensland', *Orchadian*, Vol. 5, No. 11, p. 163.

Lavarack, P. S. (1980a). '*Dendrobium antennatum* Lindl. in Australia', *Orchadian*, Vol. 6, No. 7, p. 148.

Lavarack, P. S. (1980b). '*Orchids of McIlwraith Range*', Queensland National Parks and Wildlife Service.

Lavarack, P. S. (1984). 'A Further Note on *Dendrobium johannis* Rchb.f. and *D. semifuscum* (Rchb.f.) Lavarack & Cribb', *Orchadian*, Vol. 7, No. 11, p. 252.

Lavarack, P. S. and Gray, B. (1985). *Tropical Orchids of Australia*, Thomas Nelson Australia, Melbourne.

Lawler, L. J. (1984). 'Ethnobotany of the Orchidaceae'. In: J. Arditti (ed.) *Orchid Biology, Reviews and Perspectrives, III*, Cornell University Press, Ithaca, New York, USA, pp. 27–149.

Lim, K. Y. (1985). 'The chromosomes of Orchids at Kew-2 *Dendrobium*', *American Orchid Soc. Bulletin*.

Maxwell, M. K. (1967). 'The *Dendrobium kingianum* Bidw. ex Lindl., Complex', *Australian Orchid Review*, Vol. 32, pp. 24–30, pp. 81–88 and pp. 139–141.

Millar, A. (1978). *Orchids of Papua New Guinea, an Introduction*, Australian National University Press, Canberra.

Moore, Charles (1893). *Handbook of the Flora of New South Wales*, Government Printer, Sydney.

Nicholls, W. H. (1969). *Orchids of Australia*, Thomas Nelson (Australia) Ltd, Melbourne.

Orchadian, The, Journal of the Australasian Native Orchid Society, Box C106 PO Clarence Street, Sydney 2000.

Pijl, L. van der and Dodson, C. H. (1966). *Orchid flowers, their pollination and evolution*, University of Miami Press, Coral Gables, Florida, USA.

Proceedings of the 8th Australian Orchid Conference (1983). Ed. R. Robinson, Townsville Orchid Soc., Townville, Queensland.

Rupp, H. M. R. (1943). *The Orchids of New South Wales*, Flora of New South Wales, National Herbarium, Sydney.

Rupp, H. M. R. and White, C. T. (1937). 'Review of the Genus Dendrobium in Australia', *The Queensland Naturalist*, 10:26.

Sander's List of Orchid Hybrids, The Royal Horticultural Society, London.

Schlechter, R. (1982). *The Orchidaceae of German New Guinea*, English Edition, The Australian Orchid Foundation, 107 Roberts St, Essendon, Victoria.

Schlechter, R. (1982a). *The Orchidaceae of German New Guinea*, English Edition, Group XX, *Dendrobiinae*, pp. 508–517.

Schlechter, R. (1982b). *The Orchidaceae of German New Guinea*, English Edition, Group XX, *Dendrobiinae*, p. 538.

Schlechter, R. (1982c). *The Orchidaceae of German New Guinea*, English Edition, Group XX, *Dendrobiinae*, p. 516.

Schlechter, R. (1982d). *The Orchidaceae of German New Guinea*, English Edition, Group XX, *Dendrobiinae*, pp. 627–628.

Smythe, R. S. (1970a). '*Dendrobium smilliae* Australia's First Bird Pollinated Orchid', *Orchadian*, Vol. 3, No. 8, p. 93.

Smythe, R. S. (1970b). 'A Pollinator for *Dendrobium speciosum*', *Orchadian*, Vol. 3, No. 10, p. 119.

Smythe, R. S. (1970c). 'Pollination by the Common Native Bee', *Orchadian*, Vol. 3, No. 11, p. 128.

Tierney, N. F. (1946). '*Dendrobium toftii*', *Australian Orchid Review*, Vol. 11, No. 3, p. 66.

Veitch, J. (1888). *A Manual of Orchidaceous Plants*, Part III, *Dendrobium*, James Veitch & Sons, Chelsea, UK.

Willersdorf, A. L. (1968). 'Distribution of Orchids in the Carnarvon Gorge', *Australian Orchid Review*, Vol. 33, No. 3, p. 131.

Williams, B. S. and H. (1894). *Orchid Growers Manual*, Victoria and Paradise Nurseries, London.

Withner, C. (ed.) (1959). *The Orchids—a Scientific Survey*, Ronald Press, New York.

Wooster, H. T. (1940). '*Dendrobium superbiens*', *Australian Orchid Review*, Vol. 5, No. 2, p. 65.

Year Book, Australia (1984). Australian Bureau of Statistics, Canberra.

Index

The *Dendrobium* species are listed alphabetically from p. 14, and the natural hybrids from p. 138. Figures are not indexed as they appear with the species descriptions. Artificial hybrids are indexed only when mentioned in the text; full lists are in Appendices 2 and 3. In Chapter 3 (Descriptions), artificial hybrids are listed under the sub-heading Hybrids, according to each parent species.